Post-Translational Processing

The Practical Approach Series

SERIES EDITOR

B. D. HAMES
School of Biochemistry and Molecular Biology
University of Leeds, Leeds LS2 9JT, UK

See also the Practical Approach web site at **http://www.oup.co.uk/PAS**

★ indicates new and forthcoming titles

Affinity Chromatography
Affinity Separations
Anaerobic Microbiology
Animal Cell Culture (2nd edition)
Animal Virus Pathogenesis
Antibodies I
Antibodies II
Antibody Engineering
★ Antisense Technology
Applied Microbial Physiology
Basic Cell Culture
Behavioural Neuroscience
Bioenergetics
Biological Data Analysis
Biomechanics—Materials
Biomechanics—Structures and Systems
Biosensors
Carbohydrate Analysis (2nd edition)
Cell-Cell Interactions
The Cell Cycle
Cell Growth and Apoptosis
★ Cell Separation
Cellular Calcium
Cellular Interactions in Development
Cellular Neurobiology
★ Chromatin
★ Chromosome Structural Analysis
Clinical Immunology
Complement
★ Crystallization of Nucleic Acids and Proteins (2nd edition)
Cytokines (2nd edition)
The Cytoskeleton
Diagnostic Molecular Pathology I
Diagnostic Molecular Pathology II
DNA and Protein Sequence Analysis
DNA Cloning 1: Core Techniques (2nd edition)
DNA Cloning 2: Expression Systems (2nd edition)
DNA Cloning 3: Complex Genomes (2nd edition)

DNA Cloning 4: Mammalian Systems (2nd edition)
★ Drosophila (2nd edition)
Electron Microscopy in Biology
Electron Microscopy in Molecular Biology
Electrophysiology
Enzyme Assays
Epithelial Cell Culture
Essential Developmental Biology
Essential Molecular Biology I
Essential Molecular Biology II
★ Eukaryotic DNA Replication
Experimental Neuroanatomy
Extracellular Matrix
Flow Cytometry (2nd edition)
Free Radicals
Gas Chromatography
Gel Electrophoresis of Nucleic Acids (2nd edition)
★ Gel Electrophoresis of Proteins (3rd edition)
Gene Probes 1
Gene Probes 2
Gene Targeting
Gene Transcription
★ Genome Mapping
Glycobiology
★ Growth Factors and Receptors
Haemopoiesis
★ High Resolution Chromotography
Histocompatibility Testing
HIV Volume 1
HIV Volume 2
★ HPLC of Macromolecules (2nd edition)
Human Cytogenetics I (2nd edition)
Human Cytogenetics II (2nd edition)
Human Genetic Disease Analysis
★ Immobilized Biomolecules in Analysis
Immunochemistry 1
Immunochemistry 2
Immunocytochemistry
★ *In Situ* Hybridization (2nd edition)
Iodinated Density Gradient Media
Ion Channels
★ Light Microscopy (2nd edition)
Lipid Modification of Proteins
Lipoprotein Analysis
Liposomes
Mammalian Cell Biotechnology
Medical Parasitology
Medical Virology
MHC Volume 1
MHC Volume 2
★ Molecular Genetic Analysis of Populations (2nd edition)
Molecular Genetics of Yeast
Molecular Imaging in Neuroscience
Molecular Neurobiology
Molecular Plant Pathology I
Molecular Plant Pathology II
Molecular Virology
Monitoring Neuronal Activity

Mutagenicity Testing
★ Mutation Detection
Neural Cell Culture
Neural Transplantation
Neurochemistry (2nd edition)
Neuronal Cell Lines
NMR of Biological Macromolecules
Non-isotopic Methods in Molecular Biology
Nucleic Acid Hybridisation
Oligonucleotides and Analogues
Oligonucleotide Synthesis
PCR 1
PCR 2
★ PCR 3: PCR *In Situ* Hybridization
Peptide Antigens
Photosynthesis: Energy Transduction
Plant Cell Biology
Plant Cell Culture (2nd edition)
Plant Molecular Biology
Plasmids (2nd edition)
Platelets
Postimplantation Mammalian Embryos
Preparative Centrifugation
Protein Blotting
★ Protein Expression Vol 1
★ Protein Expression Vol 2
Protein Engineering
Protein Function (2nd edition)
Protein Phosphorylation
Protein Purification Applications
Protein Purification Methods
Protein Sequencing
Protein Structure (2nd edition)
Protein Structure Prediction
Protein Targeting
Proteolytic Enzymes
Pulsed Field Gel Electrophoresis
RNA Processing I
RNA Processing II
★ RNA-Protein Interactions
Signalling by Inositides
Subcellular Fractionation
Signal Transduction
★ Transcription Factors (2nd edition)
Tumour Immunobiology

Post-Translational Processing

A Practical Approach

Edited by

S. J. HIGGINS
School of Biochemistry and Molecular Biology, University of Leeds, Leeds

and

B. D. HAMES
School of Biochemistry and Molecular Biology, University of Leeds, Leeds

OXFORD
UNIVERSITY PRESS

OXFORD
UNIVERSITY PRESS

Great Clarendon Street, Oxford OX2 6DP

Oxford University Press is a department of the University of Oxford and furthers the University's aim of excellence in research, scholarship, and education by publishing worldwide in

Oxford New York

Athens Auckland Bangkok Bogotá Buenos Aires Calcutta Cape Town Chennai Dar es Salaam Delhi Florence Hong Kong Istanbul Karachi Kuala Lumpur Madrid Melbourne Mexico City Mumbai Nairobi Paris São Paulo Singapore Taipei Tokyo Toronto Warsaw

and associated companies in Berlin Ibadan

Oxford is a registered trade mark of Oxford University Press

Published in the United States
by Oxford University Press Inc., New York

A catalogue record for this book is available from the British Library

Library of Congress Cataloging in Publication Data
(Data available)

ISBN 0–19–963794–6 (Hbk)
0–19–963795–4 (Pbk)

Typeset by Footnote Graphics,
Warminster, Wilts
Printed in Great Britain by Information Press, Ltd,
Eynsham, Oxon.

Preface

Some years ago we edited a book for The Practical Approach series entitled *Transcription and translation: a practical approach*. When the time came to consider organizing a second edition, it rapidly became clear that no one book of the desired size could include in sufficient detail the myriad of important new techniques. As a result, a decision was taken to produce a collection of books to cover this important area. *Gene transcription: a practical approach* and two volumes of *RNA processing: a practical approach* have since been published. Now, this book, *Post-translational processing: a practical approach*, and its companion volume, *Protein expression: a practical approach,* complete the 'mini-series' by providing a comprehensive and up-to-date coverage of the synthesis and subsequent processing of proteins.

Post-translational processing: a practical approach begins with a chapter on protein sequence analysis by Jeff Keen and Alison Ashcroft. Joachim Rassow then covers essential methods for the study of protein folding and import into organelles. Next follow three chapters which describe the major covalent modification events of phosphorylation (by Ivar Walaas and Anne Østvold), glycosylation (by David Ashford and Fran Platt), and lipid modification (by Nigel Hooper and Jeff McIlhinney). Another key area, proteolytic processing, is the subject of a chapter by John Hutton *et al.* The final two chapters focus on protein turnover in mammalian cells (by Aaron Ciechanover and Bernd Wiederanders) and in yeast (by Wolfgang Hilt and Dieter Wolf).

Those researchers who require practical guidance on the synthesis of proteins *in vitro* or *in vivo* for study are advised to consult the companion volume, *Protein expression: a practical approach,* which describes the expression of cloned DNA or RNA templates in all the major *in vitro* and *in vivo* systems, both prokaryotic and eukaryotic, as well as methods for monitoring expression.

The overriding goals of *Post-translational processing: a practical approach* are to describe, in precise detail, tried and tested versions of key protocols for the active researcher, and to provide all the support required to make the techniques work optimally, including hints and tips for success, advice on potential pitfalls, and guidance on data interpretation. We thank the authors for their diligence in writing such strong chapters and for accepting the editorial changes we suggested. The end-result is a comprehensive compendium of the best of current methodology in this subject area. It is a book designed both to be used at the laboratory bench and to be read at leisure to gain insight into future experimental approaches.

Leeds S.J.H.
August 1998 B.D.H.

Contents

Contributors

ALISON E. ASHCROFT
School of Biochemistry and Molecular Biology, University of Leeds, Leeds LS2 9JT, UK.

DAVID A. ASHFORD
Glycobiology: Research and Analytical, Department of Biology, University of York, PO Box 373, York YO1 5YW, UK.

ELAINE M. BAILYES
Department of Clinical Biochemistry, University of Cambridge, Addenbrooke's Hospital, Hills Road, Cambridge CB2 2QR, UK.

AARON CIECHANOVER
Department of Biochemistry, The Bruce Rappaport Faculty of Medicine and the Rappaport Institute for Research in the Medical Sciences, Technion-Israel Institute of Technology, PO Box 9649, Haifa 31096, Israel.

JOHN M. W. CREEMERS
Centre for Human Genetics, University of Leuven, Herestraat 49, B-3000 Leuven, Belgium.

WOLFGANG HILT
Institut für Biochemie, Universität Stuttgart, Pfaffenwaldring 55, D-70569 Stuttgart, Germany.

NIGEL M. HOOPER
School of Biochemistry and Molecular Biology, University of Leeds, Leeds LS2 9JT, UK.

JOHN C. HUTTON
Barbara Davis Center for Childhood Diabetes, University of Colorado Health Sciences, 4200 East 9th Avenue, Box B140, Denver, CO 80262, USA.

JEFF N. KEEN
School of Biochemistry and Molecular Biology, University of Leeds, Leeds LS2 9JT, UK.

IRIS LINDBERG
Barbara Davis Center for Childhood Diabetes, University of Colorado Health Sciences, 4200 East 9th Avenue, Box B140, Denver, CO 80262, USA.

R. A. JEFFREY MCILHINNEY
MRC Anatomical Neuropharmacology Unit, Mansfield Road, Oxford OX1 3TH, UK.

ANNE CARINE ØSTVOLD
Neurochemical Laboratory, University of Oslo, PO Box 1115, Blindern, N-0317 Oslo, Norway.

FRAN PLATT
Oxford Glycobiology Institute, Department of Biochemistry, University of Oxford, South, Parks Road, Oxford OX1 3QU, UK.

JOACHIM RASSOW
Institut für Biochemie und Molekularbiologie, Universität Freiburg, Hermann Herder Strasse 7, D-79104 Freiburg, Germany.

S. IVAR WALAAS
Neurochemical Laboratory, University of Oslo, PO Box 1115, Blindern, N-0317 Oslo, Norway.

BERND WIEDERANDERS
Institut für Biochemie, Klinikum der Friedrich-Schiller-Universität Jena, Nonnenplan 2, D-07740 Jena, Germany.

DIETER H. WOLF
Institut für Biochemie, Universität Stuttgart, Pfaffenwaldring 55, D-70569 Stuttgart, Germany.

Abbreviations

1,10-P	1,10-*ortho*phenanthroline
A_{275}	absorbance (at 275 nm)
AA	arylamine
AAC	ATP/ADP carrier
Ac	acetyl
amc	amino methylcoumarin
AMC	7-amido-4-methylcoumarine
APS	ammonium persulfate
ATP	adenosine triphosphate
ATP-γ-S	adenosine 5′-*O*-(3-thiotriphosphate)
ATZ	anilinothiazolinone
βNA	β-naphthylamide
BSA	bovine serum albumin
Bz	benzoyl
CA-074Me®	*N*-(L-3-*trans*-propylcarbamoyloxirane-2-carbonyl)-L-isoleucyl-L-prolyl-methyl ester
CaM	calcium calmodulin
cAMP	adenosine 3′,5′-cyclic monophosphate
CAPS	3-(cyclohexylamino)-1-propanesulfonic acid
Cbz	carboxybenzoyl
Cdk	cyclin-dependent kinase
CE	capillary electrophoresis
C_F	concentration of fluorophore
CHAPS	3-[(3-cholamidopropyl)dimethylammonio]-1-propane sulfonate
CON	control
CRD	cross-reacting determinant
CTP	cytidine 5′-triphosphate
DCI	dichloroisocoumarin
DE	delayed extraction
DEAE	diethylaminoethyl
DGG-U-GEE	des-Gly-Gly-ubiquitin-Gly-ethyl ester
DHFR	dihydrofolate reductase
DITC	phenyl diisothiocyanate
DMP	dimethyl pimelimidate
DMSO	dimethyl sulfoxide
DSS	disuccinimidyl suberate
DTT	dithiothreitol
ϵ_{405}	molar extinction coefficient (at 405 nm)
E-64®	*N*-(L-3-*trans*-carboxirane-2-carbonyl)-L-leucyl-agmatine

ECL	enhanced chemiluminescence
EDC	1-ethyl-3-(3-dimethylamino propyl)carbodiimide
EDTA	ethylenediaminetetraacetic acid
EGTA	ethylene glycol-*O*,*O′*-bis(2-aminoethyl)-*N*,*N*,*N′*,*N′*-tetraacetic acid
ELISA	enzyme-linked immunosorbent assay
ERK	extracellular-regulated kinases
ES	electrospray
FPLC	fast performance liquid chromatography
FT-ICR	Fourier transform-ion cyclotron resonance
GEE	glycine-ethyl ester
GPI	glycosyl phosphatidylinositol
Hepes	*N*-(2-hydroxyethyl)piperazine-*N′*-(2-ethanesulfonic acid)
HIV	human immunodeficiency virus
HMG	high mobility group
HPLC	high performance liquid chromatography
HPV	human papillomavirus
IgG	immunoglobulin class G
kDa	kilo Dalton
KRB	Krebs Ringer bicarbonate (medium)
KRP	Krebs Ringer phosphate (buffer)
λ_{em}	emission wavelength
λ_{ex}	excitation wavelength
LC	liquid chromatography
LDH	lactate dehydrogenase
LM	lysosomal fraction
MALDI	matrix-assisted laser desorption ionization
MALDITOF-MS	matrix-assisted laser desorption ionization time-of-flight mass spectrometry
MAPK	mitogen-activated protein kinase
MARKS	myristoylated alanine-rich C-kinase substrate
MCS	multiple cloning site
MeUb	methylated ubiquitin
MES	2-(morpholino)ethanesulfonic acid
MHC	major histocompatibility complex
$(M + H)^+$	protonated molecular ion
$(M - H)^-$	deprotonated molecular ion
Mo	methoxy
MoβNA	4-methoxy-β-naphthylamide
MOPS	3-(*N*-morpholino)propanesulfonic acid
MS	mass spectrometry
MS-MS	tandem mass spectrometry
MTX	methotrexate
MV	mineral-vitamin

m/z	mass-to-charge ratio
NEM	*N*-ethylmaleimide
NMT	myristoyl-CoA:protein *N*-myristoyl transferase
OD_{578}	optical density (at 578 nm)
PAGE	polyacrylamide gel electrophoresis
PBS	phosphate-buffered saline
PC	prohormone convertase
PCR	polymerase chain reaction
PDGF	platelet-derived growth factor
PEG	polyethylene glycol
PGPH	peptidylglutamyl-peptide hydrolysing activity
Pipes	piperazine-*N,N'*-bis(2-ethanesulfonic acid)
PI-PLC	phosphatidylinositol-specific phospholipase C
PITC	phenyl isothiocyanate
PKA	protein kinase A (cyclic AMP-dependent protein kinase)
PKC	phospholipid-dependent protein kinase
PMSF	phenylmethylsulfonyl fluoride
pNA	*p*-nitroanilide
PP2A	protein phosphatase 2A
PPO	2,5-diphenyloxazole
p.s.i.	pounds per square inch
P-Ser	phosphoserine
P-Thr	phosphothreonine
PTH	phenylthiohydantoin
PTP	phosphotyrosine-specific phosphatase
P-Tyr	phosphotyrosine
PVDF	polyvinylidene difluoride
Q-TOF	quadrupole-time-of-flight
SDS	sodium dodecyl sulfate
SDS–PAGE	SDS–polyacrylamide gel electrophoresis
Sulfo-MBS	*m*-maleimidobenzoyl-*N*-hydroxy-sulfosuccinimide ester
STI	soybean trypsin inhibitor
TBS	Tris-buffered saline
TCA	trichloroacetic acid
TFA	trifluoroacetic acid
TGN	*trans* Golgi network
TH	thiohydantoin
TLC	thin-layer chromatography
TMPD	tetramethyl phenylene diamine
TOF	time-of-flight
TPCK	*N*-tosyl-L-phenylalanine chloromethyl ketone
Trypsin-TPCK	trypsin treated with *N*-tosyl-L-phenylalanine chloromethyl ketone
UbA1	ubiquitin aldehyde

UBC	ubiquitin-carrier protein or ubiquitin-conjugating enzyme
UV	ultraviolet
V_{max}	maximal velocity
vol	volume
VSG	variant surface glycoprotein
YPD	yeast extract, peptone, dextrose
Z	benzyloxycarbonyl-
Z-F-F-CHN_2	Z-L-phenylalanyl-L-phenylalanyl-diazomethylketone

1

Sequence analysis of expressed proteins

JEFF N. KEEN and ALISON E. ASHCROFT

1. Introduction

The generation of recombinant proteins may be considered a relatively facile process nowadays. The wide range of expression systems available, and the marketing of specialized molecular biological kits by a number of manufacturers, virtually ensure that reasonable quantities of recombinant proteins can be produced in most biochemistry laboratories, although there are notable exceptions such as some integral membrane proteins. The engineering of appropriate 'tags' (e.g. hexa-His, StrepTag) into the DNA sequence encoding any particular protein frequently enables single-step affinity purification of the expressed polypeptide. Similarly, the creation of a fusion protein, with the desired polypeptide linked to glutathione *S*-transferase for example, allows ready expression and purification of the protein. The incorporation of a unique protease-sensitive site (e.g. for Factor X_a) into the linking region between the two polypeptide domains enables the desired product to be removed from the fusion protein. However, unequivocal proof of the identity of the expressed protein and detailed analysis of the synthetic fidelity are essential prerequisites for its use in experimental systems, particularly in structural and functional analysis, and are absolutely critical prior to release of the protein into the environment in numerous biotechnological applications. The confirmation of N- and C-termini is of paramount importance, and amino acid sequence analysis remains the most unambiguous technique for this. This is complemented by mass spectrometric methods, particularly utilizing electrospray ionization, for accurate mass measurement of the expressed protein. This in turn enables a comparison to be made with the theoretical value, which may indicate the presence of post-translational modifications and lead to their identification.

Although biochemical research has encountered major changes over the years, the basics of protein sequencing methodology have barely changed since Pehr Edman introduced his degradative chemistry for N-terminal sequencing in the 1950s (1, 2). However, relatively minor improvements to the overall chemistry, coupled with extensive automation, have led to greatly enhanced sensitivity and consequently the routine determination of long sequences. Only the relatively recent introduction of biological mass spectrometry has offered a better alternative strategy in certain cases.

It has been suggested that the extremely rapid determination of DNA sequences will eliminate the requirement for routine protein sequencing, but in fact the two approaches have become increasingly complementary. Protein sequencing has remained central to modern molecular biological research, although the emphasis of the work has necessarily changed. Today, rather than sequencing entire proteins, the requirements are much more as prerequisites for DNA cloning work, providing the information required for the design of oligonucleotide probes and PCR primers. Sequencing also provides data for the manufacture of synthetic peptides for antibody production, for protein identification, and for the study of post-translational modifications. In the quality control of recombinant proteins, a few residues of N-terminal sequence are sufficient to confirm identity and the correctness of the reading frame. The rapid rise in the manufacture of recombinant proteins has also led to the perfection of C-terminal sequencing chemistry (3, 4) and the development of automated C-terminal analysis instrumentation to complement N-terminal sequencers and thus assist in the routine characterization of expressed proteins.

The introduction of fully-automated instrumentation to perform multiple cycles of the Edman chemistry reproducibly has eliminated much labour-intensive and inefficient manipulation of samples. Manual methods are virtually unused nowadays, as most researchers can obtain access to core facilities operating automated sequencers. Procedures for the removal of contaminants from samples prior to sequencing, the means by which the sample is exposed to the chemistry, the optimization of that chemistry, the reduction of side-product formation, and identification methods for the product have all been improved to enable the routine determination of sequences from ever-smaller amounts of material. Thus, the sensitivity of the technique has been reduced from nanomolar to sub-picomolar levels. In many cases sensitivity is not critical, as expressed proteins are not limited in supply, but in others this is an issue and minimization of usage is then important, e.g. for integral membrane proteins expressed from baculovirus vectors in insect cells.

In contrast to N-terminal sequencing, the determination of the C-terminal sequence of a protein has proved to be extremely difficult. Traditional enzymic approaches have been both laborious and relatively ineffective, sometimes producing only a few ambiguous residues. Recently, however, successful automated chemical approaches have been developed, allowing several residues to be identified reliably from a few hundred picomoles of material. The implementation of these approaches has been driven by the need to fully characterize expressed proteins.

The introduction of mass spectrometry (MS) into the field of protein sequence analysis has been a significant change. The technique has become an almost essential complementary approach to chemical protein sequencing. Accurate mass information aids the interpretation of sequence data, particularly with respect to post-translational modifications. MS analysis of either N-terminal or C-terminal sequencing 'ladders' generated by incomplete reactions can provide sequence information. MS approaches are also used for the direct determination of amino acid sequences. Spontaneous or directed fragmentation products of peptides from protein digests can be analysed to generate sequence information, which is particularly useful for the identification and characterization of known sequences.

In this chapter, we will describe the essential details of modern chemical N- and C-terminal protein sequencing. The importance of sample preparation and some recommended methods to produce good samples will be provided. The use of MS in sequence analysis and its prospects for the future will also be discussed.

2. N-terminal sequence analysis

Modern automated protein sequencers utilize the degradative chemistry developed in the 1950s by Pehr Edman (1, 2) for the determination of the N-terminal sequences of proteins (*Figure 1*). The protein sample is first made alkaline by exposure to a volatile amine, then exposed to the Edman reagent, phenyl isothiocyanate (PITC), which reacts with the N-terminal amino group (and some side chains). Excess reagents are then washed away from the sample using a variety of organic solvents and the modified N-terminal residue is cyclized and cleaved from the polypeptide chain using anhydrous trifluoroacetic acid (TFA). The released N-terminal residue is washed into a second reaction chamber for conversion from the relatively unstable anilinothiazolinone (ATZ)-derivative to a more stable phenylthiohydantoin (PTH)-amino acid which can be identified subsequently by reverse-phase HPLC. The truncated protein remains in the original reaction chamber where it can undergo further rounds of Edman degradation, leading eventually to the generation of a sequence of residues.

2.1 Automated sequencing

Commercial automated sequencers utilize various approaches for subjecting the sample to the Edman chemistry, and differ slightly in the details of the chemistry. In solid phase instruments (e.g. the MilliGen ProSequencer) the sample is attached covalently to a support membrane, whereas in gas phase or liquid pulse equipment (e.g. PE Applied Biosystems; Hewlett-Packard) the sample is simply adsorbed non-covalently onto the support. The difference between covalent and non-covalent attachment can govern the exact cycle programme used and the efficiency of the washing steps; covalent attachment allows much more stringent washing. The vast majority of work, however, is

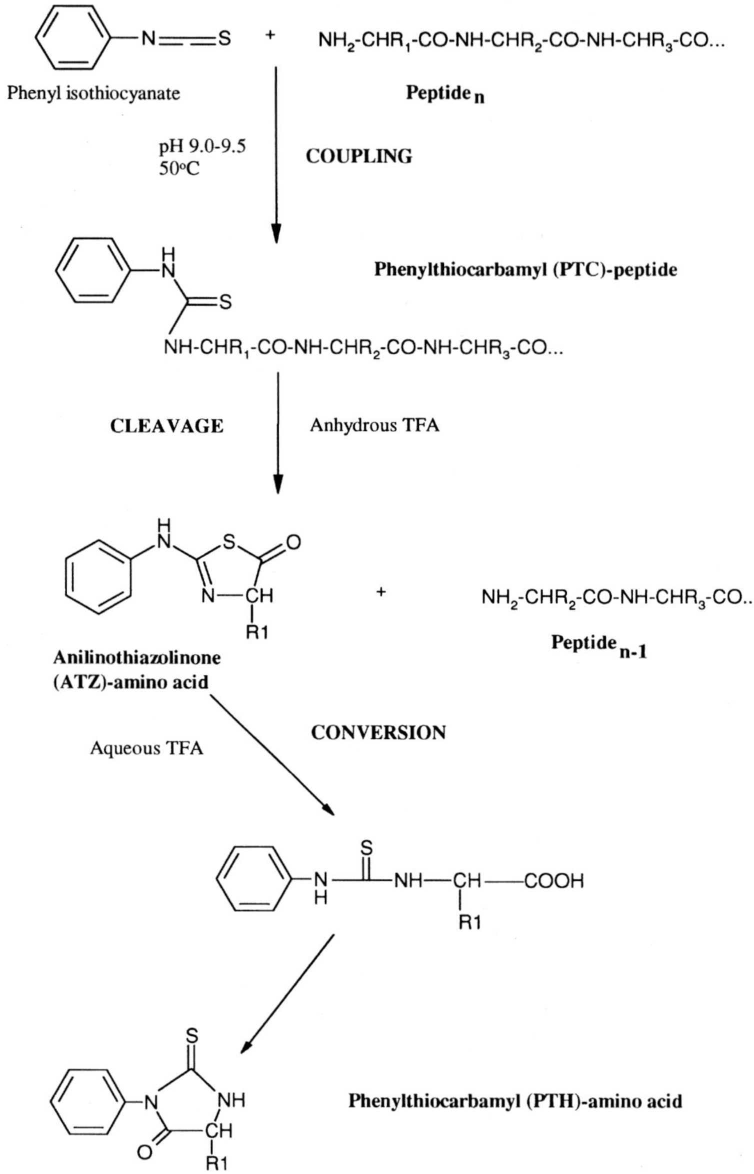

Figure 1. Edman degradation procedure. At pH > 8, phenyl isothiocyanate (PITC) reacts with the free N-terminal amino group of the protein, forming a phenylthiocarbamyl (PTC) derivative. Excess reagent is removed by washing with solvent and the modified residue is cyclized and cleaved from the protein using anhydrous trifluoroacetic acid (TFA), leaving a truncated protein with a new N-terminal amino group for the subsequent cycle. The cleaved residue, an anilinothiazolinone (ATZ)-amino acid, is converted using aqueous TFA into a stable phenylthiohydantoin (PTH)-amino acid and identified using reverse-phase HPLC.

Table 1. Comparison of adsorptive and solid phase sequencing[a]

Adsorptive sequencing	Solid phase sequencing
Sample adsorbed to PVDF membrane or glass fibre disc	Sample attached covalently to DITC membrane (via Lys), AA membrane (via Asp/Glu), or polyamino polymer (via Lys) on PVDF
Tolerant of small amounts of contaminants	Coupling intolerant of primary amines/thiol groups (polyamino polymer, DITC) or acidic groups (AA) Low yield of attached residues
Sample washout problem	No washout; stringent washing provides clean background and allows long runs
Loss of charged residues in reaction chamber	Charged residues (e.g. phosphorylated amino acids) recovered
Long cycle times	Reduced cycle times, due to high flow rates
SDS a problem in line blockage and sample washout	SDS solubilization of samples possible

[a] Abbreviations: PVDF polyvinylidene difluoride; DITC diisothiocyanate; AA arylamine.

carried out using adsorptive technology, which is convenient for most samples. *Table 1* compares and contrasts these two approaches.

2.1.1 Adsorptive sequencing

The sample may be applied to a glass fibre disc, usually pre-treated with a polycationic carrier (polybrene) to aid in the entrapment of the protein using ionic and H bond interactions (5). Alternatively, and preferably, the sample may be dried onto a polyvinylidene difluoride (PVDF) membrane, either by direct spotting, by electrophoretic transfer following SDS–PAGE, or by using ProSpin or ProSorb cartridges (see Section 3.3). As the protein is only adsorbed and not bound covalently to the support, care must be taken to ensure that the sample is not washed too quickly out of the system during the various sequencing reactions. Thus, adsorptive sequencers use gaseous delivery and/ or small pulses of liquid to supply the solubilizing reagents (PITC and TFA) and to reduce the exposure of the sample to liquid. Solvent washes are also kept low to minimize washout whilst still removing excess reagents and by-products.

2.1.2 Biphasic column technology

Hewlett-Packard developed an alternative strategy to subject the sample to the Edman chemistry. The sample is applied to a biphasic column, comprising reverse-phase material in one-half and ion exchange material in the other. The sample, which may be several millilitres of protein solution contaminated with high levels of buffer salts and/or detergents, is applied to the reverse-phase material and washed thoroughly with a polar solvent (e.g. 2% (v/v) aqueous TFA), causing inorganic ions to elute. The reverse-phase segment of the column (now containing the protein) is then attached to the ion exchange part and placed in the sequencer. Small organic impurities are washed to waste during the initial stages of sequencing, but the proteinaceous material becomes trapped at the interface between the two resins as the organic solvent washes encounter the sample and leach it from the reverse-phase resin.

2.1.3 Solid phase sequencing

Solid phase sequence analysis, pioneered by Richard Laursen (6), has generally not been implemented in core sequencing facilities. However, it has proven to be a particularly effective approach for the sequencing of extremely hydrophobic samples, such as integral membrane proteins, which tend to elute rapidly from instruments utilizing adsorptive procedures. In solid phase instruments the sample is attached covalently to the support matrix, usually a chemically modified PVDF membrane, activated with either phenyl diisothiocyanate (DITC) (*Protocol 1*) or arylamine (AA) (*Protocol 2*). Once attached covalently to the support membrane, the delivery of reagents and solvent washes can be greatly increased for efficient reaction and for thorough removal of excess reagents and by-products with minimal sample washout. The sample must be chemically clean prior to covalent attachment to the membrane, otherwise attachment may be severely compromised. Alternatively, following electroblotting onto a PVDF membrane after SDS–PAGE, the protein can be immobilized by cross-linking it to an overlying polymeric matrix, the membrane entrapment procedure (*Protocol 3*), which limits washout (7).

For methods of preparation for the protein samples used in *Protocols 1–3*, see Section 3.

Protocol 1. Solid phase attachment of protein to Sequelon-DITC membrane discs

Equipment and reagents

- Lyophilized protein sample for analysis[a]
- 0.2 M 4-methylmorpholine, 0.1% (w/v) SDS[b]
- Sonicator bath (optional, Jencons)
- Heating block at 56°C
- Sequelon-DITC membrane disc (PerSeptive Biosystems)
- 0.2 M 4-methylmorpholine, 50% (v/v) propan-2-ol

Method

1. Dissolve 10–1000 pmol of the protein sample in 35 μl 0.2 M 4-methylmorpholine, 0.1% SDS. Warm the solution to 56°C and sonicate it if necessary to aid solubilization.
2. Wet the membrane in 0.2 M 4-methylmorpholine, 50% propan-2-ol and place it in the cap of a microcentrifuge tube.
3. Apply the solubilized protein sample to the wetted membrane and allow it to dry at 56°C.[c]

4. Wet the membrane with 5 μl 0.2 M 4-methylmorpholine, 50% propan-2-ol and allow it to redry.[d]

[a] The sample must be free of contaminating primary amines and thiols which will seriously compromise coupling (see Section 3 for purification methods).
[b] The SDS concentration can be increased to 2% (w/v) for very hydrophobic proteins.
[c] Attachment to the membrane occurs via the N-terminus and lysine side chains.
[d] Following this coupling step, the membrane may be washed in 0.2 M 4-methylmorpholine, 50% propan-2-ol to remove excess salts and detergent.

Protocol 2. Solid phase attachment of protein to Sequelon-AA membrane discs

Equipment and reagents

- Lyophilized protein sample for analysis[a]
- 50% (v/v) aqueous acetonitrile
- Coupling buffer: 10 mg/ml 1-ethyl-3-(3-dimethylaminopropyl) carbodiimide (EDC) in 2-(morpholino)ethanesulfonic acid (MES) buffer pH 5[b]
- Sequelon-AA membrane disc (PerSeptive Biosystems)[c]
- Heating block at 56°C

Method

1. Dissolve 10–1000 pmol of the protein sample in 30 μl 50% acetonitrile.
2. Apply the solution to the Sequelon-AA membrane disc in 10 μl aliquots and allow it to dry at 56°C between additions.
3. Transfer the disc to room temperature and add 10 μl coupling buffer.
4. Allow it to dry at room temperature (about 30 min).[d]

[a] The sample must be free of contaminating organic acids (e.g. acetate), phosphate, and detergents (e.g. SDS) which will prevent attachment of the protein to the membrane (see Section 3 for purification methods).
[b] Supplied as part of the kit; the concentration of the MES buffer is unspecified.
[c] Sequelon-AA discs are supplied in a kit containing solid EDC and MES buffer.
[d] The protein attaches to the membrane covalently via the C-terminus and acidic side chains.

Protocol 3. Membrane entrapment procedure[a]

Equipment and reagents

- PVDF membrane onto which the protein sample has been deposited by spotting or by electroblotting
- Fine forceps
- 0.1% (v/v) PITC in ethyl acetate
- Heating block at 55°C
- 2% (v/v) triethylamine in 50% (v/v) aqueous methanol
- 0.1% (w/v) DITC in ethyl acetate
- Poly(allylamine) hydrochloride solution: 0.1% (w/v) poly(allylamine) hydrochloride (Aldrich) in 2% (v/v) triethylamine, 50% (v/v) aqueous methanol

Protocol 3. *Continued*

Method

1. Hold the PVDF membrane with fine forceps and pipette 5 μl 0.1% PITC in ethyl acetate onto each side. Allow it to air dry (15–20 sec).[b]
2. Place the PVDF membrane on the heating block at 55 °C and add 30 μl 2% triethylamine in 50% methanol. Allow it to dry (7–8 min).
3. Return the PVDF membrane to room temperature and add 5 μl 0.1% DITC in ethyl acetate to each side. Allow it to air dry (15–20 sec).[c]
4. Place the PVDF membrane at 55 °C and add 30 μl poly(allylamine) hydrochloride solution. Allow it to dry (5–6 min).[d]
5. Add 20 μl 2% triethylamine in 50% methanol and allow it to dry (10 min).
6. Wash the membrane thoroughly with methanol, then with water, and finally with methanol again. Allow it to dry.

[a] From ref. 7.
[b] A proportion of the primary amino groups are modified, allowing their later identification.
[c] The remaining amino groups are modified for cross-linking.
[d] The modified amino groups cross-link to the polymer, immobilizing the protein on the membrane.

2.1.4 Amino acid identification

Most automated sequencers use on-line HPLC to identify the modified amino acids recovered at each cycle of sequencing. In N-terminal sequencers the released amino acid is first converted to the stable PTH-derivative and then injected onto a reverse-phase C_{18} resin equilibrated with acetate buffer and eluted with an increasing gradient of acetonitrile. The exact buffer components and compositions vary between instruments, but most use essentially the same approach. The PTH-amino acids are generally detected at 269 nm, the absorbance maximum of the PTH moiety. Sensitivity is around 1 pmol using UV detection, but attempts have been made by various manufacturers to enhance this, e.g. by the use of diode array detectors (ratio of 269/293 nm; Beckman/Porton) or previous chromatogram subtraction (MilliGen; PE Applied Biosystems). The recent introduction of a capillary LC system (PE Applied Biosystems Procise cLC) improves sensitivity to 50–100 fmol. The use of radiolabelled samples or fluorescent derivatives may also enhance sensitivity, but these approaches have not been incorporated successfully into routine analysis. However, in the case of recombinant proteins which generally are available in relatively large quantities, sensitivity is rarely an issue and thus identification of the amino acids is facile.

Amino acid identification may be a problem in cases where the protein has been modified post-translationally. The modified residue may not be recovered from the reaction cartridge (e.g. in the case of highly charged phosphoamino acids, which are insoluble in the organic transfer solvent) or may display modified behaviour during HPLC and not be recognized (e.g. fast eluting glycosylated residues), leading to a 'blank' cycle in the sequencer run. Alternatively, an amino acid may be particularly prone to degradation during the Edman reaction and its presence not detected (e.g. cysteine, tryptophan). Cysteine can be positively identified if it is first modified to protect the labile side chain (e.g. by *S*-pyridylethylation; *Protocol 4*). Tryptophan can usually be identified at very low yield, but at times can be overlooked quite easily. Phosphoserine can be identified if it is converted to *S*-ethylcysteine prior to analysis (8), although alternative means of identification (e.g. utilizing MS) may be preferable.

In the case of recombinant proteins, a 'blank' cycle may not be a problem, since the flanking sequences can be used to confirm the identity of the expressed polypeptide. If absolute identification of the 'missing' residue is required, then analysis using MS may be more appropriate. An accurate mass determined by MS may be sufficient to identify a modification of the protein, by comparison to the mass predicted from the DNA sequence, or tandem MS sequencing of a peptide containing the modified amino acid may be required for absolute confirmation (see Section 6.1.3).

Protocol 4. Modification of cysteine using 4-vinyl-pyridine[a]

Equipment and reagents

- Lyophilized protein sample for analysis
- 0.2 M Tris–HCl pH 8.5, 6 M guanidinium chloride
- 0.5 M DTT
- 4-Vinyl-pyridine (Aldrich)[b]

Method

1. Dissolve 10–1000 pmol of the protein sample in 50 μl 0.2 M Tris–HCl pH 8.5, 6 M guanidinium chloride.
2. Add 0.5 M DTT to a final concentration of 10 mM. Incubate the mixture for 1 h at 37 °C.
3. Add 2 μl 4-vinyl-pyridine. Incubate for 1 h at 37 °C.
4. Desalt the sample by appropriate means.[c]

[a] From ref. 9.
[b] 4-Vinyl-pyridine is unstable. Use recently purchased reagent that is stored under argon at –20 °C.
[c] Gel filtration or solvent precipitation are suitable methods. Alternatively the sample may be transferred to a PVDF membrane using ProSpin or ProSorb cartridges (see Section 3.3).

3. Sample preparation

For efficient sequence analysis, protein samples must be extremely clean. Substances which compromise the Edman chemistry (e.g. primary amines and thiols) or solid phase coupling procedures such as the attachment of protein to DITC membranes (e.g. primary amines and thiols) or AA membranes (e.g. organic acids and anionic detergents) must be avoided. Other compounds can cause machine problems, e.g. detergents may froth and block reagent delivery lines, or prevent recovery of the cleaved amino acid from the reaction chamber. Many of the potential problems can be avoided by using PVDF membrane (e.g. for electroblotting from SDS gels) which binds proteins very strongly and consequently can be exhaustively washed after protein adsorption to remove contaminants. Typically, SDS–PAGE or HPLC are used for the final stages of sample purification.

3.1 Sample preparation by SDS–PAGE

Polyacrylamide gel electrophoresis in the presence of the surfactant SDS (SDS–PAGE) offers a simple route to the purification of a specific protein from rather impure samples. As long as the protein of interest can be identified against the background of protein bands present following gel separation of the protein components (e.g. based on molecular weight), it is relatively straightforward to use electroelution (followed by dialysis) or preferably electroblotting to generate a pure sample for sequence analysis. Generally, the overexpression of recombinant proteins allows their facile identification on a gel, but care must be taken that other cellular proteins are not overexpressed in response to the presence of the recombinant protein. Experience has often demonstrated that the major polypeptide produced during expression of a recombinant protein may not be the desired protein, emphasizing the absolute necessity for ensuring the product is what is expected.

Protocol 5. Electroblotting from SDS–PAGE gels

Equipment and reagents

- 3MM chromatography paper (Whatman)
- 10 mM 3-(cyclohexylamino)-1-propanesulfonic acid (CAPS) pH 11, 10% (v/v) methanol
- PVDF transfer membrane (preferably sequencer grade, reduced pore size)
- SDS–PAGE gel on which the protein of interest has been separated
- Semi-dry electroblotter (Hoefer Semiphor)
- Power supply (capable of very low voltage, moderate current; Pharmacia EPS600)[a]
- Staining solution: 0.005% (w/v) sulforhodamine B (Sigma) in 30% (v/v) aqueous methanol, 0.2% (v/v) acetic acid

Method

1. Soak two sets of two pieces of 3MM chromatography paper in 10 mM CAPS, 10% methanol.
2. Rinse the PVDF membrane in methanol, then soak it in 10 mM CAPS, 10% methanol.[b]

3. Remove the SDS–PAGE gel from the electrophoresis plates, slice off the stacking gel and discard it. Cut off one corner from the separating gel to allow its orientation to be maintained in subsequent manipulations. Rinse the gel in 10 mM CAPS, 10% methanol.
4. Build up a sandwich comprising two pieces of wet 3MM chromatography paper (step 1), the PVDF membrane (step 2), the gel, and two further pieces of wet 3MM chromatography paper.
5. Place the sandwich on the electroblotter such that the PVDF membrane lies between the gel and the anode.
6. Assemble the blotter and blot the gel at 1–1.5 mA/cm^2 for 1–2 h.[c]
7. Stain the gel in staining solution after blotting to establish the efficiency of protein transfer.[d]
8. Wash the PVDF blot extensively in water and dry it thoroughly.[e]
9. Soak the blot in the staining solution until the protein bands are visible.[f] Rinse the blot in water and allow it to dry. Store the membrane in a sealed bag at –20 °C for subsequent analysis (*Protocol 3*).

[a] There is an absolute requirement for a low voltage power supply to achieve this. Many power supplies have a minimal output of a few volts (rather than zero volts) and are incapable of operating in these conditions.
[b] Ensure the membrane wets thoroughly. Any area which dries out will not allow the protein to be transferred to it.
[c] Scorching of the blot may occur after longer blotting times as the buffer becomes exhausted.
[d] Protein transfer is usually effective, but high molecular weight proteins and hydrophobic proteins may not always transfer quantitatively, hence it is advisable to check by staining the gel after blotting has been carried out.
[e] This is absolutely critical for the staining procedure in the next step.
[f] Pink bands appear almost immediately against a white background (as long as the blot has been washed and dried thoroughly beforehand).

Electroblotting is the method of choice for preparing protein samples for sequence analysis, as it involves transfer of the protein from the gel to a support matrix which can be placed directly into the sequencer (*Protocol 5*). Several membranes are available specifically for this purpose, based on the generic PVDF membrane, but comprising double or triple layers with reduced pore size to eliminate the over-transfer of protein through the membrane. These membranes are marketed by various companies: PE Applied Biosystems (ProBlott), Millipore (Immobilon-P^{SQ}), and Pall (Fluorotrans). They adsorb protein by hydrophobic interaction, so instrument cycle conditions must be chosen to minimize sample washout. They are suited to the mild washing conditions of adsorptive sequencers, but can also be used in solid phase instruments following cross-linking of the protein to an overlying polyamino compound as in the membrane entrapment procedure described in *Protocol 3*.

3.2 Sample preparation by HPLC

Following conventional purification, or as an integral part of the purification procedure, a final reverse-phase HPLC 'polishing' step offers a simple approach to the production of a sequencer grade sample, effectively desalting the sample into a volatile solvent system. The type of resin and the column dimensions required will depend on the amount, concentration, and properties of the protein or peptide to be purified. For many peptides, a resin with C_{18} functionality is preferred, although for larger and more hydrophobic peptides, a C_8 or even a C_4 resin may be more suitable. For proteins, a C_4 or phenyl-resin may be particularly applicable. An analytical column (3.9–4.6 mm i.d.) is typical for the majority of applications, but for small amounts of sample it may be preferable to resort to narrow-bore (2.1 mm i.d.), microbore (1 mm i.d.), or even capillary ($<$ 1 mm i.d.) columns. The most important consideration for sample clean-up is the choice of the mobile phase. A gradient elution system based on water and acetonitrile and incorporating a volatile ion-pairing agent (usually TFA) is generally successful. For very hydrophobic proteins or peptides, the organic phase can be supplemented with propan-2-ol to reduce the interaction between sample and column and allow easier elution.

3.3 Other procedures for sample preparation

Many recombinant protein samples are most readily produced by conventional chromatographic procedures, but as a result are frequently generated in solutions containing high concentrations of buffers, salts, or other compounds which may interfere with sequence analysis, particularly once they are concentrated into small volumes for loading onto the instrument. It is often necessary, therefore, to clean-up pure samples which may not be particularly suited to purification by SDS–PAGE or HPLC. In these cases, one of the three methods below may be used successfully.

3.3.1 Use of ProSpin cartridges

This useful procedure involves the low speed centrifugation of samples through a disc of PVDF membrane inside a microcentrifuge tube (ProSpin cartridges, from PE Applied Biosystems) as described in *Protocol 6*. During centrifugation the protein adsorbs onto the PVDF membrane and subsequently can be decontaminated by passing further aliquots of water or aqueous solvent through the membrane. The disc containing the adsorbed protein can then be excised and placed directly into the sequencer. This approach is suitable for protein samples containing many common interfering compounds, such as high inorganic salt concentrations, detergents, thiol reagents, and glycerol. The only prerequisite is that the desired protein is sufficiently pure for sequencing (at least 80% of the total protein), since all proteins will bind to the PVDF membrane and hence be exposed to the subsequent sequencing chemistry.

Protocol 6. Sample preparation using ProSpin cartridges

Equipment and reagents

- ProSpin cartridge (PE Applied Biosystems)[a]
- Methanol
- Protein sample for analysis (in solution resulting from a purification procedure)
- 0.1% (v/v) aqueous TFA
- Low speed microcentrifuge (5000 *g*)
- Punch (PE Applied Biosystems)

Method

1. Wet the PVDF membrane of the ProSpin cartridge with 10 μl methanol.
2. Add 10–1000 pmol of the protein sample in 50–400 μl of solution. Dilute smaller volumes of the sample with 0.1% TFA to a final volume of 50–400 μl.
3. Centrifuge the cartridge at low speed (5000 *g*) until the membrane appears dry.[b]
4. Centrifuge an additional 200 μl 0.1% TFA through the cartridge to wash the PVDF membrane.
5. Allow the PVDF membrane to air dry and punch out the membrane from the cartridge for sequencing. Store the membrane in a microcentrifuge tube at –20 °C until analysed.

[a] The ProSpin cartridge comprises a microcentrifuge tube into which is inserted a sample reservoir containing a disc of PVDF (ProBlott) membrane.
[b] The protein adsorbs strongly to the membrane during this process. This may take more than 30 min if the sample has a high viscosity (e.g. high salt concentration or detergent).

3.3.2 Use of ProSorb cartridges

A similar result can be achieved even more simply using ProSorb cartridges (from PE Applied Biosystems) as described in *Protocol 7*. Rather than using centrifugation, the solution is drawn through a disc of ProBlott from beneath using an absorptive wick, such that the protein adheres to the PVDF.

Protocol 7. Sample preparation using ProSorb cartridges

Equipment and reagents

- ProSorb cartridge (PE Applied Biosystems)[a]
- Methanol
- Protein sample for analysis (in solution resulting from a purification procedure)
- 0.1% (v/v) aqueous TFA
- Punch (PE Applied Biosystems)

Protocol 7. *Continued*

Method

1. Wet the PVDF membrane of the ProSorb cartridge with 10 μl methanol.
2. Add 10–1000 pmol of the protein sample in 50-400 μl of solution. Dilute smaller volumes of the sample with 0.1% TFA to a final volume of 50–400 μl.
3. Draw the liquid through the PVDF by pressing the wick to the underside of the membrane.[b]
4. Draw through an additional 200 μl 0.1% TFA to wash the membrane.
5. Allow the PVDF membrane to air dry and punch out the membrane. Store the membrane in a microcentrifuge tube at –20 °C until analysed.

[a] The ProSorb cartridge comprises a polypropylene tube into which is inserted a sample reservoir containing a disc of PVDF membrane. An absorbent wick can be pressed to the underside of the PVDF disc.

[b] The sample is drawn through within a few minutes and adheres strongly to the PVDF membrane.

3.3.3 Use of capillary blotting

PE Applied Biosystems have recently launched the 173A MicroBlotter, which combines reverse-phase capillary liquid chromatography of compounds with direct elution onto a strip of PVDF membrane. The simplicity and sensitivity of this technology, eliminating any problems (particularly sample losses) which may originate from sample transfer between HPLC, fraction collector, and sequencer, suggest that it may become an important strategy for preparing samples for sequencing, particularly where amounts of available material are rather limited.

4. N-terminal blocking

The most common and aggravating problem encountered in protein sequencing is N-terminal modification (blocking of the free N-terminal amino group), which prevents reaction with PITC and thus precludes Edman degradation. However, compared with proteins purified from native systems, recombinant proteins are only infrequently not amenable to direct N-terminal sequence analysis. In the case of eukaryotic proteins expressed in bacterial systems, for example, the absence of the eukaryotic processing machinery which leads to N-terminal acetylation allows the recombinant protein to retain a free N-terminus and sequencing is not compromised. Determination of the N-terminal sequence of a protein may be critical, and since the method of expression and the isolation and purification procedures for proteins are so variable, N-terminal blockage may still be a significant problem. Thus, the predominant causes of blockage, and methods of overcoming the problem, are presented below:

(a) Proteins expressed in prokaryotes may retain the initiating formyl-methionine residue, thus preventing sequencing. Removal of the formyl group may be achieved by treating the sample with HCl in methanol. The concentration, temperature, and reaction time can be varied to effect unblocking, but typically overnight incubation at room temperature in 0.5 M HCl in anhydrous methanol is suitable (10).
(b) A rather less common blocking mechanism is the cyclization of an N-terminal glutamine, either *in vivo* or in acidic isolation conditions, to pyroglutamic acid. The removal of this amino acid derivative can be partially effected (at 10–15% efficiency) using the enzyme pyroglutamate aminopeptidase (EC 3.4.19.3) available from Boehringer Mannheim. This generates a free N-terminus which represents the second residue of the original protein (11).
(c) The most common modification is acylation, particularly acetylation, occurring in an estimated 80–90% of eukaryotic proteins. No procedure has yet been devised for the routine removal of these groups. Partial success has been reported in cases of acetyl-Ser or acetyl-Thr N-termini, where heating for extended periods in TFA vapour (e.g. 65 °C for 16 h) may be effective (12). However, exposure of peptide bonds to such harsh conditions frequently results in non-specific partial hydrolysis of the protein. A modification of this approach, using TFA:methanol (1:1, v/v), is reported to be more effective and also to minimize internal bond cleavage (13). Acylamino acid-releasing enzyme (acetylamino acid hydrolase; EC 3.4.19.1) available from Boehringer Mannheim can be used to remove the acetylamino acid from short peptides (< 20 residues), thereby allowing the remainder of the peptide to be sequenced. A simple application of this procedure to obtain sequence information is presented in *Protocol 8*. Basically, the protein is digested thoroughly with a protease (e.g. trypsin) and then the N-termini of the resultant peptides are irreversibly blocked by incubating the digest with succinic anhydride. Acylamino acid-releasing enzyme is then used to remove the acetyl group from the N-terminal peptide, allowing it to be sequenced without any requirement for purification (14). However, the frequent use of bacterial systems for the overexpression of recombinant eukaryotic proteins circumvents the problem of blockage by eliminating the exposure of the nascent polypeptide chain to the acetylation machinery normally encountered in eukaryotic cells.
(d) The removal of other acyl groups, e.g. myristoyl groups, is also difficult. The purification of an enzyme capable of achieving demyristoylation of peptides has been described (15). A peptide *N*-fatty acylase is now commercially available (Wako BioProducts).

Protocol 8. Sequence analysis of N-terminally acetylated protein[a]

Equipment and reagents

- Protein for analysis (e.g. separated in an SDS–PAGE gel)
- Lyophilizer (Heto-Holten Maxi Dry Plus vacuum concentration system)
- Microcentrifuge tubes (1.5 ml)
- Succinic anhydride
- 12% (v/v) aqueous triethylamine
- 20% (v/v) aqueous TFA
- Ether (diethyl ether)
- Dry nitrogen gas
- 0.2 M NaOH
- 50 mM NH_4HCO_3 pH 7, 1 mM EDTA, 2 mM $MgCl_2$
- Acylamino acid hydrolase (Boehringer Mannheim)
- 20% (v/v) acetic acid

Method

1. Thoroughly digest a known quantity of the protein with trypsin into short peptides (*Protocol 9*). Lyophilize the mixture of peptides generated in a 1.5 ml microcentrifuge tube.
2. Redissolve the peptides in 50 μl water.
3. Add solid succinic anhydride (5000 pmol/pmol protein) in small portions over the course of 1 h, mixing vigorously on a vortex stirrer. Maintain a pH of 9.5–10 by the addition of 12% triethylamine solution.
4. Leave the reaction mixture at pH 10 overnight, then acidify it to pH 2–2.5 with 20% TFA. Lyophilize the sample.
5. Extract the sample with 1 ml ether (six to ten times) to remove succinic acid and triethylamine salts.[b]
6. Dry off any residual ether from the residue (which will be invisible) with a stream of nitrogen.
7. Redissolve the sample in 500 μl water and adjust to pH 7 with 2–6 μl 0.2 M NaOH. Lyophilize the sample.
8. Redissolve the sample in 50 μl 50 mM NH_4HCO_3 pH 7, 1 mM EDTA, 2 mM $MgCl_2$.
9. Add acylamino acid hydrolase to an enzyme:substrate molar ratio of 1:80. Incubate the mixture at 37 °C for 12 h.
10. Heat the mixture at 100 °C for 5 min, acidify the mixture with 20% acetic acid, and lyophilize the sample.
11. Sequence the digestion mixture containing the deacetylated N-terminal peptide to determine its sequence.

[a] From ref. 14.
[b] This volume of ether is suitable if 1.5 ml microcentrifuge tubes are used. Adjust the volume of ether accordingly if other tubes are used.

4.1 Protein fragmentation

In most situations, N-terminal blockage cannot be easily overcome, thus it is necessary to resort to the production of peptides by chemical cleavage or enzymic digestion of the protein. The purified peptides can then be used to generate internal sequence information. This is sufficient to confirm the reading frame, but obviously does not confirm the integrity of the N-terminus. The conditions used for proteolysis in solution are infinitely variable (ratio of chemical reagent or enzyme to protein, temperature, time, presence of denaturants). So one must first decide on the extent of proteolysis required, e.g. partial digestion to generate large fragments for SDS–PAGE and electroblotting, or more thorough digestion to generate short peptides for HPLC purification or MS analysis. *Table 2* lists some of the common enzymes and chemicals used, together with their specificities.

If the protein has been purified to homogeneity, e.g. by chromatography, digestion can be achieved in solution, an approach which is generally very effective. More often, however, the protein of interest is not sufficiently pure and it is necessary to use SDS–PAGE to generate a band of the pure protein for digestion. The piece of gel containing the band can be excised for treatment, or the gel may be electroblotted and the piece of membrane treated.

It is possible to digest a protein within a piece of polyacrylamide gel in several ways. One approach involves placing the gel piece in a sample well of a second gel, together with an appropriate enzyme (e.g. endoproteinase Glu-C)

Table 2. Some common enzymes and chemical reagents for the fragmentation of proteins

Reagent	Specificity[a]	Reference(s)
A. Enzymes		
Endoproteinase Glu-C (*Staphylococcus aureus* V8 protease)	C-terminus of Glu/Asp	16, 17
Endoproteinase Lys-C	C-terminus of Lys	18
Endoproteinase Arg-C (Clostripain)	C-terminus of Arg	19, 20
Endoproteinase Asp-N	N-terminus of Asp/Cys acid	21
Trypsin	C-terminus of Lys/Arg	
Chymotrypsin	C-terminus of Phe/Tyr/Trp	
B. Chemical reagents		
Cyanogen bromide	C-terminus of Met	22
o-Iodosobenzoic acid	C-terminus of Trp	23, 24
Hydroxylamine	Between Asn–Gly	25
Formic acid	Between Asp–Pro	26

[a] The desired cleavage specificity can be achieved by varying parameters such as the substrate: reagent ratio, time, temperature, and buffer conditions. Many of the enzymes are able to tolerate moderate amounts of denaturants, including SDS and urea.

and allowing digestion to take place during the stacking process (27). Electrophoresis is usually halted for a period during the stacking phase to allow adequate digestion, then the run is continued to separate the resultant peptide fragments. More often, the gel piece is treated with reagent to generate small peptide fragments which can be eluted from the gel for subsequent analysis. In this *in situ* approach (see *Protocol 9* for a typical method using trypsin), the gel pieces are dehydrated by allowing them to dry in air, then infused with buffer containing the enzyme of choice (1:50–1:200 enzyme:protein weight ratio), and incubated for several hours at 30 °C. The resultant peptides are subsequently eluted with 60% (v/v) acetonitrile for fractionation by HPLC or analysis by MS. Variations on this theme have been published, so trial and error is recommended; each protein presents its own unique problems!

Protocol 9. Tryptic digestion of protein within acrylamide gel pieces

Equipment and reagents

- SDS–PAGE gel containing the proteins to be analysed
- Fixing and staining solution: 0.25% (w/v) Coomassie Brilliant Blue R250 (Sigma) in 45% (v/v) aqueous methanol, 7% (v/v) aqueous acetic acid
- Destaining solution: 25% (v/v) aqueous methanol, 7% (v/v) aqueous acetic acid
- Scalpel or razor blade
- Microcentrifuge tubes (1.5 ml and 0.5 ml)
- 0.2 M NH_4HCO_3 pH 8.9
- Parafilm
- 0.2 M NH_4HCO_3 pH 8.9, 0.02% (w/v) Tween 20
- Trypsin (modified sequencing grade; Boehringer Mannheim)
- 10% (v/v) aqueous TFA
- 60% (v/v) aqueous acetonitrile
- Lyophilizer (Heto-Holten Maxi Dry Plus vacuum concentration system)

Method

1. Briefly fix and stain the SDS–PAGE gel just sufficiently to detect the protein bands.[a] To do this, immerse the gel in the staining solution for 15–30 min, then wash it in the destaining solution for 30–60 min until protein bands are visible.
2. Cut out the desired bands from the gel using a scalpel or razor blade. Place each band in a separate 1.5 ml microcentrifuge tube.
3. Cover each of the gel pieces with 0.2 M NH_4HCO_3 pH 8.9. Incubate them at 30 °C for 20 min with occasional mixing.[b] Remove the liquid with a micropipettor and discard it. Repeat this washing twice more with fresh buffer.
4. Air dry the gel pieces on Parafilm for about 10 min until they are thin and curl up.
5. Rehydrate each of the gel pieces in 2–5 μl 0.2 M NH_4HCO_3 pH 8.9, 0.02% Tween 20, containing 0.1–1 μg trypsin.[c]
6. Reswell each gel piece to its original size with repeated additions of 2–5 μl 0.2 M NH_4HCO_3 pH 8.9, 0.02% Tween 20.

7. Cut each gel piece into 1 mm^3 pieces with a scalpel on Parafilm. Transfer the pieces to a 0.5 ml microcentrifuge tube and cover them with 0.2 M NH_4HCO_3 pH 8.9, 0.02% Tween 20.
8. Incubate the tubes at 30 °C for 3–24 h. Add more 0.2 M NH_4HCO_3 pH 8.9, 0.02% Tween 20 as necessary to ensure the pieces remain covered.
9. Add 0.1 vol. 10% TFA to stop the reaction.
10. Remove the liquid to a fresh 0.5 ml microcentrifuge tube using a micropipettor.
11. Wash the gel pieces at 30 °C with two 50 μl lots of 60% acetonitrile. Add these washes to the original (step 10).
12. Concentrate the combined extract by lyophilization to about 20 μl.
13. Either analyse the concentrated extract directly by MS or purify the peptides by HPLC and analyse them individually by MS or Edman degradation.

[a] Digestion is best achieved if the gel is minimally fixed and stained. Overstaining may reduce recoveries of peptides.
[b] This washing procedure removes detergent and excess stain from the gel pieces.
[c] Use a trypsin:protein weight ratio of 1:50–1:200, estimated from the intensity of the stained band.

In cases where a sample has been applied to the sequencer in electroblotted format, but turns out to be N-terminally blocked, the lack of additional sample may make it necessary to use this material for further analysis. The following procedures have all been used for *in situ* proteolysis on the membrane but produce relatively low amounts of cleavage:

(a) Cyanogen bromide (0.15 M) in 70% (v/v) formic acid overnight at room temperature in the dark (28).
(b) BNPS-skatole (1 μg/μl) in 75% (v/v) acetic acid for 1 h at 47 °C (29).
(c) Trypsin in 1% (v/v) hydrogenated Triton X-100, 10% (v/v) acetonitrile, 100 mM Tris–HCl pH 8, for 24 h at 37 °C (30).

Due to steric problems, enzymes are particularly poor at digesting proteins on membranes. Small peptides may be eluted from the membrane (e.g. with 60% (v/v) aqueous acetonitrile at 30 °C for 1 h) for purification by HPLC and subsequent sequencing, whereas large and hydrophobic peptides tend to remain bound to the membrane. The strong affinity of PVDF membrane for protein effectively precludes the elution from it of proteins or large peptides for further analysis.

However, the advent of peptide mass fingerprinting and tandem mass spectrometric sequencing has necessitated the further development of procedures for the routine digestion of proteins electroblotted onto PVDF membrane, e.g.

from two-dimensional gels. The general approach is to use a minimal volume (< 5 μl) of digestion buffer containing trypsin (1:50 relative to protein) and a detergent such as *n*-octyl-β-D-glucopyranoside (1%, w/v) to wet the membrane surface and 'release' the protein for subsequent cleavage. A cleavage and recovery efficiency of over 80% for a few pmol of starting material has been reported. *Protocol 10* describes a procedure for the digestion of proteins on PVDF membranes.

Protocol 10. Digestion of proteins on a PVDF membrane[a]

Equipment and reagents

- PVDF membrane with proteins of interest attached (e.g. from an electroblotting procedure)
- Microcentrifuge tubes (0.5 ml) or microtitre plates
- Trypsin (modified sequencing grade; Boehringer Mannheim)
- Digestion buffer: 0.2 M NH_4HCO_3 pH 8.9, 1% (w/v) *n*-octyl-β-D-glucopyranoside, 2% (w/w) trypsin (relative to protein)
- Formic acid:ethanol (1:1, v/v)
- Lyophilizer (Heto-Holten Maxi Dry Plus vacuum concentration system)

Method

1. Place the pieces of PVDF blot containing the protein in a 0.5 ml microcentrifuge tube or in a well of a microtitre plate.
2. Cover the pieces with 5 μl digestion buffer.[b] Incubate the mixture at 30 °C for from 3 h to overnight.
3. Remove the liquid to a fresh microcentrifuge tube or microtitre plate.
4. Extract the pieces of blot with two 10 μl lots of formic acid:ethanol to release the peptides from the membrane. Add these washes to the original extract (step 3).
5. Concentrate the extract by lyophilization if necessary.
6. Analyse the extract by MS or purify the peptides and sequence them as with gel extracted peptides (*Protocol 9*).

[a] Based on ref. 31.
[b] The detergent reduces the interaction between the protein and the membrane, allowing the trypsin to access the protein and digest it.

In addition to protein fragmentation for the determination of internal sequence information, proteolysis can be used under very mild conditions to provide information on the domain structure of the protein of interest. Protein domains are generally well-folded regions of the polypeptide chain, but are often separated by 'linker' regions of extended structure which may be prone to enzymic attack. Mild proteolysis can provide information relating to the correct folding of the protein and delineation of domains. Sequence analysis of the resulting fragments (separated by SDS–PAGE) pin-points the actual points of proteolytic attack.

5. C-terminal sequencing

The unambiguous determination of the C-terminal sequence of a protein has not proven to be as straightforward as N-terminal analysis. Frequently, the generation of N-terminal or internal peptide sequences using Edman degradation has been sufficient for the system under investigation, providing information to allow the design of oligonucleotide probes or confirmation of identity of expressed proteins, for example. Consequently, less emphasis has been paid to the analysis of the C-terminus of a protein. However, the elucidation of the C-terminal sequence is an important part of the structural characterization of a protein in a number of situations. Post-translational processing of a precursor polypeptide may be an important step in creating the biologically significant form of the protein. Identification of sites of C-terminal processing, coupled with functional analysis may suggest mechanisms of action of a protein. The ability to obtain some sequence information from the C-terminus of a protein may be important if the N-terminus is blocked and refractive to Edman degradation, since it may provide information confirming identity or allowing the generation of a suitable oligonucleotide probe for cloning approaches.

The rapid expansion in the generation of recombinant proteins for numerous therapeutic and other biotechnological applications has stimulated the further development of C-terminal sequencing techniques to aid the complete characterization of these proteins, a critical requirement prior to testing and release into the environment.

5.1 Chemical analysis of the C-terminus

The development and implementation of a routine chemical procedure to degrade proteins from the C-terminus eluded protein chemists for many years, although various approaches showed partial success, allowing the identification of certain residues in some proteins (for a review see ref. 32). Dedication eventually paid dividends, with the publication of methodology for the stepwise chemical removal and identification of all 20 amino acids, whilst leaving the protein in a condition suitable for continued sequential analysis (3, 4). This chemistry has been incorporated into a fully automated instrument by Hewlett-Packard. PE Applied Biosystems have developed an alternative chemistry, but this is halted by the presence of a proline residue. Although all 20 amino acids can now be identified, at least with the Hewlett-Packard system, the main problem relates to the amounts of material required and the extent of obtainable sequence. Currently, at least 100 pmol of protein are required and only five to ten residues can be obtained, significantly less well developed than the Edman N-terminal chemistry. Consequently, neither C-terminal procedure is sufficiently well-developed for high sensitivity protein sequencing applications, but both are very useful for the analysis of recombinant proteins, where amounts of material are not usually limited and only a few residues are required for the necessary quality controls.

The overall C-terminal reaction scheme utilized by Hewlett-Packard in the HP 241 combined N- and C-terminal sequencer involves three stages (see *Figure 2*). First, the protein is treated with TFA to generate the fully protonated carboxylic acid at the C-terminus. This then undergoes a coupling step with diphenyl phosphoroisothiocyanatidate, and cyclization in the presence of pyridine, forming the peptidyl thiohydantoin. The addition of TFA and water vapour allows the derivatization of proline (notoriously difficult), without affecting other amino acids. Cleavage of the thiohydantoin (TH)-amino acid with potassium trimethylsilanolate generates a shortened polypeptide in activated form for the subsequent cycle, as well as providing the TH-amino acid for identification by reverse-phase HPLC.

The samples are supplied to the sequencer adsorbed non-covalently onto porous Teflon (Zitex) membranes. PVDF membranes are not suitable for the basic conditions encountered during the cleavage step. Proteins and relatively short peptides are all suitable for analysis. Proteins can also be electroblotted directly onto the Zitex membranes, although this procedure is quite inefficient compared to using PVDF membranes. Samples prepared in this way have been analysed sequentially by both N- and C-terminal methods, generating sequence information from both ends of the same sample. However, N-terminal analysis must precede C-terminal analysis, as the conditions used in the latter lead to modification of the N-terminal amino group, preventing N-terminal analysis.

5.2 Enzymic analysis of the C-terminus

Prior to the successful implementation of fully-automated procedures for chemical C-terminal sequence analysis, only manual enzymic methods were both readily available and reasonably successful for the sequential degradation of proteins from the C-terminus. A number of carboxypeptidases with differing specificities could be used to determine the C-terminal sequence, either by amino acid analysis of aliquots of the reaction mixture taken at different time intervals or by MS analysis of the residual protein mixture (ladder sequencing).

In the case of recombinant proteins, the expected C-terminal sequence may dictate which particular carboxypeptidases can be used successfully to generate a C-terminal sequence. The known amino acid specificity of each enzyme allows the selection of an appropriate enzyme or mixture of enzymes for the stepwise liberation of amino acids from the C-terminal end of the protein. However, it is not uncommon for proteins to undergo C-terminal proteolytic

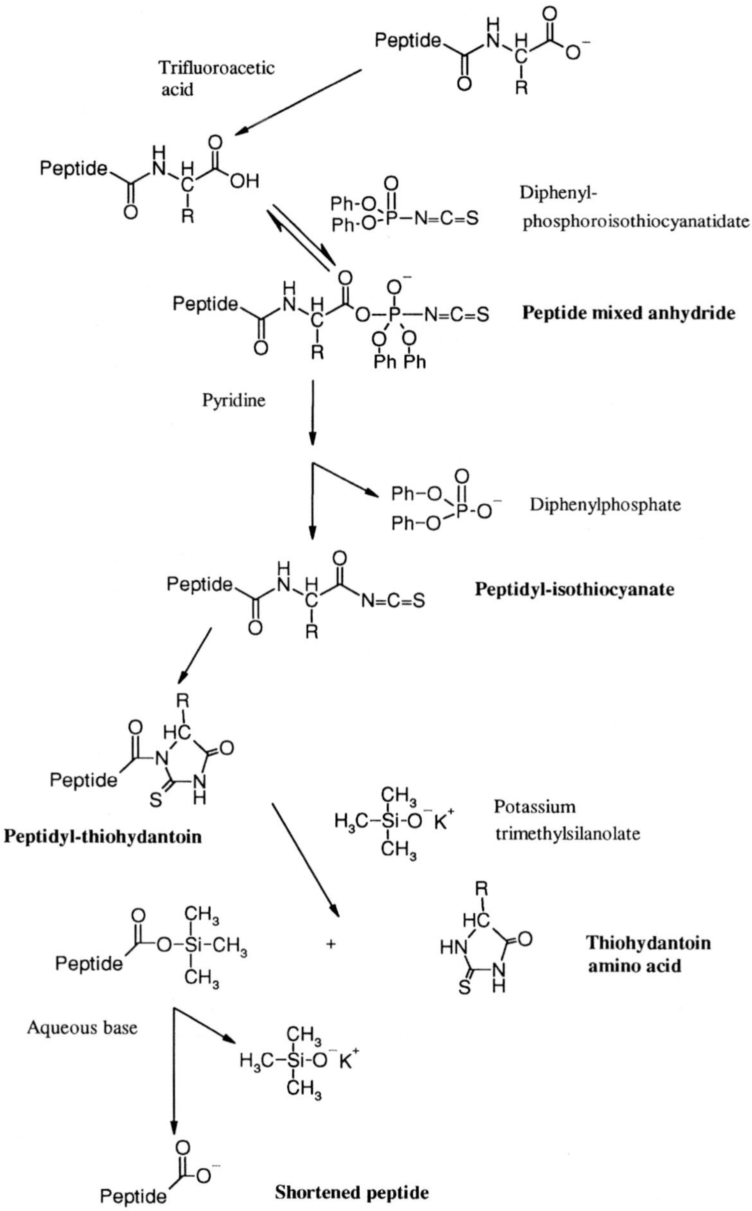

Figure 2. C-terminal sequencing chemistry. The polypeptide is treated with trifluoroacetic acid (TFA) to generate a protonated C-terminal carboxylic acid. Reaction of the C-terminus with diphenyl phosphoroisothiocyanatidate followed by pyridine results in the formation of the peptidyl-thiohydantoin. TFA is then applied to stabilize any peptidyl-thiohydantoin formed from C-terminal proline. Treatment with potassium trimethyl-silanolate generates a shortened peptide for the subsequent round of sequencing and a thiohydantoin-amino acid which can be identified by reverse-phase HPLC. (Ph = phenyl)

events, leading to an unexpected sequence, or worse a 'ragged' C-terminus. In most cases, carboxypeptidase Y from yeast is appropriate, working on all amino acids (including proline) with similar activity (33). Generally, the amounts of recombinant protein normally available are ample (tens to hundreds of micrograms) and the results generated are unambiguous. However, carboxypeptidases can yield rather complex and variable results, due to the dependence of the reaction kinetics on the identity of the amino acid being acted upon and the inefficiency of the exopeptidase activity. Some endopeptidase activity also occurs, resulting in several fragments which are digested to generate alternative C-terminal sequences. This is further complicated if the true protein C-terminus is 'ragged' due to post-translational processing or degradation. Thus, results must be interpreted with caution and preferably confirmed by complementary approaches.

In the approach based on amino acid analysis, aliquots of digested material are removed from the reaction mixture at appropriate time intervals (usually over the course of several minutes to a few hours), the target protein and enzyme are precipitated by treatment with acid (usually acetic or TCA), and the amino acid composition of the supernatant is determined. The appearance (and subsequent disappearance) of particular amino acids during the time course of the experiment is interpreted to provide the C-terminal sequence of the protein. This is described in *Protocol 11*.

Protocol 11. C-terminal analysis by cleavage with carboxypeptidases

Equipment and reagents

- Peptide or protein for analysis[a]
- Digestion buffer: 0.1 M pyridine–acetate buffer pH 5.5
- Carboxypeptidase Y (Sigma): make an appropriate stock solution in water (see step 2)
- Ice-cold 20% (v/v) acetic acid
- Lyophilizer (Heto-Holten Maxi Dry Plus vacuum concentration system)
- Access to an amino acid analyser

Method

1. Dissolve the protein (usually hundreds of pmol to tens of nmol) in 200 μl digestion buffer.
2. Add carboxypeptidase Y to the protein sample (1% to 5% molar ratio). Incubate the mixture at 25 °C.
3. Remove 20 μl samples at intervals over the time course of the digestion. Take time points within the first few minutes and over a longer period, e.g. 0, 1, 2, 5, 10, 20, 30, 60, 120 min.
4. Immediately precipitate the protein in each sample by adding 5 μl 20% acetic acid. After 10 min centrifuge the sample at 13 000 *g* to pellet the protein.

5. Remove and lyophilize the supernatants. Analyse them for amino acid content using an amino acid analyser.

[a] Although digestion proceeds rapidly with peptides, protein conformation may severely restrict the effectiveness of this approach with proteins, necessitating prior denaturation and reduction plus alkylation of the protein. This may severely compromise sample solubility, requiring supplementation of the digestion buffer (step 1) with up to 5 M urea, which itself may interfere with the subsequent amino acid analysis.

When using the MS approach (see Section 6.3.2 and *Protocol 15*), the ladder of sequentially shorter proteins generated over the time course of C-terminal analysis is interpreted by examining the mass differences between each component. Owing to the high mass accuracy required for proteins, electrospray-MS or delayed extraction MALDITOF-MS is essential. Elucidation of the identity of the amino acid corresponding to each mass difference provides the sequence.

Alternatively, after thorough proteolytic fragmentation of the protein, the C-terminal peptide can be specifically isolated and subjected to N-terminal sequencing in order to elucidate the C-terminus of the intact protein. This can be achieved effectively if the recombinant protein has been engineered to contain a C-terminal affinity tag, thus enabling single step purification of the required peptide.

6. Mass spectrometric analysis of proteins

Mass spectrometry has played a critical role in the characterization of recombinant proteins for several years (34), providing an accurate molecular weight to enable the recognition of post-translational modification of the expected gene product and allowing the characterization of any modifications, in addition to providing direct sequence information. Precise characterization of the product and its consistency from batch to batch are of prime importance, and the accuracy, reproducibility, and ability to detect minor impurities associated with MS have ensured its widespread usage.

6.1 A brief guide to mass spectrometers

Mass spectrometers can be divided simply into three basic components: namely the ionization source into which the sample is introduced and ionized, the analyser where the ions are separated according to their mass-to-charge (m/z) ratios, and finally the detector where the signal from the separated ions is detected, amplified, and sent to a data system for processing and printing. The variety and complexity of these instruments merits a brief introduction to the ionization methods and analysers available.

6.1.1 Ionization techniques

The reader is referred to ref. 35 for detailed information. Currently the two most popular techniques used to ionize samples of virtually any type, but particularly biomolecules, are electrospray (ES) (36) and matrix-assisted laser desorption ionization (MALDI) (37).

ES is a liquid inlet system whereby samples are introduced in solution into the ionization source of the mass spectrometer. ES is thus a useful technique for analysing proteins in their native forms, for on-line reaction monitoring, and for direct coupling to HPLC or capillary electrophoresis (CE). Ionization of the solute and solvent takes place in solution and is brought about by the application of a high voltage (*c.* 3–4 kV) to the tip of the capillary through which these liquids are spraying. The majority of the solvent is pumped away at this stage, leaving the sample ions to be extracted into the analyser of the mass spectrometer. Samples with molecular weights in excess of *c.* 1200 Da form multiply charged ions, thus appearing at relatively low m/z values in the mass spectrum, and so ES can be coupled to analysers with standard (i.e. $< m/z$ 3000) mass-to-charge ranges such as quadrupoles. The associated technique of nanospray, or nano-electrospray (38) is frequently used when sample quantities are limited, as this enables high quality mass spectra to be obtained from small volumes of samples.

MALDI on the other hand requires the sample to be crystallized first with a vast excess of an appropriate chemical matrix on a specially designed target, and then the combined solid mixture to be inserted into the ionization source of the mass spectrometer where it is bombarded and ionized by a laser. The matrix transforms the energy delivered from the laser into excitation energy for the sample molecules, and in this way the sample is ionized but spared any excessive energy which may otherwise lead to decomposition. MALDI produces predominantly protonated $(M + H)^+$ or deprotonated $(M - H)^-$ singly charged molecular ions, with little fragmentation. Hence, a mass spectrometer with a large mass range analyser, such as a time-of-flight (TOF) instrument, is required for biomolecular analyses. MALDI is more tolerant than ES to the presence of involatile components such as buffers and salts.

6.1.2 Analysers

The analyser refers to the portion of the mass spectrometer that is responsible for separating the ions formed in the ionization source. It is generally of the quadrupole or time-of-flight type, or less frequently, of magnetic sector or ion trap (either Fourier transform-ion cyclotron resonance (FT-ICR) or quadrupole ion trap) technology. Each one of these resolves, or separates, ions on the basis of their m/z ratios. All mass spectrometers generate molecular weight information for samples.

The most popular mass analysers in biomolecular laboratories are of the quadrupole and TOF type due to their relative ease of use, comparatively low cost and, perhaps above all, their ability to perform well and produce high quality, meaningful, reproducible data for the majority of biomolecules. However, Fourier transform technology is gaining in usage, and should make a greater contribution in the future.

6.1.3 Tandem mass spectrometry

Whereas MS generates predominantly molecular weight information, tandem mass spectrometry (MS-MS) generates useful information regarding the structure, or sequence in the case of proteins, of the sample under investigation. In order to carry out authentic MS-MS analyses, the mass spectrometer should be of the multisector type, comprising more than one analyser, or of the ion trap type where individual ions can be selected uniquely. In multisector mass spectrometers, the analysers can be of the same or different types.

Mass analysers, single or multiple, can be coupled to different ionization methods. *Table 3* shows a summary of the more popular commercial instruments together with their capabilities. As this chapter is primarily concerned with proteins and associated biomolecules, *Table 3* has been limited to include ES and/or MALDI ionization methods, these being by far the most effective methods for biomolecular analyses. There are other, earlier ionization techniques still in use such as fast atom bombardment ionization, liquid secondary ion mass spectrometry, and field desorption/ionization.

For MS-MS studies, the tandem quadrupole is the most commonly used instrument. Large mass spectrometers of the magnet-quadrupole and four-sector types have declined in popularity due to the manufacturing expense involved in the production of magnetic sector instruments and also the higher degree of complexity required for their operation and usage. Quadrupole-

Table 3. Specifications of typical commercially available mass spectrometers

Mass analyser	**Standard ionization methods**	**MS capability**	**MS-MS capability**	**Cost**
Quadrupole (Q)	Electrospray	Yes	No	Low
Tandem quadrupole (QQ)	Electrospray	Yes	Yes	Moderate
Time-of-flight (TOF)	MALDI/electrospray	Yes	No	Low–moderate
Quadrupole-time-of-flight (Q-TOF)	Electrospray	Yes	Yes	High
Magnet (double focusing) (B, BE, EB)	Electrospray	Yes	No	High
Magnet/magnet (4 sector)	Electrospray	Yes	Yes	Very high
Magnet/quadrupole (M/Q)	Electrospray	Yes	Yes	High
Quadrupole ion trap	Electrospray/MALDI	Yes	Yes	Low
Fourier transform -ion cyclotron resonance (FT-ICR)	Electrospray/MALDI	Yes	Yes	Very high

orthogonal time-of-flight (Q-TOF) (39) analysers are generating interest not only due to their novelty, but also more deservedly to their high resolution capabilities, good sensitivity, and ease of use.

Structural studies can be performed using a number of MS-MS experiments, which are listed in *Table 4*. The Q-TOF type of instrument is used solely in the product ion scanning mode. The ion trap type of mass spectrometer, including both quadrupole ion traps and FT-ICR mass spectrometers, can perform not only MS-MS product ion scanning experiments, but further studies described as MS^n experiments. For the MS-MS product ion studies, the ions of interest (usually the molecular-related ions) are selected uniquely in the ion trap and encouraged to fragment; these fragment ions are then detected. If any of the fragment ions are of particular interest, these can be selected in turn, and the fragment ions from these detected. This second generation of fragment ions represents an MS-MS-MS study, or an MS^3 experiment.

Other methods for generating and monitoring fragmentation processes, which are not strictly MS-MS techniques, are also widely used. Although these rely on a single analyser, and hence are not as specific as MS-MS experiments, they can produce useful structural information.

Amongst these techniques are ES ionization in-source fragmentation, during which ions are encouraged to fragment in the ionization source soon after their formation by the input of extra energy. The fragments produced tend to be of the same nature as would be expected from low energy MS-MS product ion scanning experiments, the difference being that as this experiment is not selective, then all the ions within the ionization source will frag-

Table 4. Types of tandem mass spectrometric experiments

	1st analyser	**Collision cell**	**2nd analyser**	**Uses**
Product (daughter) ion scan	Selects ions of interest	Fragmentation of selected ions	Analysis of fragment ions	Structural elucidation, e.g. peptide sequencing
Precursor (parent) ion scan	Analyses all ions	Fragmentation of all ions	Selects ions of interest	Detection of groups of compounds with common fragment ions, e.g. glycopeptide mapping
Constant neutral loss	Analyses all ions	Fragmentation of all ions	Analyses ions resulting from a specified mass loss	Detection of groups of compounds that fragment by the same mass loss
Multiple reaction monitoring	Selects ions of interest	Fragmentation of selected ions	Selects ions of interest	Highly specific monitoring of a known compound at very low detection levels

ment, not just the ions of interest. However, if ES is coupled to on-line liquid chromatography, then there is a good chance that individual components of a mixture are separated and hence enter the source individually.

Another method of fragmentation which has been used successfully is the recently introduced post-source decay technique (40), used with MALDITOF-MS, where specific ions exiting the source are selected and given time to fragment. The resulting metastable ions are analysed and detected, and have been found to correspond to low energy MS-MS product ions.

6.2 Molecular mass determination

The primary objective for analysing samples by MS is to obtain a molecular weight for the sample under investigation. The molecular weight determination will in most cases confirm the existence of an expected peptide, protein, or glycoprotein. It will also give an indication of the purity of the sample, because any coexisting compounds will also be analysed and their molecular weights determined simultaneously. The number of disulfide bonds can be verified by molecular weight before and after reduction with DTT, as the change from disulfide bridge to dithiol increases the molecular weight by 2 Da for each bridge broken.

The accuracy of the molecular weight determination will vary from one type of mass spectrometer and one particular ionization method to another. ES coupled to a quadrupole mass analyser, for example, should provide a molecular weight accuracy of 0.01% of the mass for samples up to *c.* 30 kDa. Therefore, a protein of molecular weight 20 kDa should be confirmed to within 2 Da using this type of MS analysis. MALDI, coupled with a TOF analyser and incorporating a reflectron lens and delayed extraction, provides a similar accuracy. This high degree of accuracy, unique to MS, will highlight minor changes in a protein, such as the replacement of one amino acid for another, or a post-translational modification. It is the first step in protein identification; if the molecular weight is incorrect then the amino acid sequence almost certainly is not as expected.

For samples of molecular weight less than 1200 Da, a modern mass spectrometer should be able to measure the mass to within 0.1 Da, or much more accurately if calibrated carefully with an internal calibrant (41).

6.2.1 Electrospray sample analysis

To perform a molecular weight analysis by ES, the sample should be in solution, ideally at a concentration between 1–30 pmol/μl, in an appropriate solvent. Acceptable solvents can be formulated from the following: water, acetonitrile, methanol, propan-2-ol, chloroform, and any miscible mixtures thereof. Less volatile solvents such as DMSO and *N,N*-dimethylformamide should be avoided. Less stable ones such as tetrahydrofuran should be treated with caution.

Often it is necessary to change the pH of the mobile phase to help the ionization process. Most proteins and peptides produce the most unambiguous and sensitive results when analysed by positive ionization MS and the protonation process involved is aided by the addition of a trace of a suitable acid to the solvent. Formic acid and acetic acid are best used at concentrations up to *c.* 2% (v/v), although a concentration of between 0.05–1% (v/v) is usually sufficient to promote the production of protonated ions. TFA can be used, although if the concentration in the solvent system is higher than 0.1% (v/v), the MS sensitivity will be reduced considerably. Inorganic acids such as sulfuric and phosphoric acids should be avoided. Only occasionally will a protein generate superior results by negative ionization analysis. In such cases, it is more beneficial to raise the pH of the solution by the addition of a small percentage (less than 1%, v/v) of a volatile amine, such as aqueous ammonia solution or triethylamine.

Great care should be taken to ensure that the sample is as free as possible from involatile buffers (including Tris, CHAPS, Hepes, citrate, borate, sulfate, phosphate, and perchlorate), inorganic cations (Na^+, K^+), detergents, and other additives. The involatile components will be deposited on the surface of the ionization source quickly leading to significant losses in sensitivity. Inorganic salts tend to generate sample–adduct ions which complicate the spectrum and obscure sample ions arising from other components in the mixture. Most additives lead to reduced sensitivity and many produce ions which preclude sample ions and dominate the mass spectrum. Surfactants—be they anionic, cationic, zwitterionic, or non-ionic—will preclude any protein signal when present at a level of 1% (42). At a level of 0.01%, the protein signal may be present albeit at reduced sensitivity, but not in the cases of such ionic detergents as SDS and sodium taurocholate. Buffers which can be used successfully are volatile ones such as ammonium acetate, ammonium hydrogen carbonate, and ammonium carbonate. These can all be used at concentrations up to 100 mM, in the presence of the volatile organic acids listed above if necessary.

It is of great importance to ensure that the sample and any additives are fully dissolved, as any particulate matter will block the narrow capillaries used to transport the samples to the ionization source, thus impeding any sample analyses.

The preparation of samples for ES-MS is described in *Protocol 12*. With the sample in an appropriate solution, the ES analysis can then be carried out. Standard ES is the most common method of analysing proteins, and the sample can either be infused into the ES ionization source at a flow rate between 2–300 μl/min, or a discrete injection (10–100 μl) can be made into a solvent system flowing between 2–300 μl/min into the ionization source. In the latter case, the solvent system should preferably be the same as the solvent used for the analysis, or at least if different the two should be totally miscible.

Protocol 12. Preparation of samples for electrospray mass spectrometry

Equipment and reagents

- Protein or peptide sample for analysis, free from salts, detergents, and surfactants (see text)
- HPLC grade solvents, preferably degassed (e.g. water, acetonitrile, methanol, propan-2-ol)
- Formic acid (BDH, Analar grade)
- 1:1 mixture (v/v) of 0.1% (v/v) aqueous formic acid:acetonitrile
- 1:1 mixture (v/v) of 0.1% (v/v) aqueous formic acid:methanol
- Triethylamine (BDH, Analar grade)
- 1:1 mixture (v/v) of 0.1% (v/v) aqueous triethylamine:propan-2-ol
- Access to ES-MS

Method

1. Dissolve the protein in water to a concentration of 100–500 pmol/μl.
2. (a) For positive ionization analyses,[a] dilute this solution with 0.1% aqueous formic acid:acetonitrile or with 0.1% aqueous formic acid:methanol, depending on the solubility characteristics of the protein, to a concentration of 1–30 pmol protein/μl.[b]
   (b) For negative ionization analyses,[c] dilute the aqueous protein solution with 0.1% aqueous triethylamine:propan-2-ol to a concentration of 1–30 pmol protein/μl.[b]
3. Mix the final solution thoroughly using a vortex mixer.
4. Centrifuge at 13 000 *g* to remove any remaining insoluble material.
5. Analyse the supernatant solution by ES-MS (see the text following this protocol).

[a] Positive ionization mode can be used for the majority of proteins and peptides.
[b] A minimum of 1 μl of this final solution is required, but preferably 10–100 μl.
[c] Negative ionization mode can be used for the occasional protein or peptide whose analysis is unsuccessful by positive ionization.

If nano-electrospray is to be used, an aliquot (1–10 μl) of the sample solution is used to load a small glass sample vial, taking great care to avoid trapping any air bubbles or damaging the delicate vial tip. The use of methanol as the organic portion of the solvent mixture is better than acetonitrile for sustaining a steady flow of solvent into the ionization source, and sample concentrations of less than 10 pmol/μl usually produce the optimum results. Good data have been obtained with sample concentrations in the low fmol range, depending on the sample and its purity.

Some sample purification can be accomplished on-line to the ionization source using HPLC techniques. As ES requires a constant flow of solvent into the source, it is an ideal technique for coupling directly to HPLC. Columns of any size can be coupled. Flow rates of 300 μl/min or less are directed entirely into the source, while higher flow rates can be split prior to entering the mass spectrometer. Desalting cartridges (such as the ones supplied by Jones Chromatography) which allow the sample to be washed and the washings directed to waste before eluting the peptide or protein into the mass spectrometer, are particularly useful for 'salty' protein samples.

After analysis a mass spectrum will be generated, which in the case of ES will show a series of multiply charged ions all relating directly to the intact protein, but with varying numbers of protons attached (in positive ionization mode). The data are processed automatically or semi-automatically and a molecular weight profile is generated from which the molecular weight of the sample(s) can be read directly and unambiguously. An estimate of the purity and composition of the sample can be made.

6.2.2 Matrix-assisted laser desorption ionization analysis

MALDI requires the protein to be crystallized on an MS target with a vast excess of a suitable, light absorbing matrix (43, 44). A good, standard method for preparing samples is given in *Protocol 13*, but other methods are also available (45).

After sample preparation, the target on which the dried mixture has been applied is inserted into the ionization source of the mass spectrometer and bombarded with a laser beam. The matrix transforms the laser energy into excitation energy for the sample and also enhances sample ion formation by participating in the proton transfer mechanism. An appropriate matrix for proteins, especially those with molecular weights above 20 kDa, is 3,5-dimethoxy-4-hydroxycinnamic acid (sinapinic acid). Peptides are usually successfully analysed with this same matrix, or alternatives such as α-cyano-4-hydroxycinnamic acid, 2,5-dihydroxybenzoic acid, or 6,7-dihydroxycoumarin.

Protocol 13. Sample analysis by matrix-assisted laser desorption mass spectrometry

Equipment and reagents

- Protein or peptide sample for analysis
- 0.1% (v/v) aqueous TFA
- Protein sample buffer[a] (for protein samples): 10 mg/ml 3,5-dimethoxy-4-hydroxycinnamic acid in 7:3 mixture (v/v) of 0.1% (v/v) aqueous TFA:acetonitrile
- Peptide sample buffer[a] (for peptide samples): 10 mg/ml α-cyano-4-hydroxycinnamic acid in 1:1 mixture (v/v) of 0.1% (v/v) aqueous TFA:acetonitrile
- MALDI-MS sample target (for the instrument to which access is available)
- Access to MALDI-MS

Method

1. Dissolve the protein or peptide in water or 0.1% TFA at a concentration of 10–50 pmol/μl.[b]

2. Dilute an aliquot (1–2 μl) of the protein solution with an equal volume of protein sample buffer or peptide sample buffer.[c]
3. Mix thoroughly using a vortex mixer and then allow the saturated solution to settle.
4. Apply an aliquot (1–2 μl) of the final solution to the MALDI-MS sample target. Allow it to dry at room temperature for several minutes.
5. Analyse by MALDI-MS (positive ionization mode).

[a] Prepare fresh solution daily due to the light-sensitive nature of the matrix.
[b] Choice of water or 0.1% aqueous TFA (or other volatile solution) is determined by solubility of sample.
[c] The matrix should be in a molar ratio of at least 5000:1 to the sample.

The *m/z* mass spectrum obtained by MALDI generally depicts the singly charged, protonated molecular ions $(M + H)^+$ in positive ion mode or the deprotonated molecular ions $(M - H)^-$ in negative ionization mode. These ions may be accompanied by weaker doubly charged ions $(M + 2H)^{2+}$ and/or dimeric ions $(2M + H)^+$, as well as (sample + matrix) adducts, and also ions arising from any other components or contaminants present in the sample. The spectra are usually straightforward to interpret.

MALDI copes reasonably well, and invariably better than ES, with the contaminants often associated with protein and peptide samples, i.e. inorganic and organic salts, buffers, and detergents. For example, detergents can be tolerated at the 0.1% level, and alkali metal salts at levels as high as 1 M. *Table 5* summarizes the maximum amounts tolerated in MALDI analyses before the sensitivity and spectral integrity is compromised (35). More detailed lists have been compiled (46).

Table 5. Concentrations of common contaminants tolerated by MALDI protein analysis[a]

Contaminant	**Maximum concentration tolerated**[b]
Phosphate buffer	20 mM
Tris buffer	50 mM
Detergents[c]	0.1%
SDS	0.01%
Alkali metal salts	1 M
Glycerol	2% (w/v)
Ammonium hydrogen carbonate	30 mM
Guanidine	1 M
Sodium azide	1%

[a] From ref. 35.
[b] In many cases, potentially problematic concentrations of contaminants can be overcome simply by diluting the sample further.
[c] Such as Tween, Triton.

6.2.3 Electrospray or MALDI mass spectrometry?

The choice, if any, will usually be dictated by the instrumentation available. The major difference between the two techniques is that ES analyses samples in solution and so can be coupled to on-line separation techniques such as HPLC-MS, whereas MALDI analyses samples in a solid form. Traditionally, ES has been coupled to quadrupole mass analysers and so has had the advantage of mass accuracy and resolution. More recently nano-electrospray has improved the sensitivity of this ionization method (38). On the other hand MALDI, which has the advantage of being more tolerant to additives and contaminants, has been closely associated with TOF analysers. These provide the advantage of sensitivity, but until the relatively recent advent of TOF enhancements such as reflectron lenses and delayed extraction (47, 48), the resolution and mass accuracy had been less satisfactory. Recent developments in instrumentation have seen the coupling of ES (and nano-electrospray) with TOF mass spectrometers to produce a range of products with high sensitivity, good resolution, mass accuracy, and a large mass range. The future is looking extremely bright for biochemists!

6.3 Sequencing by mass spectrometry

Once a molecular weight has been measured for a particular protein, the next step is to glean information regarding its amino acid sequence. This requires breaking down the protein into smaller components, and there is a number of ways of achieving this:

(a) One successful and accepted approach is to digest the protein enzymically and then to analyse all of the resulting components by MS; first to obtain molecular weights for all of the peptide components, and then to generate structural information by the use of MS-MS. Each of the peptides will fragment by MS-MS (see Section 6.3.1), and from the pieces of the jigsaw some structural information can be constructed. Sometimes a complete sequence can be assembled for a peptide, more often a partial one, which may or may not be adequate to search a library and substantiate the composition of the original protein.

(b) Another approach for protein structural determination is to subject the entire protein to MS-MS (49, 50), but this requires a great deal of expertise and is generally not a routine procedure. It is most useful for confirming suspected sequences, rather than establishing new sequences.

(c) A third approach is to monitor the chemical or enzymic removal of individual amino acid residues sequentially from the N- or the C-terminus of either the intact protein or a peptide formed from enzymic digest of the protein. This technique is known as 'ladder sequencing' (see Section 6.3.2).

6.3.1 Tandem mass spectrometry of peptides and protein digests

Early studies on MS-MS experiments with peptides indicated that a degree of predictability existed concerning their modes of fragmentation. A standard nomenclature was suggested and has been adapted (51, 52). Very briefly, the peptide backbone bonds have the possibility of fragmenting between the N–C atoms, the C–C(O) atoms, or the C(O)–N atoms. It is the latter bonds which are the weakest of these covalent bonds and so *usually* fragment most easily to give rise predominantly (but not necessarily!) to the y″ and/or the b ions. Briefly, the a, b, and c″ fragment ions are formed with the charge retained at the N-terminus of the peptide, while the x, y″ and z ions retain the charge at the C-terminus. The immonium (or i) ions arise from cleavage of an amino acid from the peptide chain, and so confirm the presence of a certain amino acid, but do not give any indication of its position in the peptide backbone. These ions all arise from what are classified as low energy collision-induced decompositions, and are achieved with the widely-used tandem mass spectrometers with analysers of the type quadrupole-quadrupole or more recently Q-TOF. Experiments have been devised to differentiate between the isobaric leucine and isoleucine residues using such instruments (53).

Other fragmentations are possible, but only under high energy collision conditions achievable with certain mass spectrometers. These generate the d, v, and w ions, all resulting from side chain cleavages. These ions coexist with the low energy fragments, and the spectra are correspondingly more complex and hence difficult to interpret.

If the protein under investigation is digested with the aim of further MS analyses in mind, then the digest conditions should be adjusted accordingly. For example, a tryptic digest will work well if ammonium hydrogen carbonate is used as the buffer. As this is a volatile buffer, the whole reaction mixture can be injected directly into the mass spectrometer without compromising sensitivity. After the digest has reached completion, there are two approaches to the analysis: the whole mixture can be analysed directly by ES or MALDI to determine the molecular weights of all of the components from a single spectrum. Alternatively, if ES is available, then on-line HPLC-MS (54) using nanobore packed capillary columns (55) or standard columns $\leq$ 1 mm with flow rates between 0.1–50 μl/min, should generate a clean spectrum for each separated component in the mixture.

The molecular weights obtained from the digest mixture may be in agreement with those expected for the theoretical digest fragments, if the protein was of the expected sequence, and this evidence may be sufficient to characterize the protein. If further evidence is required, or if the molecular weights are not in agreement with those expected, then MS-MS fragment or daughter ion scanning is required. Again the MS-MS studies can be performed on either the whole mixture (56), or combined with an on-line HPLC-ES-MS run if the mixture is very complex. Each peptide in the mixture will generate an

MS-MS spectrum from which a sequence can be verified, or deduced at least partially. Software is available to aid interpretation, although experience is extremely useful!

More elaborate and specific MS-MS experiments can be carried out to identify certain classes of compounds by monitoring fragments relating specifically to these components, e.g. the detection of glycopeptides in a peptide mixture is achieved admirably by such principles (57, 58), as is the identification of phosphorylated sites (59).

The analysis of protein digest mixtures by on-line HPLC-ES using mass spectrometers such as tandem quadrupole and Q-TOF type is a highly sensitive—fmol quantities are achievable (60)—rapid, and accurate means of analysing proteins routinely. In some cases on-line CE-MS may prove a useful alternative (61), but to date this is not routine.

6.3.2 Ladder sequencing by mass spectrometry

Rather than the fragmentation of peptides inside the mass spectrometer, an alternative approach to sequencing utilizes the sequential removal of the C- or N-terminal amino acid residues by enzymic or chemical methods followed after each step by MS analysis of the residual truncated peptide. This is known as 'ladder sequencing'. MALDI-MS has been the analytical method of choice for monitoring such reactions due to its high sensitivity and uncomplicated spectra. The resulting spectrum contains molecular-related ions from several components—the original peptide and all of the reaction intermediates formed by removal of one or more residues. The difference in mass between the original peptide and the first reaction product (following the removal of one amino acid residue) will be specific to a particular amino acid residue removed, except in the cases of the isobaric pairs leucine and isoleucine (which would both show a mass loss of 113 Da) and glutamine and lysine (which would both show a mass loss of 128 Da). An example mass spectrum of a ladder sequence is shown in *Figure 3*. When carrying out such analyses, it is important to remember that as the reaction mixture is sampled in its entirety without any purification, then all of the reagents, solvents, and buffers should be compatible with the MS conditions being used.

N-terminal ladder sequencing has been accomplished successfully by Edman techniques using partial blocking (62), or by adding fresh peptide at the start of each Edman cycle (63). A rather elegant example of the latter methodology, using a volatile and hence MS-compatible degradation reagent, is described in *Protocol 14*. The degradation cycle is repeated $n - 1$ times, where n is the total number of amino acid residues in the peptide. At the start of each cycle, the intact peptide is added, so that the final reaction mixture contains the intact peptide, the peptide less one amino acid residue, the peptide less two amino acid residues, etc. The mass spectrum of the final solution characterizes components representing each stage of the sequence, starting with the intact peptide and finishing with the peptide less $n - 1$ residues.

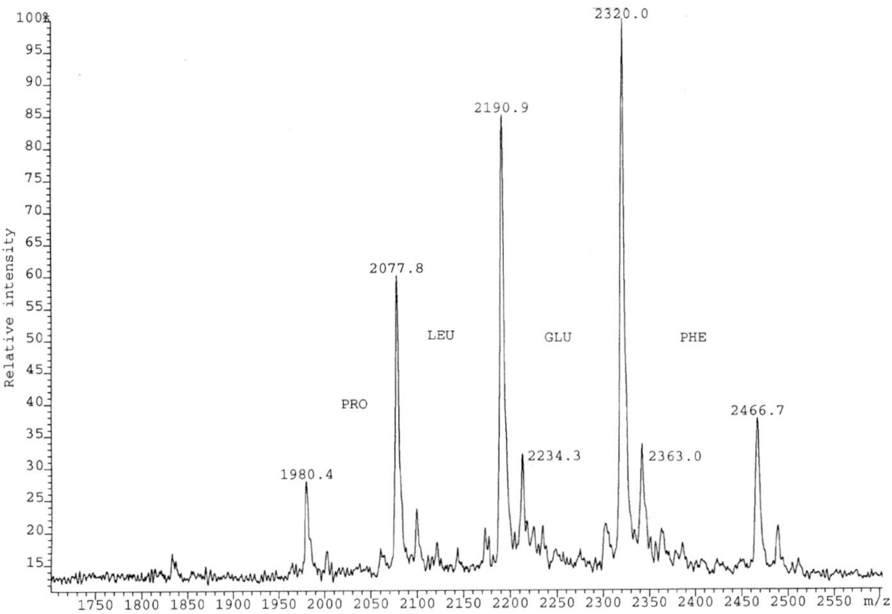

Figure 3. MALDI mass spectrum showing the 'ladder sequence' generated by digestion of a peptide with carboxypeptidase Y. A fragment (residues 18–39) of adrenocorticotrophic hormone (ACTH 18–39) was subjected to digestion with carboxypeptidase Y as described in *Protocol 15*. An aliquot was taken from the digest after 2 min and analysed by MALDI-MS. The mass differences between the ions in the 'ladder' produced can be interpreted to generate the C-terminal sequence of the peptide.

Protocol 14. N-terminal ladder sequencing by matrix-assisted laser desorption mass spectrometry[a]

Equipment and reagents

NB: all reagents must be Analar (or similar) grade; solvents must be HPLC grade.

- Peptide sample
- Microcentrifuge tubes (0.5 ml)
- Trifluoroethanol buffer: 5:4:1 mixture (by vol.) of trifluoroethanol:water:12.5% (w/v) aqueous trimethylammonium hydrogen carbonate pH 8
- Trifluoroethylisothiocyanate solution: 10% (v/v) trifluoroethylisothiocyanate in acetonitrile
- Heptafluorobutyric acid
- Lyophilizer (Heto-Holten Maxi Dry Plus vacuum centrifugation system)
- 1:1 mixture (v/v) of 0.1% (v/v) aqueous TFA:acetonitrile
- MALDI-MS sample target (for the instrument to which access is available)
- Matrix solution:[b] 10 mg/ml α-cyano-4-hydroxycinnamic acid in 1:1 mixture (v/v) of 0.1% (v/v) aqueous TFA:acetonitrile
- Access to MALDI-MS

Method

1. In a microcentrifuge tube, dissolve the peptide in trifluoroethanol buffer.

Protocol 14. *Continued*

2. Remove a 2 μl aliquot of this solution into a fresh microcentrifuge tube.
3. To this aliquot, add 2 μl trifluoroethylisothiocyanate solution.
4. Seal the tube and heat at 80 °C for 5 min.
5. Open the tube and dilute the contents with 5 μl water. Evaporate to dryness under vacuum.
6. Add 2 μl heptafluorobutyric acid to cleave the peptide N-terminal residue.
7. Seal the tube and heat at 80 °C for 5 min.
8. Open the tube and evaporate the contents under vacuum for 15 min.
9. Add another 2 μl aliquot of the original peptide solution to the contents of the tube.
10. Add 2 μl trifluoroethylisothiocyanate solution to the contents of the tube.
11. Repeat steps 4–10 for a total of $n - 1$ cycles, where n = total number of amino acid residues in the peptide.
12. Dissolve the final product in 3–5 μl 0.1% aqueous TFA:acetonitrile and apply 1–2 μl to the MALDI target.
13. Add 0.3 μl of the matrix solution to the target and allow it to dry in air.
14. Analyse by MALDI-MS (positive ionization mode).

[a] From ref. 64.
[b] Prepare fresh solution daily.

For C-terminal sequencing, carboxypeptidase Y is a useful enzyme for digestion as it cleaves all residues from the C-terminus (proline included), and has been used for MS ladder sequencing (65). Briefly, the peptide and enzyme are mixed together and aliquots removed at regular time intervals. The aliquots are analysed separately, or pooled together, and the mass differences between subsequent reaction products measured. From this information the amino acid sequence can be deduced, at least for a good number (about five on average) of residues (65), although the isobaric pairs remain indistinguishable. *Protocol 15* describes the procedure. Although digestion proceeds rapidly with peptides, protein conformation may severely restrict the effectiveness of this approach (see *Protocol 11*). The use of ES-MS precludes the use of urea to solubilize proteins unless the proteins are to be precipitated and washed prior to analysis. However, analysis by MALDI-MS may allow the use of urea.

Protocol 15. C-terminal cleavage using carboxypeptidases and ladder sequencing by mass spectrometry[a]

Equipment and reagents

- Peptide or protein sample
- 0.1 M pyridine–acetate buffer pH 5.5
- Carboxypeptidase Y (Sigma)
- 1:1 mixture (v/v) of 0.1% (v/v) aqueous formic acid:acetonitrile
- Peptide sample buffer[c] (for peptides): 10 mg/ml α-cyano-4-hydroxycinnamic acid in 1:1 mixture (v/v) of 0.1% (v/v) aqueous TFA:acetonitrile
- Protein sample buffer[c] (for proteins): 10 mg/ml 3,5-dimethoxy-4-hydroxycinnamic acid in 7:3 mixture (v/v) of 0.1% (v/v) aqueous TFA:acetonitrile
- MALDI-MS sample target for analysis by MALDI-MS (for the instrument to which access is available)
- Access to a mass spectrometer[b]
- Ice-cold 20% (v/v) aqueous acetic acid (for ES-MS, see footnote *f*)

Method

1. Dissolve the protein or peptide sample (100–5000 pmol) in 200 μl pyridine–acetate buffer.
2. Add an aliquot of carboxypeptidase Y to the sample (1–5% molar ratio)[d] and incubate the mixture at 25 °C.
3. Remove 20 μl samples at intervals over the time course of the digestion.[e]
4. (a) For ES-MS analysis, dilute the aliquots 1:1 with aqueous formic acid:acetonitrile solution.[f]

   (b) For MALDI-MS analysis, mix 10% of each aliquot with an equal volume of peptide sample buffer or protein sample buffer (as appropriate) and apply to a MALDI-MS sample target. Allow the sample to dry at room temperature for several minutes.
5. Analyse each sample by MS (positive ionization mode) for accurate M_r determination.
6. Compare the spectra from successive samples, checking for new reaction products.[g] Calculate the mass difference between successive products to elucidate the C-terminal sequence.

[a] From ref. 65.
[b] Electrospray-MS or MALDI-MS (preferably with delayed extraction for enhanced mass accuracy) is required to provide the necessary accuracy.
[c] Prepare fresh solution daily due to the light-sensitive nature of the matrix.
[d] The specific activity of the enzyme varies with the source. The effects of protein conformation also vary, necessitating pilot experiments being done with each protein and enzyme source.
[e] Time points within the first few minutes and over a longer period are generally suitable.
[f] Alternatively, the protein may be precipitated using 20% aqueous acetic acid and redissolved in aqueous formic acid:acetonitrile solution for ES-MS analysis.
[g] If the reaction is proceeding too slowly, increase the temperature from 25 °C to 37 °C in step 2.

Chemical methods have also been used for C-terminal ladder sequencing, successfully cleaving after all types of amino acid residue (66). The chemistry involved includes activation of the C-terminal end-group to a mixed anhydride, followed by cyclization with a thiocyanate to form a peptidylthiohydantoin, and finally cleavage to the C-terminal shortened peptide together with the severed and derivatized amino acid residue. In this way, C-terminal sequences of up to eight amino acids have been determined, although the enzymic method generally has the dual advantages of speed and sensitivity. All of the reagents involved (acetyl anhydride, acetic acid, aqueous ammonium hydroxide) were chosen because of their volatility and hence compatibility with MS.

References

1. Edman, P. (1950). *Acta Chem. Scand.*, **4**, 283.
2. Edman, P. (1960). *Ann. N. Y. Acad. Sci.*, **88**, 602.
3. Bailey, J. M., Nikfarjam, F., Shenoy, N. R., and Shively, J. E. (1992). *Protein Sci.*, **1**, 1622.
4. Bailey, J. M., Tu, O., Issai, G., Ha, A., and Shively, J. E. (1995). *Anal. Biochem.*, **224**, 588.
5. Hewick, R. M., Hunkapillar, M. W., Hood, L. E., and Dreyer, W. J. (1981). *J. Biol. Chem.*, **256**, 7990.
6. Laursen, R. A. and Bonner, A. G. (1970). *Fed. Proc.*, **29**, 727.
7. Pappin, D. J. C., Coull, J. M., and Köster, H. (1990). *Anal. Biochem.*, **187**, 10.
8. Meyer, H. E., Hoffmann-Posorske, E., Korte, H., and Heilmeyer, L. M. G. Jr. (1986). *FEBS Lett.*, **204**, 61.
9. Friedman, M., Krull, L. H., and Cavins, J. F. (1970). *J. Biol. Chem.*, **245**, 3868.
10. Hurley, C. K. and Stout, J. T. (1980). *Biochemistry*, **19**, 410.
11. Podell, D. N. and Abraham, G. N. (1978). *Biochem. Biophys. Res. Commun.*, **81**, 176.
12. Wellner, D., Panneerselvam, C., and Horecker, B. L. (1990). *Proc. Natl. Acad. Sci. USA*, **87**, 1947.
13. Bergman, T., Gheorghe, M. T., Hjelmqvist, L., and Jornvall, H. (1996). *FEBS Lett.*, **390**, 199.
14. Krishna, R. G., Chin, C. C. Q., and Wold, F. (1991). *Anal. Biochem.*, **199**, 45.
15. Kimura, Y. and Yasuda, N. (1989). *Agric. Biol. Chem.*, **53**, 497.
16. Drapeau, G. R., Boily, Y., and Houmard, J. (1972). *J. Biol. Chem.*, **247**, 6720.
17. Drapeau, G. R. (1976). In *Methods in enzymology* (ed. L. Lorland), Vol. 45, p. 469. Academic Press, New York.
18. Doonan, S., Doonan, H. J., Hanford, R., Vernon, C. A., Walker, J. M., Airoldi, P. da S., Bossa, F., Bara, D., Carloni, M., Fasella, P., and Riva, F. (1975). *Biochem. J.*, **149**, 497.
19. Mitchell, W. M. and Harrington, W. F. (1968). *J. Biol. Chem.*, **243**, 4683.
20. Schenkein, I., Levy, M., Franklin, E. C., and Frangione, B. (1977). *Arch. Biochem. Biophys.*, **182**, 64.
21. Noreau, J. and Drapeau, G. R. (1979). *J. Bacteriol.*, **140**, 911.
22. Gross, E. and Witkop, B. (1961). *J. Am. Chem. Soc.*, **83**, 1510.
23. Mohoney, W. C., Smith, P. K., and Hermondson, M. A. (1981). *Biochemistry*, **20**, 443.

24. Fontana, A., Dalzoppo, D., Grandi, C., and Zambonin, M. (1981). *Biochemistry*, **20**, 6997.
25. Enfield, D. L., Ericsson, L. H., Fujikawa, K., Walsh, K. A., Neurath, H., and Titani, R. (1980). *Biochemistry*, **19**, 659.
26. Inglis, A. S., McKern, N. M., and Stike, P. M. (1979). *Proc. Aust. Biochem. Soc.*, **12**, 12.
27. Cleveland, D. W., Fischer, S. G., Kirschner, M. W., and Laemmli, U. K. (1977). *J. Biol. Chem.*, **252**, 1102.
28. Scott, M. G., Crimmins, D. L., McCourt, D. W., Tarrand, J. J., Eyerman, M. C., and Nahm, M. H. (1988). *Biochem. Biophys. Res. Commun.*, **155**, 1353.
29. Crimmins, D. L., McCourt, D. W., Thoma, R. S., Scott, M. G., Macke, K., and Schwartz, B. D. (1990). *Anal. Biochem.*, **187**, 27.
30. Fernandez, J., DeMott, M., Atherton, D., and Mische, S. M. (1992). *Anal. Biochem.*, **201**, 255.
31. Sutton, C. W., Pemberton, K. S., Cottrell, J. S., Corbett, J. M., Wheeler, C. H., Dunn, M. J., and Pappin, D. J. (1995). *Electrophoresis*, **16**, 308.
32. Inglis, A. S. (1991). *Anal. Biochem.*, **195**, 183.
33. Hayashi, R., Moore, S., and Stein, W. H. (1973). *J. Biol. Chem.*, **248**, 2296.
34. Clerc, F. F., Monegier, B., Faucher, D., Cuine, F., Pourcet, C., Holt, J. C., Tang, S. Y., van Dorsselaer, A., Becquart, J., and Vuilhargne, M. (1994). *J. Chromatogr. B (Biomed. Appl.)*, **662**, 245.
35. Ashcroft, A. E. (1997). *Ionization methods in organic mass spectrometry*. The Royal Society of Chemistry, Cambridge.
36. Fenn, J. (1984). *J. Phys. Chem.*, **88**, 4451.
37. Hillenkamp, F., Karas, M., Beavis, R. C., and Chait, B. T. (1991). *Anal. Chem.*, **63**, 1193.
38. Wilm, M. and Mann, M. (1996). *Anal. Chem.*, **68**, 1.
39. Verentchikov, A. N., Ens, W., and Standing, K. G. (1994). *Anal. Chem.*, **66**, 126.
40. Kaufmann, R., Spengler, B., and Lutzenkirchen, F. (1993). *Rapid Commun. Mass Spectrom.*, **7**, 902.
41. Tyler, A. N., Clayton, E., and Green, B. N. (1996). *Anal. Chem.*, **68**, 3561.
42. Ogorzalek Loo, R. R., Dales, N., and Andrews, P. C. (1994). *Protein Sci.*, **3**, 1975.
43. Karas, M., Bachmann, D., Bahr, U., and Hillenkamp, F. (1987). *Int. J. Mass Spectrom. Ion Processes*, **78**, 53.
44. Carroll, J. A. and Beavis, R. C. (1996). In *Laser desorption and ablation* (ed. J. C. Miller and R. F. Hagulund), Ch. 7. Series: Experimental methods in the physical sciences.
45. Chait, B. T. Rockefeller University, New York. http://chait-sgi.rockefeller.edu/recipes/sproteas.htm
46. Vorm, O., Chait, B. T., and Roepstorrf, P. (1994). In *Mass spectrometry of protein samples containing detergents*. Proc. 41st. ASMS Conf., 621.
47. Wiley, W. C. and McLaren, I. (1955). *Rev. Sci. Inst.*, **26**, 1150.
48. Lennon, J. J. and Brown, R. S. (1995). *Anal. Chem.*, **67**, 1988.
49. Witkowska, H. E., Green, B. N., Morris, M., and Shackleton, C. H. L. (1996). *J. Mass Spectrom.*, **S111**.
50. Loo, J. A., Edmonds, C. G., and Smith, R. D. (1991). *Anal. Chem.*, **63**, 2488.
51. Roepstorrf, P. and Fohlmann, J. (1984). *Biomed. Mass Spectrom.*, **11**, 601.
52. Johnson, R. S. and Biemann, K. (1989). *Biomed. Environ. Mass Spectrom.*, **18**, 945.

53. Hulst, A. G. and Kjentz, C. E. (1996). *J. Mass Spectrom.*, **31**, 1188.
54. Covey, T. (1996). In *Methods in molecular biology* (ed. J. R. Chapman), Vol. 61, p. 83. Humana Press Inc., Totowa, NJ.
55. Kassel, D. B., Shushan, B., Sakuma, S., and Salzman, J. P. (1994). *Anal. Chem.*, **66**, 236.
56. Wilm, M., Neubauer, G., and Mann, M. (1996). *Anal. Chem.*, **68**, 527.
57. Huddleston, M. J., Bean, M. F., and Carr, S. A. (1993). *Anal. Chem.*, **65**, 877.
58. Medzihradszky, K. F., Maltby, D. A., Hall, S. C., Settineri, C. A., and Burlingame, A. L. (1994). *J. Am. Soc. Mass Spectrom.*, **5**, 350.
59. Allen, M., Anacleto, J., Bonner, B., Bonnici, P., Shushan, B., and Nuwaysir, L. (1997). *Rapid Commun. Mass Spectrom.*, **11**, 325.
60. Wilm, M., Shevchenko, A., Houthaeve, T., Breit, S., Schweigerer, L., Fotsis, T., and Mann, M. (1996). *Nature*, **379**, 466.
61. Major, H. J. and Ashcroft, A. E. (1996). *Rapid Commun. Mass Spectrom.*, **10**, 1421.
62. Chait, B. T., Wang, R., Beavis, R. C., and Kent, S. B. H. (1993). *Science*, **262**, 89.
63. Bartlet-Jones, M., Jeffery, W., Hansen, H. F., and Pappin, D. J. C. (1994). *Protein Chem.*, **13**, 455.
64. Bartlet-Jones, M., Jeffery, W. A., Hansen, H. F., and Pappin, D. J. C. (1994). *Rapid Commun. Mass Spectrom.*, **8**, 737.
65. Patterson, D. H., Tarr, G. E., Regnier, F. E., and Martin, S. A. (1995). *Anal. Chem.*, **67**, 3971.
66. Thiede, B., Salnikow, J., and Wittmann-Liebold, B. (1997). *Eur. J. Biochem.*, **244**, 750.

2

Protein folding and import into organelles

JOACHIM RASSOW

1. Introduction

How do proteins find their distinct functional location within the cell after they have been synthesized? In the last 15 years this basic biological question has opened a new field of molecular cell biology. The sites of protein synthesis and function are often separated by membranes. Therefore, the structures and mechanisms which drive proteins across membranes are of fundamental importance in the correct distribution of newly synthesized proteins in the cell (1). In this chapter, several of the techniques will be described which are applied to characterize intracellular protein traffic. All the protocols involved are easy to perform and do not require special equipment so they can be readily adopted by non-specialists. More sophisticated protocols can be obtained from the literature cited in the references.

The key role of protein folding in this context becomes evident in considering the different stages of protein translocation. All proteins leave the ribosome in an extended conformation. Therefore, newly synthesized proteins not only have to find their correct location within the cell, but also their correct three-dimensional structure. Some proteins are synthesized directly at the target membrane and are translocated co-translationally. These are especially the proteins of the mammalian endoplasmic reticulum. They adopt their correct folding state after translocation across the membrane. However, most proteins, for example all mitochondrial proteins, are translocated post-translationally. After synthesis they bind to specific factors in the cytosol and are kept in a non-native, but translocation-competent, conformation. Translocation across the target membrane again requires complete unfolding of the polypeptide in transit. After translocation, the proteins fold into their native state, often after interaction with several chaperone proteins or protein folding enzymes which interfere with this process. Only nuclear and peroxisomal proteins may be imported into their organelle without major structural transformations. In the endoplasmic reticulum, in mitochondria, and in chloroplasts, the translocating polypeptides transiently bind to members of the Hsp70 family of chaperone proteins at the *trans* side of the membrane. This interaction drives the translocation of the protein.

In this chapter, I concentrate on methods for studying the import of proteins into mitochondria and the endoplasmic reticulum. Protocols for the study of special aspects of other protein transport systems can be obtained from the literature given in Section 5. The chapter starts by describing the synthesis of preprotein substrates for studying import into organelles (Section 2). It then describes the preparation of mitochondria and endoplasmic reticulum and the basic protocols for the uptake of preproteins into these organelles (Sections 3 and 4). The final section deals with methods for analysing protein import and association of proteins with the import machinery (Section 6).

2. Preparation of preprotein substrates for *in vitro* import into organelles

Intracellular protein transport can be studied *in vivo*, for example by expressing the desired preprotein in the yeast *Saccharomyces cerevisiae*. Several genetic screening procedures have taken advantage of this possibility (2–5). However, for *in vitro* import studies, organelles are first isolated and then incubated with preproteins, which can be obtained from various sources, usually by translation of the appropriate mRNA in an *in vitro* protein synthesis system.

2.1 Preparation of mRNA for *in vitro* translation

To obtain mRNA for *in vitro* protein synthesis, the corresponding cDNA sequence is normally cloned into an expression vector that allows expression of the cDNA from bacteriophage T7 or SP6 promotors. Before translation in cell-free systems the mRNA is usually synthesized in a separate reaction using purified T7 or SP6 RNA polymerase. The yield and integrity of the mRNA can easily be tested by agarose gels containing ethidium bromide which are commonly used to resolve DNA fragments. In comparing samples prior to and after the incubation with the RNA polymerase, the mRNA should become visible as a band additional to the plasmid DNA. The mRNA may be precipitated by ethanol in the presence of 250 mM LiCl and dissolved in water (keep the ethanol pellet wet as dried RNA is difficult to redissolve in water) and stored at –70 °C or –80 °C (where it is stable for very long times). For the relevant protocols, see ref. 6.

2.2 Synthesis of preproteins in the reticulocyte lysate

Translation of the appropriate mRNA in rabbit reticulocyte lysate is the most common way to obtain radiolabelled preproteins for *in vitro* studies. The lysate is a diluted preparation of reticulocyte cytosol (7) and is commercially available (e.g. from Amersham, Boehringer Mannheim, or Promega). ^{35}S-radiolabelled amino acids can be obtained from Amersham or NEN. The concentration of the radiolabelled protein which can be achieved by translation in the reticulocyte lysate is about 1–20 pmol/ml. This low concentration is not sufficient to detect the protein in Western blots. Advantages of the reticulocyte lysate are that it is very easy to use and (if obtained from commercial suppliers) very reliable. Moreover, preproteins are synthesized in an environment which often closely resembles their environment *in vivo*. The lysate contains all of the cytosolic factors necessary to keep preproteins in a translocation-competent state. Protocols for use with reticulocyte lysates are normally provided by the supplier. Alternatively, consult ref. 7 if laboratory prepared lysates are being used. If translation gives only poor results, it is sometimes an advantage to use a coupled transcription and translation system, e.g. the TNT lysate from Promega.

At the end of the translation, a reticulocyte lysate which is to be used for import of preproteins into isolated organelles should be supplemented with non-labelled amino acids to chase the ^{35}S-labelled amino acids. As appropriate, methionine and/or cysteine are added from tenfold concentrated stock solutions to final concentrations of 5.8 mM. This allows the completion of translation of incompletely synthesized proteins. Without this treatment, incompletely synthesized proteins could cause a contamination in the pattern of labelled proteins. Additionally, it reduces the background of non-specific binding of the labelled amino acids to amino acid-binding proteins. In experiments with mitochondria or chloroplasts, it also prevents the use of labelled amino acids in mitochondrial protein synthesis. Sucrose should also be added to stabilize the organelles of the import assay; for mitochondria, a final concentration of 250 mM is required, most conveniently added from a 1.5 M sucrose stock solution.

Preproteins which are sufficiently soluble can be separated from most of the non-incorporated radioactivity and the haemoglobin by ammonium sulfate precipitation without loss of import competence (see Section 2.6.2, *Protocol 2*). In addition, this depletes reticulocyte lysate of aberrant translation products formed for example if the ribosomes recognize an internal AUG as a start codon. Sometimes the specific recognition of the start codon can be improved by titrating the concentration of magnesium ions (up to 5 mM).

Since successful use of labelled preproteins requires a sufficient specific radioactivity in the reaction products, it is often helpful to test for the amount of incorporated radioactivity after translation in the reticulocyte lysate. For instance, only reticulocyte lysates which show at least 75 000 c.p.m./μl should be used for subsequent cross-linking studies (Section 6.2.3). *Protocol 1* describes a simple and quick assay. In parallel, the purity (the absence of additional artefact proteins) can be tested by SDS–PAGE.

Protocol 1. Determination of the incorporated radioactivity after mRNA translation in a reticulocyte lysate

Equipment and reagents

- Reticulocyte lysate containing the radio-labelled protein
- Filter paper (thin paper commonly used for example for fluted filters)
- 100 ml beaker, and device for heating
- 5% trichloroacetic acid (TCA)
- Ethanol
- Scintillation counter plus vials and scintillation cocktail
- Fume-hood
- Safety spectacles to protect your eyes

Method

1. Separately pipette 1 μl aliquots of the reticulocyte lysate in a grid pattern onto a piece of filter paper. Leave the paper until it is completely dry.
2. Cut up the filter paper so each piece contains a reticulocyte lysate sample. Use a piece of the filter of the same size without reticulocyte lysate as a background control.
3. Add the filter pieces to about 15 ml 5% TCA in a beaker at room temperature and incubate for about 5 min.
4. Heat the beaker until the TCA starts boiling. Caution: do this in a fume-hood and protect your eyes!
5. Remove the beaker from the heat (do not boil the TCA for longer) and wait until the TCA has cooled down.
6. Carefully discard the TCA solution (which contains the non-incorporated radioactivity). Rinse the filters with ethanol.
7. Incubate the filters in fresh ethanol for 5 min at room temperature.
8. Remove the filters from the ethanol and let them air dry.
9. Add the filter pieces to scintillation vials containing scintillation cocktail and determine radioactivity in a scintillation counter.

2.3 Synthesis of preproteins in wheat germ lysate

In principle, synthesis of preproteins in wheat germ lysate is an alternative to synthesis in reticulocyte lysate. However, it is only used in special cases, for example using preproteins from plant sources. One reason is that preproteins are often far less import-competent when synthesized in wheat germ lysate as compared to the reticulocyte lysate, probably due to a lower concentration of Hsp70 and/or other cytosolic factors. Wheat germ lysate is available commercially from Boehringer and Promega, or can be prepared following the protocols of Erickson and Blobel (8). These authors also describe the use of the lysate in cell-free translation.

2.4 Synthesis of preproteins in yeast cytosol

The use of a cell-free lysate from the yeast *Saccharomyces cerevisiae* for translation is helpful especially where a homologous system is required for use with yeast organelles. Several protocols have been published; one of the best is described by Garcia *et al.* (9) with modifications in ref. 10. A disadvantage of translations in yeast cytosol is that the efficiency is only about 10% as compared to reticulocyte lysate. Moreover it is difficult to use; even in the hands of experts it often requires several attempts to obtain a suitable preparation.

2.5 Synthesis of preproteins in *Escherichia coli*

The synthesis of preproteins in *E. coli* enables preproteins to be obtained pure and in chemical amounts. For detailed protocols, see ref. 11.

2.6 Dihydrofolate reductase as a model protein for import studies

2.6.1 Advantages of DHFR as a domain of preproteins

In studying intracellular protein transport and chaperone-mediated protein folding, the dihydrofolate reductase (DHFR) of the mouse is one of the most often used substrate proteins. The amino acid sequence has been published by Stone and Phillips (12) (but note that the start methionine is missing in the sequence in the paper), and the nucleotide sequence has been determined by Nunberg *et al.* (13). By fusion to appropriate presequences, DHFR can be imported into different organelles. Several properties of DHFR are of advantage:

(a) The molecular weight of DHFR is only 21.5 kDa. Therefore, it is easy to construct hybrid proteins of a reasonable size. DHFR is a monomeric protein and readily soluble.

(b) The native folding state of DHFR can be stabilized by addition of methotrexate, a ligand which binds reversibly with a binding constant in the nanomolar range (described in detail in Section 3.3.3). The native folding state of DHFR can be further stabilized by addition of NADPH (at a concentration of about 1 mM) which is the second substrate of DHFR in the cell (besides dihydrofolate). Translocation of a DHFR domain across the endoplasmic reticulum membrane and across the mitochondrial membranes is reversibly inhibited in the presence of methotrexate because the ligand inhibits unfolding of DHFR (14, 15). Aminopterin can be used as a membrane permeable analogue of methotrexate (16). By parallel import of DHFR fusion proteins into mitochondria and glycosomes, it has been demonstrated that unfolding of proteins is not required for transport into glycosomes (17).

(c) In the native folding state, DHFR is highly resistant to proteinase K, trypsin, and other proteases. On the other hand, it is extremely protease-sensitive after unfolding. Therefore, DHFR is an excellent substrate protein for monitoring protein folding reactions in intracellular protein traffic (18, 19).

(d) DHFR can be bound to different Hsp70s and to the chaperonin of *E. coli*, GroEL. Thus, DHFR can be used to investigate reactions which depend on chaperone proteins. Mouse DHFR also binds to the mitochondrial Hsp60 of the fungus *Neurospora crassa* but it does not bind to the homologous Hsp60 in the yeast *S. cerevisiae* (18, 20, 21).

(e) Mouse DHFR is a substrate for the peptidyl-prolyl *cis/trans* isomerase cyclophilin both *in vivo* and *in vitro* (22, 23).

(f) The crystal structure of mouse DHFR has been resolved (24).

2.6.2 Denaturation of a DHFR hybrid protein for use as a preprotein substrate

Protocol 2 describes the precipitation and urea denaturation of a hybrid protein (Su9-DHFR) containing the presequence of the subunit 9 of the mitochondrial ATP synthase (residues 1–69) of *N. crassa* fused to the entire DHFR of the mouse. This construct can be imported into isolated mitochondria from *N. crassa* as well or *S. cerevisiae* by diluting the urea solution directly into the import assay. In the same way the protein can be bound to the chaperone protein GroEL. Hybrid proteins containing other presequences can be treated similarly. DHFR fusion preproteins have been particularly useful in the identification of intermediates involved in the import process in mitochondria (Section 3.3) and the analysis of protein folding *in vivo* (Section 6.3).

Protocol 2. Denaturation of Su9-DHFR

Reagents

- Reticulocyte lysate in which Su9-DHFR mRNA has been translated in the presence of [^{35}S]methionine (Section 2.2)
- Dithiothreitol (DTT) 100 mM stock solution
- 1 M MOPS pH 7.2 (with KOH) stock solution[a]
- Saturated ammonium sulfate (ice-cold)
- 0.5 M $MgCl_2$
- Urea or guanidinium hydrochloride
- *Either* 8 M urea solution: 8 M urea, 2 mM DTT, 5 mM $MgCl_2$, 50 mM MOPS pH 7.2 (add DTT, $MgCl_2$, and MOPS from stock solutions, in total 600 μl, to 480 mg solid urea to prepare 1 ml 8 M urea)[b]
- *Or* 6 M guanidinium chloride solution: 6 M guanidinium chloride, 2 mM DTT, 5 mM $MgCl_2$, 50 mM MOPS pH 7.2[c]

Method

1. Place 25 μl reticulocyte lysate in a test-tube on ice. Add 1 μl 100 mM DTT and 1 μl 1 M MOPS pH 7.2 and mix.
2. Add 5 μl saturated ammonium sulfate, immediately mix carefully, and

incubate on ice for 2 min. Repeat this four times with further 5 μl aliquots.

3. Incubate for an additional 20 min on ice, pellet the precipitated protein by centrifugation for 10 min at 16000 *g* at 2°C. Remove the supernatant, and centrifuge the pellet again for 1 min, remove residual supernatant.
4. Store the pellet at –20°C, or dissolve it in 8 M urea solution or 6 M guanidinium chloride solution, and proceed to import studies (Section 3.2.3).

[a] The MOPS buffer may be replaced by other buffers as required for subsequent import studies.
[b] Urea solutions are only of limited stability and should be freshly made.
[c] 6 M guanidinium chloride solution is stable at –20°C.

In the procedure, the hybrid protein Su9-DHFR is first synthesized in the reticulocyte lysate, for example in the presence of [^{35}S]methionine, and then precipitated by 45% saturated ammonium sulfate. Other preproteins may require higher concentrations of ammonium sulfate (up to 66%). The precipitated protein can be denatured either in 8 M urea or in 6 M guanidinium chloride. However, guanidinium salts should be used with caution since they may preclude subsequent import studies due to their detrimental effect on cell organelles (see Section 3.2.3). The time required for complete denaturation depends on the preprotein. Often a few seconds at low temperature are sufficient, but sometimes even an hour at room temperature may be necessary. Denaturation for only a few seconds retains the conformations of peptidyl-prolyl *cis/trans* conformers; after longer denaturation they partition between *cis* and *trans* conformations.

3. Import of proteins into mitochondria

To set up an *in vitro* system for mitochondrial protein import is usually very simple and does not require special equipment or specialist knowledge. All that is required are isolated mitochondria and the purified preprotein in a standard buffer or, alternatively, the preprotein synthesized in the reticulocyte lysate or one of the translation systems described in Section 2. As soon the preprotein is added to the mitochondria and the sample is incubated at an appropriate temperature, the preprotein will be imported. The protocols in this section cover the basic technique and are intended to act as a starting point. For additional methods see refs 25 and 26.

Most experiments on protein import into mitochondria are currently carried out with the yeast *S. cerevisiae*. Therefore, this section starts with a protocol for the isolation of mitochondria from this organism. Heterologous import (i.e. import of preproteins into mitochondria from different species) is often possible. In general, heterologous import can be carried out with preproteins and mitochondria from the fungus *N. crassa* and the yeast *S. cerevisiae.* Mammalian mitochondria and preproteins may show different specificities and do not always match with those from *Neurospora* and yeast. In some cases, efficient import has been observed, in other cases import could not be achieved (27). Note that mitochondria isolated from any organism never retain their original structure. Inside cells, mitochondria are worm-like structures which are several micrometres in length (28). During isolation, the mitochondria form small spherical bodies of about 1 μm diameter (15). By an unknown mechanism the isolated mitochondria retain their content, including the soluble proteins of the intermembrane space and the matrix, and they seem to retain all their functions. Biochemical data which refer to 'a single mitochondrion' usually refer to these isolated mitochondria.

3.1 Isolation of mitochondria for import studies

3.1.1 Isolation of mitochondria from *Saccharomyces cerevisiae*

The isolation of mitochondria from yeast is usually performed in the following steps:

- digestion of the cell wall by Zymolyase
- breaking of the cells by osmotic shock
- isolation of the mitochondria by differential centrifugation

To isolate mitochondria, the cells are usually grown in a non-fermentable medium which induces the growth of mitochondria. The classical medium of this type is the YPEG medium which contains glycerol and ethanol as carbon sources. However, it is also possible to grow the cells on substrates such as lactate, glucose, or galactose and get a sufficient yield of mitochondria. In working with mitochondria from different yeast strains, all preparations should be carried out using the same growth conditions to allow a direct comparison of their mitochondria. Ideally the mitochondria should be isolated in parallel. Growth and storage of yeast strains has been described in detail in ref. 29. *Protocol 3* gives standard conditions for growth on glycerol/ethanol. Yeast is usually grown either in large Erlenmeyer flasks or in fermenters. Growth in fermenters can give poor results when the medium is autoclaved too long. This can be avoided by autoclaving the fermenter containing only water and separately autoclaving a fivefold concentrated medium which is diluted afterwards into the sterilized water in the fermenter.

Protocol 4 is a modified version of a method which was first published by Daum *et al.* (30). The preparation of the mitochondria takes about 5 h. The isolated mitochondria still contain considerable amounts of other membranes such as microsomes (from the endoplasmic reticulum). However, the purity is sufficient for protein import experiments. Separation of the mitochondria from residual microsomes is difficult and does not improve the import efficiencies of preproteins. An additional purification step by a Nycodenz gradient has been described by Glick and Pon (31).

Protocol 3. Growth of *Saccharomyces cerevisiae* for the isolation of mitochondria

Equipment and reagents

- Appropriate yeast strain
- YP medium: 1% (w/v) yeast extract (Difco), 2% (w/v) peptone (Bacto-Peptone, Difco); adjust to pH 4.7–5 with conc. HCl, autoclave 1.5 litres in a 5 litre flask at 121 °C for 20 min
- Shaker with 5 litre Erlenmeyer flasks for growth at 30 °C
- Photometer for determining the OD_{578} or OD_{600}
- 30% (v/v) glycerol, autoclaved separately
- Ethanol (absolute)

Method

1. Prepare a starter culture by inoculating 100 ml YP medium with the yeast strain and growing it overnight at 30 °C to an OD_{578} = 1.5–5.0.[a]
2. Add the following to 800 ml autoclaved YP medium:
   (a) 100 ml autoclaved 30% glycerol.
   (b) 20 ml absolute ethanol.
   (c) 10 ml (approx.) of the starter culture.[b]
3. Shake the flasks (about 150 r.p.m.) at 30 °C for 16–18 h and monitor the growth by OD_{578}.
4. When the OD_{578} reaches 1–4 proceed to the isolation of mitochondria as in *Protocol 4*.

[a] Temperature-sensitive mutant strains are usually grown at 23 °C. Growth can also be monitored at OD_{600}.
[b] The exact volume depends on the yeast strain and has to be determined individually.

Protocol 4. Isolation of mitochondria from *Saccharomyces cerevisiae*

Equipment and reagents

- 5 litre culture of *S. cerevisiae* (*Protocol 3*)
- Centrifuge with a large rotor for isolation of the cells up to 3000 *g* and appropriate bottles
- Rotor for isolation of the mitochondria at 17 500 *g* and suitable tubes
- 100 mM Tris pH 9.4 with H_2SO_4 (store at 4 °C)
- 1 M DTT (freshly made)
- 100 mM Tris–H_2SO_4 pH 9.4, 10 mM DTT; make this fresh using stock solutions
- Shaking water-bath at 30 °C
- 2.4 M sorbitol stock solution (store at 4 °C)—use this to make a 1.2 M solution
- 100 mM potassium phosphate pH 7.4 stock solution (store at 4 °C)
- 1.2 M sorbitol, 20 mM potassium phosphate pH 7.4—make this using stock solutions
- Zymolyase 20T from *Arthrobacter luteus* (30 U/mg; Seikagaku Corporation)[a]
- 1 M Tris–HCl pH 7.4 stock solution

Protocol 4. *Continued*

- 250 mM EDTA pH 8 stock solution
- BSA (Sigma, 9003T; fatty acid-free)
- Homogenizing buffer: 0.6 M sorbitol, 10 mM Tris–HCl pH 7.4, 1 mM EDTA, 0.5% BSA. Use the stock solutions to prepare this just before use (see step 8). Add BSA, adjust the pH to 7.4 after solubilization of the BSA, and cool on ice.[b]
- 100 mM PMSF in ethanol (caution: very toxic)
- Tight-fitting Dounce homogenizer
- SEM buffer: 250 mM sucrose, 1 mM EDTA, 10 mM MOPS–KOH pH 7.2 (store at 4 °C)
- Reagents for protein determination
- Liquid nitrogen and vials for freezing samples

Method

1. Weigh the empty centrifuge bottles and fill them with the yeast culture.
2. Pellet the cells by centrifugation for 5 min at 2800 *g* at room temperature.
3. Wash the cells by resuspending them in distilled H_2O and pellet them for 5 min at 2800 *g* at room temperature. Determine the weight of cells. A culture of 1000 ml should yield a wet weight of 2–8 g yeast cells.
4. Resuspend the cells in 100 mM Tris–H_2SO_4 pH 9.4, 10 mM DTT (1 ml buffer/0.5 g wet weight of cells). Incubate the suspension for 10 min at 30 °C (ts mutants at 23 °C) with gentle shaking.
5. Pellet the cells at 4 °C by centrifugation at 2800 *g* for 5 min.
6. Wash the cells once by resuspension in 1.2 M sorbitol and centrifugation for 5 min at 2800 *g* and 4 °C.
7. Resuspend the cells in 1.2 M sorbitol, 20 mM potassium phosphate pH 7.4 (1 ml buffer/0.15 g wet weight of cells).
8. Add Zymolyase 20T (3 mg enzyme/g of wet cells) and incubate for 30 min at 30 °C (ts mutants at 23 °C) with gentle shaking. Prepare the homogenization buffer while the cells are incubating.
9. Test for conversion of yeast cells into spheroplasts by dilution of 10 μl of the suspension into 1 ml distilled H_2O. The suspension should become clear after a few seconds. If this is not observed add the same amount of Zymolyase again (see step 8) and continue the incubation.[c]
10. Pellet the spheroplasts by centrifugation for 5 min at 2800 *g* and 4 °C.
11. Wash the spheroplasts twice in 1.2 M sorbitol at 4 °C. The spheroplasts are labile and should be handled with care.
12. Resuspend the spheroplasts in cold homogenization buffer (1 ml/0.15 g wet weight of cells). Add PMSF from the 100 mM stock solution to a final concentration of 1 mM.
13. Break the spheroplasts by 10–20 strokes in a Dounce homogenizer with continuous cooling in ice. If the homogenizer pestle is a tight fit, ten strokes should be sufficient.

14. Add an additional volume of homogenization buffer (as in step 12) together with 1 mM PMSF. Mix, and centrifuge the suspension at 4 °C for 5 min at 1500 *g*. Discard the pellet.
15. Isolate the mitochondria from the supernatant by centrifugation at 4 °C for 12 min at 17 500 *g*.
16. Resuspend the pelleted mitochondria in SEM buffer and centrifuge for 5 min at 1500 *g*.
17. Pellet the mitochondria from the supernatant by centrifugation at 4 °C for 12 min at 17 500 *g*.
18. Resuspend the mitochondria in SEM (1 ml/1000 ml original culture) and determine the protein concentration.
19. Adjust the protein concentration to 5 mg/ml and freeze the mitochondria in aliquots in liquid nitrogen.
20. Store the mitochondria at –70 °C.[d]

[a] The source of this enzyme may be critical for the successful formation of spheroplasts and the intactness of the mitochondria after lysis of the spheroplasts.
[b] Homogenization medium can be stored frozen in aliquots at –20 °C.
[c] If formation of spheroplasts is still not achieved, this may be due to thick cell walls formed in cultures which have been grown too long. In this case the cells should be harvested at a lower OD (OD_{578} = 0.8–1.5).
[d] If stored at –70 °C, the mitochondria remain competent for import of preproteins for more than a year without loss of activity. They may be stored at –20 °C for some weeks.

3.1.2 Isolation of rat liver mitochondria

If mitochondria from mammalian tissue are required they are usually isolated from rat liver since it is relatively easy to obtain pure mitochondria in a high yield from this source. *Protocol 5* is a modified version of a method published in ref. 32. The tissue is broken by a Teflon homogenizer, lysosomal membranes are lysed by digitonin at a low concentration, and the mitochondria are isolated by differential centrifugation. Mitochondria isolated from young rats show better import efficiencies than mitochondria from adult animals.

Protocol 5. Isolation of mitochondria from rat liver

Equipment and reagents

- Rat (Wistar, 90–120 g)
- Teflon homogenizer (motorized, Potter-type)
- HMS buffer: 220 mM mannitol, 70 mM sucrose, 20 mM Hepes pH 7.4
- Centrifuge, rotors, and tubes for centrifugations at 3000 *g* and 14 500 *g*
- Digitonin (Sigma, D-1407; used as delivered from the manufacturer, do *not* recrystallize as described in *Protocol 15*). Dissolve 1.75 mg/ml in HMS buffer at about 40 °C and then cool it on ice. Make this up just before use.
- Small Dounce homogenizer with glass pestle (for resuspension of mitochondria)
- Reagents for protein determination

Protocol 5. *Continued*

Method

1. Kill the rat by decapitation.[a]
2. Remove the liver and cool it on ice.[b]
3. Chop the liver into small cubes and transfer the pieces into a pre-cooled Teflon Potter homogenizer. Add 50 ml HMS buffer.
4. Homogenize the tissue by three up and down strokes of the pestle rotating at 300 r.p.m.
5. Centrifuge 1 min at 3000 *g* (or 10 min at 1000 *g*) at 4 °C.
6. Transfer the supernatant into a new tube and pellet the mitochondria at 14 500 *g* for 5 min at 4 °C.
7. Discard the supernatant and resuspend the mitochondria in 1 ml HMS buffer.
8. Add 1 ml of digitonin solution, mix, and incubate for 5 min at 0 °C.
9. Add 50 ml cold HMS buffer and pellet mitochondria for 5 min at 4 °C and 14 500 *g*.
10. Resuspend the mitochondria in 1 ml HMS buffer using a Dounce homogenizer.
11. Determine the protein concentration and adjust the concentration of the suspension to 14 mg protein/ml.[c]
12. Use the isolated mitochondria for protein import immediately (see Section 3.2).

[a] Starvation of the rat overnight is recommended by some laboratories, but this is not obligatory.
[b] Perfusion of the liver with HMS buffer is recommended in some protocols but is not necessary.
[c] It is often useful to have a relatively high concentration of the mitochondria (8 mg/ml) in protein import assays.
[d] It has been reported that the isolated mitochondria stay import competent for several hours. Most laboratories find that rat liver mitochondria lose their import competence during a single freeze–thaw cycle.

3.1.3 Isolation of mitochondria from other sources

A more general discussion of the isolation of mitochondria from different sources together with detailed protocols has been given by the authors of ref. 33. Additional protocols also can be obtained from ref. 25.

3.1.4 Assessment of the mitochondrial membrane potential

The membrane potential is essential in the translocation of mitochondrial presequences across the mitochondrial inner membrane. Therefore, a reduced membrane potential can be the reason for decreased efficiencies in the import of preproteins. The membrane potential can be assessed by a quenching assay using the fluorescence dye $DiSC_3(5)$. The dye is positively charged and taken up by the mitochondria. Inside the mitochondria the fluorescence of the dye is quenched by an unknown mechanism. Upon dissipation of the membrane potential the dye is quickly released. The difference between fluorescence before and after uncoupling of the mitochondria is taken as an indicator of the mitochondrial membrane potential. The assay can be carried out at higher temperatures, for example for expression of the temperature-sensitive phenotype of mutant mitochondria. The assay is described in *Protocol 6*. Valinomycin can be used in place of cyanide to dissipate the membrane potential.

Protocol 6. Assessment of the mitochondrial membrane potential

Equipment and reagents

- Fluorescence spectrophotometer and cuvette (3 ml volume): the fluorescence is determined at 25°C, excitation at 622 nm, emission at 670 nm, slit width 5
- BSA (fatty acid-free; Sigma, 9003T)
- Phosphate buffer: 0.6 M sorbitol, 0.1% (w/v) BSA, 10 mM $MgCl_2$, 0.5 mM EDTA, 20 mM potassium phosphate pH 7.2
- 3,3′-Dipropylthiadicarbocyanine iodide [$DiSC_3(5)$] (Molecular Probes; Cat. No. C-306): make a 2 mM solution in DMSO[a]
- Mitochondria, isolated as described in *Protocols 4* or *5*
- 1 M KCN (or NaCN) (caution: toxic)

Method

1. Add 3 ml phosphate buffer to the cuvette and place the cuvette in the fluorescence spectrophotometer for incubation at 25°C.
2. Add 3 μl of the 2 mM $DiSC_3(5)$ solution (final concentration in the cuvette 2 μM) and mix.
3. Add 20 μl mitochondria suspension (final concentration of mitochondrial protein in the cuvette 33 μg/ml) and mix immediately. The dye is taken up by the mitochondria which leads to a decrease in the fluorescence.
4. After 2 min (or later) add 3 μl 1 M KCN to dissipate the membrane potential. Measure the increase of the fluorescence. This indicates the membrane potential (see *Figure 1*).
5. If the membrane potential is low, repeat the measurement using a higher concentration of mitochondria (final concentration of mitochondrial protein in the cuvette 66 μg/ml). Alternatively, if under the standard conditions (steps 1–4) the fluorescence decreases completely, the concentration of mitochondria can be reduced.

[a] The Sigma reagent D-8658 is *not* identical to $DiSC_3(5)$ and does not work in the assay.

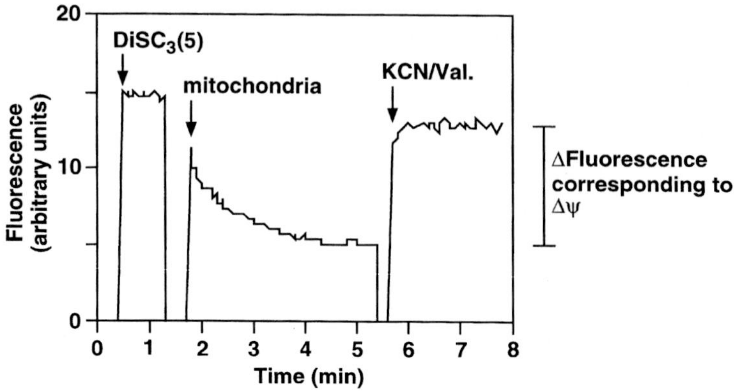

Figure 1. Assessment of the mitochondrial membrane potential. See *Protocol 6* for explanations.

Note that this assay does *not* allow the calculation of the absolute values of the membrane potential. However, it provides a means of determining the *relative* membrane potential in a comparison of different batches of mitochondria or different mutant strains. The relative values obtained by this assay correlate with the relative efficiencies in mitochondrial protein import experiments.

3.2 Import of preproteins into isolated mitochondria

3.2.1 The standard protocol

Nearly all mitochondrial preproteins can be imported using the standard conditions of *Protocol 7*. The conditions of this protocol can be used to import preproteins into mitochondria isolated from *S. cerevisiae*, *N. crassa*, and rat liver. The same conditions can be used to import *soluble* proteins in the absence of reticulocyte lysate. Factors affecting the import and modifications to the standard protocol are described in subsequent sections.

Protocol 7. Standard protocol for import of preproteins from reticulocyte lysate into isolated mitochondria

Equipment and reagents

- Centrifuge and 1.5 ml tubes to run at 2°C and 16000–17500 *g*
- BSA (fatty acid-free; Sigma, 9003 T)
- BSA buffer: 3% (w/v) BSA, 250 mM sucrose, 80 mM KCl, 5 mM $MgCl_2$, 10 mM MOPS–KOH pH 7.2
- 100 mM NADH (Boehringer Mannheim), freshly made in H_2O
- 0.2 M ATP (Sigma): adjust to pH 7.2 with KOH or NaOH, and store at −20°C
- 100 μM valinomycin (Sigma, V-0627; caution: extremely toxic): dissolve in ethanol and store at −20°C
- Ethanol (absolute)

- Reticulocyte lysate containing radiolabelled preprotein (or several preproteins), supplemented with 0.1 vol. 58 mM methionine and 58 mM cysteine (see Section 2.2), and 0.2 vol. 1.5 M sucrose[a]
- Isolated yeast mitochondria as prepared in *Protocol 4* (suspension in SEM buffer; 5 mg/ml mitochondrial protein)
- 1 mg/ml proteinase K (Boehringer Mannheim): dissolve in H_2O immediately before use[b]
- Shaker (e.g. Eppendorf mixer 5432)
- 100 mM PMSF (Sigma, P-7626; caution: very toxic by inhalation, contact with skin, or if swallowed): dissolve in ethanol and store the solution on ice[c]
- SEM buffer: 250 mM sucrose, 1 mM EDTA, 10 mM MOPS–KOH pH 7.2 (see *Protocol 4*)
- SDS–PAGE sample buffer: 62.5 mM Tris–HCl pH 6.8, 2% (w/v) SDS, 5% 2-mercaptoethanol, 10% glycerol, 0.002% (w/v) bromophenol blue

Method

1. Set up two series of 1.5 ml microcentrifuge tubes on ice.
2. To each tube add 86 μl BSA buffer, 2 μl 100 mM NADH, 1 μl 200 mM ATP pH 7.2, and mix.
3. To half of the tubes add 1 μl 100 μM valinomycin, add 1 μl ethanol to the other tubes. Vortex each tube immediately after the addition.
4. Add 5 μl reticulocyte lysate containing radiolabelled preprotein to all the tubes and mix.
5. Add 5 μl mitochondria (containing 25 μg mitochondrial protein). Vortex 1–2 sec at medium speed.
6. Incubate the tubes at 25°C for 15 min.
7. Cool the samples on ice. Divide each sample into two halves.
8. Add 2 μl proteinase K (1 mg/ml) to one-half of each sample. Vortex 1–2 sec at medium speed.
9. Incubate the samples at 0°C for 15 min.
10. Add 1 μl 100 mM PMSF to all samples. Vortex every sample immediately after addition of the PMSF.
11. Incubate for 5 min at 0°C to achieve complete inactivation of the proteinase K.
12. Pellet the mitochondria by centrifugation for 10 min at 16 000–17 000 *g*, 2°C.
13. Remove and discard the supernatants.
14. Remove residual BSA buffer and reticulocyte lysate from the pellets by adding about 150 μl SEM buffer and spinning for 5 min at 16 000 *g* (complete resuspension is *not* necessary). Discard the supernatants.
15. Add 25 μl SDS–PAGE sample buffer to each sample and shake vigorously for 7 min at room temperature to dissolve the mitochondria. Heat the tubes for 3 min at 95°C. Analyse the samples by SDS–PAGE and fluorography.[d]

[a] Sucrose is added to adjust the osmolarity of the reticulocyte lysate to that of the mitochondria.
[b] Alternatively the proteinase K solution can be stored in aliquots at –20°C.
[c] PMSF solution is stable on ice for a few hours.
[d] See ref. 73.

Homologous import in a cell-free system using yeast mitochondria in a yeast translation system has been reported by Hönlinger *et al.* (10).

The import of mitochondrial preproteins is monitored by several criteria. Most mitochondrial preproteins carry presequences which are cleaved upon transfer into the matrix. This is easily detectable by the increased mobility of the shortened protein in SDS–PAGE. However, cleavage of the presequence does not necessarily indicate complete translocation across the mitochondrial membranes; part of the mature protein may still be located outside the organelle. This can be tested by treatment of the mitochondria with proteinase K. Only those protein molecules which have been completely translocated across at least the outer membrane will be resistant to this treatment. In this context it is very useful to include a sample containing valinomycin to dissipate the mitochondrial membrane potential. Initial experiments on mitochondrial protein import should always include this control. It allows one to distinguish between membrane potential-dependent and independent processes. The membrane potential is particularly required for the insertion of proteins into and the translocation across the inner membrane. Only after translocation into the mitochondrial matrix should presequences become accessible to the matrix-localized processing enzyme. 'Leaky' mitochondria will release their processing enzyme and cause specific cleavage of presequences outside the mitochondria. This will only become visible by means of the valinomycin control. With intact mitochondria neither processing nor resistance against externally added proteases should occur. After dissipation of the membrane potential preproteins usually accumulate at the outer surface of the organelle.

Cleavage of the presequence and resistance against externally added proteases are the most important means to characterize the import of mitochondrial proteins. However, some preproteins do not carry a cleavable presequence. Examples are the ADP/ATP carrier of the mitochondrial inner membrane and all proteins of the mitochondrial outer membrane. Furthermore, many outer membrane proteins will remain exposed to the outer surface of the organelle even after arrival at their functional location. In this case the import reaction can be characterized by extraction with carbonate to test for insertion into the membrane (Section 6.1), by co-immunoprecipitations to demonstrate insertion into a protein complex (Section 6.2.1), or by testing for the formation of a characteristic pattern of proteolytic fragments after treatment with very low concentrations of proteases, especially by more specific proteases such as trypsin or V8 protease.

3.2.2 Factors affecting protein import: modification of the standard import conditions

i. Buffer conditions

The BSA buffer described in *Protocol 7* has been applied successfully in the import of preproteins into mitochondria from yeast, *N. crassa*, and rat liver. However, use of several other buffers has been reported in the literature.

- MOPS can be substituted by Hepes or by phosphate; Tris is not recommended.
- The sucrose is used to stabilize the mitochondria osmotically; it can be substituted by mannitol or sorbitol at a concentration of 500 mM.
- BSA considerably increases import efficiencies but can be omitted from the import buffer if required for specific reasons; it is important that the BSA is fatty acid-free because free fatty acids may uncouple the mitochondria. BSA buffers are not only suitable for import assays but also for subsequent protease treatment (as described in Section *vii*).
- The optimum salt concentration is different for different preproteins; 80 mM KCl is usually sufficient for preproteins containing cleavable presequences. Other preproteins (for example the ADP/ATP carrier and other preproteins which bind to mitochondrial import receptors by hydrophobic interactions) show better import at higher KCl concentrations (up to 200 mM) or in the presence of an additional 5 mM $MgCl_2$.
- For import into mitochondria from *S. cerevisiae*, it is of advantage to include about 20 mM potassium phosphate; the phosphate ions inhibit a proton-conducting channel in the inner membrane and thereby indirectly stabilize the mitochondrial membrane potential (34).

ii. Membrane potential

Translocation of preproteins across the mitochondrial membranes is driven by the membrane potential across the inner membrane ($\Delta\psi$ = about 160 mV) and by the Hsp70 of the mitochondrial matrix (mtHsp70) which in yeast is encoded by the *SSC1* gene and is ATP-dependent in its reaction cycle. This is the reason for the addition of NADH and ATP in the standard import buffer. NADH is added to the import assay to supply electrons indirectly to the respiratory chain. As an alternative, electrons can be provided by a combination of ascorbate and *N,N,N',N'*-tetramethyl-1,4-phenylenediamine (TMPD; Sigma, T-3134). In this case, the electrons are fed directly into cytochrome *c* and from there to complex IV of the respiratory chain. If this alternative is to be used, dissolve potassium ascorbate in H_2O at a concentration of 400 mM and adjust to pH 7 with KOH. The solution is stable at –20 °C. Use the ascorbate in the import assay at a final concentration of 8 mM. Freshly dissolve the TMPD in H_2O at a concentration of 10 mM and use it at a final concentration of 0.2 mM. If required, the respiratory chain can be blocked by antimycin A (although this only partially works with mitochondria from *S. cerevisiae*) and the membrane potential can subsequently be re-established by ascorbate and TMPD. Dissolve the antimycin A (Sigma, A-8674) in ethanol as a stock solution of 0.8 mM and use it at a final concentration of 8 μM. Usually oligomycin is added together with antimycin A to prevent the mitochondrial ATP synthase establishing the membrane potential. Dissolve the oligomycin (Sigma, O-4876) in ethanol at a concentration of 1 mM and use it at a final concentration of

20 μM. If several inhibitors are combined in one sample, take into account that the sample should not contain more than 2% ethanol during the import reaction. However, after completion of the import reaction (*Protocol* 7, step 7) more ethanol may be added, for example with the PMSF (*Protocol* 7, step 10). Higher concentrations of ethanol in the import reaction can be avoided by combining two inhibitors in the same stock solution (e.g. antimycin A and oligomycin) or by using higher concentrations of inhibitors in the stock solutions.

iii. ATP requirement

ATP is contained in the reticulocyte lysate and further addition of ATP can be omitted if sufficient reticulocyte lysate is contained in the samples for the import reactions. At lower concentrations of reticulocyte lysate the addition of 1–4 mM ATP may be of advantage. It facilitates release of preproteins from cytosolic chaperone proteins and it fuels the mtHsp70 motor of the protein import machinery. In some cases, especially at higher concentrations, ATP may inhibit the import. Therefore, it is worth testing the effect of different concentrations of ATP on the efficiency of the import reaction in a separate assay. The ATP dependence of mitochondrial protein import has been described in more detail in ref. 35.

iv. Divalent cations

Inside the mitochondrial matrix, presequences of precursor proteins are cleaved by the enzyme MPPα/MPPβ. The activity of this enzyme is dependent on divalent cations. Thus the enzyme is inhibited in the presence of chelators and can be reactivated by the addition of manganese ions. This provides an opportunity to inhibit cleavage of the presequence in the *in vitro* import system. EDTA can act as a chelator but it does not penetrate the mitochondrial membranes. To inhibit processing, add 5 mM EDTA and 0.1 mM 1,10-phenanthroline (*ortho*-phenanthroline; Sigma, P-9375) to the import assay. Freshly dissolve the 1,10-phenanthroline in water at a concentration of 20 mM. After dilution into the import buffer, it quickly diffuses across the mitochondrial membranes and inactivates the MPP enzyme. Add the EDTA from a stock solution of 250 mM which has been adjusted to pH 8. Treatment with chelators may reduce the import efficiency. Therefore, the optimal concentration of 1,10-phenanthroline should be determined by titration to achieve the optimal conditions.

v. Concentration of mitochondria

The amount of mitochondria in the import reaction may be reduced to 5 μg mitochondrial protein per sample or even less. Import reactions have been performed using only 50 ng mitochondrial protein in a standard assay (15). Higher concentrations of mitochondria (up to 200 μg/100 μl) may be used if the import efficiency of a particular preprotein is very poor.

vi. Import kinetics

In comparing import efficiencies under different conditions, or with mitochondria which have been isolated from different strains, it is essential to compare not only the absolute amount of protein imported at a given time point but also the import kinetics. The import reactions should be linear with respect to time. For example, if one reaction reaches 50% import efficiency after 2 min of import and the other reaction requires 10 min to reach the same import efficiency, this difference would not be visible if the reactions are only compared after 15 min. Besides comparing import efficiencies after short import times, it is possible to reduce the amount of mitochondria to slow down the import reaction and thus facilitate the comparison of different samples.

vii. Protease treatment

Treatment of mitochondria with proteinase K (*Protocol 7*, steps 8 and 9) is intended to digest all the preprotein outside the membranes. Therefore, it is necessary first to determine the protease sensitivity of the preprotein. At 0°C mitochondria can be treated with up to 200 μg/ml proteinase K, but 40 μg/ml are usually sufficient. Authentic mitochondrial preproteins are bound to chaperone proteins and are easily degraded by different proteases. Mitochondrial presequences typically form amphiphilic α-helices which remain protease-sensitive even if they are linked to a protease-resistant protein domain. Mitochondria can be treated with trypsin at similar concentrations to proteinase K. In this case, proteolysis is stopped by adding (and thoroughly mixing!) the sample with a 30-fold weight excess of soybean trypsin inhibitor relative to the amount of trypsin. Incubate the samples at 0°C for 5 min to allow complete binding of the trypsin to the inhibitor. The advantage of trypsin is that it can be combined with a treatment with proteinase K in a later step of the experiment. Proteinase K is not inhibited by residual trypsin inhibitor whereas trypsin is partially inhibited by the PMSF added at the end of the experiment. For example, it is possible to pre-treat mitochondria with trypsin to digest the import receptors at the outer surface (final concentration of trypsin required is 25 μg/ml at 0°C) and subsequently to perform an import experiment including the use of proteinase K.

3.2.3 Import of preproteins from 8 M urea solutions

Import from solutions containing 8 M urea is possible for many proteins, including some preproteins which are insoluble in the absence of urea. This is of special interest if the preprotein is synthesized in chemical amounts and found to be insoluble, for example after expression in *E. coli*. In principle, preproteins can also be imported from 6 M guanidinium chloride. However, in contrast to urea, this denaturant easily destroys the mitochondria. This problem can be overcome only by sufficient dilution of the guanidinium chloride (to about 50 mM). Import from 8 M urea requires a 40-fold dilution to 200 mM which is tolerated by the mitochondria.

The method is mainly limited to proteins of less than 40 kDa, probably due to facilitated aggregation of larger proteins. The efficiency of import of urea denatured proteins can be improved by addition of reticulocyte lysate (36) or by other methods which suppress aggregation of the preprotein during refolding in the import buffer. Aggregation of chemical amounts of proteins causes light scattering which can be assayed spectrophotometrically, recording the absorption at 320 nm. Aggregation of lower amounts of proteins can be monitored by centrifugation at 100 000 *g* and then comparing the amounts of protein in the pellet and supernatant.

Protocol 8 is suitable for import of preproteins which have been precipitated from reticulocyte lysate by ammonium sulfate (as described in *Protocol 2*) and for the import of preproteins which have been expressed in *E. coli*. Isolation and import of insoluble preproteins from *E. coli* inclusion bodies has been described in detail by Wienhues *et al.* (37) and by Becker *et al.* (36).

The major differences between this procedure and the standard protocol for protein import from reticulocyte lysate (*Protocol 7*) are:

- Since it is often more difficult to achieve high efficiencies of import, higher amounts of mitochondria are used in the import assay.
- The samples are pre-incubated at 25 °C to allow uptake of the preprotein by the mitochondria immediately after dilution into the import buffer.
- Since import of high amounts of preproteins may cause a delay in processing of the imported preproteins, processing is improved by longer times of incubation or, in some cases, by addition of $MnCl_2$ (final concentration not more than 2 mM) to the samples for a second incubation after the import reaction is finished.

Denaturation of preproteins in 8 M urea is described in *Protocol 2*. The samples may be stored at –70 °C but storage of preproteins in urea solution may cause covalent modifications of the preprotein and reduced import efficiencies. Therefore, it is often better to store small aliquots of the preprotein in the absence of urea and to denature the preprotein immediately before the import experiment.

Protocol 8. Import of urea denatured preproteins into isolated mitochondria

Equipment and reagents

- Preprotein sample (see Section 2)
- BSA buffer, 100 mM NADH, 200 mM ATP, 1 mg/ml proteinase K, 100 mM PMSF, 100 μM valinomycin, ethanol, centrifuge, and tubes (see *Protocol 7*)
- 8 M urea solution, freshly made (see *Protocol 2*)
- Isolated yeast mitochondria (suspension in SEM buffer; 5 mg/ml mitochondrial protein) and SEM buffer (see *Protocol 4*)

Method

1. Denature the preprotein for 45 min in freshly made 8 M urea solution.
2. Set up the import incubations as in *Protocol 7*, steps 1–3.
3. Add 10 μl mitochondria suspension (containing 50 μg mitochondrial protein) to each sample, vortex (1–2 sec at medium speed).
4. Incubate for 3 min at 25 °C.
5. Add 2.5 μl urea denatured preprotein (step 1) and mix immediately by vortexing.
6. Incubate at 25 °C for 10 min.
7. Cool the samples on ice. Divide each sample into two halves, and proceed as described in *Protocol 7*, steps 8–15.

If chemical amounts of preprotein have been imported in *Protocol 8*, the samples can be analysed by SDS–PAGE, Western blotting, and decoration with antibodies directed against the mature part of the preprotein. Defined amounts of the preprotein serve as standards for quantification.

3.3 Generation of translocation intermediates

The analysis of many different translocation intermediates has been an important contribution in the elucidation of the mechanisms of mitochondrial protein import. Methods to arrest preproteins at distinct stages of translocation have been described by Söllner *et al.* (38), Cyr *et al.* (39), and Alconada *et al.* (40). The following sections give an introduction to methods for studying some of the most interesting intermediates. In the procedures described, the basic method of *Protocol 7* is followed, modified as indicated below.

3.3.1 Translocation intermediates spanning both mitochondrial membranes: import of preproteins at low temperature

The translocation of many preproteins can be slowed down simply by lowering the temperature during the import reaction. In general this does not apply to the import of urea denatured proteins (as described in *Protocol 8*) but is suitable in the case of import from reticulocyte lysate as described in *Protocol 7*. If the import is performed at 8–10 °C instead of 25 °C, many preproteins will accumulate as translocation intermediates spanning both mitochondrial membranes. The presequence is cleaved, indicating translocation of the presequence into the mitochondrial matrix, and the same molecules remain accessible to externally-added proteases, indicating that parts of the protein are still located at the outer surface of the mitochondria. Fragments characteristic of the translocation intermediates are usually not formed in this protease treatment.

The disadvantage of this procedure is that the intermediates are kinetically trapped and are stable for only a limited time. On the other hand, the advantage of this type of intermediate is that it can be accumulated without chemical modifications of the system. Moreover, the intermediates can easily be 'chased' across the membranes by increasing the temperature, demonstrating their location on an authentic import pathway.

3.3.2 Stable translocation intermediates: import of preproteins after depletion of ATP

A more stable translocation intermediate can be formed for many preproteins by lowering the concentration of ATP in the import assay. This can be achieved by separately pre-incubating the mitochondria and the reticulocyte lysate with the enzyme apyrase (apyrase grade VIII; Sigma, A-6160). Dissolve the enzyme in water at a concentration of 1 U/μl; aliquots can be frozen in liquid nitrogen and stored at –20 °C. Pre-incubate the mitochondria and reticulocyte lysate for about 15 min with up to 20 U per ml of the enzyme at 0–25 °C. To the samples containing mitochondria, also add 20 μM oligomycin (Sigma, O-4876) from a 100-fold concentrated stock solution in ethanol (store at –20 °C) along with the apyrase. Oligomycin suppresses the synthesis of ATP inside the mitochondria and is essential in these experiments. The optimal concentrations of apyrase have to be determined individually for each preprotein. Since commercial preparations of apyrase contain many impurities, higher concentrations of the enzyme and incubations at higher temperatures should be avoided. If mitochondrial protein import is completely blocked, the pre-incubation of the mitochondria with the apyrase can be omitted or reduced to very mild conditions.

The treatment with apyrase depletes the mitochondria of ATP and hence inactivates the mitochondrial Hsp70 which is regarded as the 'motor of mitochondrial protein import'. This implies that preproteins will proceed through the mtHsp70-independent steps of import and arrest at the mtHsp70-dependent step of the translocation, i.e. only the presequence or part of the presequence will traverse the membranes, and the mature part of the preprotein will remain outside the organelle. The effect of the ATP depeletion on different preproteins is not identical because preproteins differ in their requirement for mtHsp70. The import of some preproteins is completely ATP-independent. A preprotein which is easily handled in this system is the fusion protein Su9-DHFR containing the first 69 amino acids of subunit 9 of the mitochondrial ATP synthase from *N. crassa* fused to the DHFR of the mouse (18, 19, 41). Accumulated translocation intermediates can be chased by the addition of ATP and subsequent incubation at 25 °C. An intriguing example of this technique (together with detailed protocols) is described by Manning-Krieg *et al.* (19). A second well established system is provided by the ADP/ATP carrier (see Section 3.3.4).

3.3.3 Translocation arrest by reversible folding: import of preproteins containing a DHFR domain

Fusion proteins containing mitochondrial presequences fused to mouse DHFR have been used in many studies to address details of the mechanism of mitochondrial protein import and folding (for general properties of the enzyme, see Section 2.6.2). One advantage of such fusion proteins is the possibility of reversibly arresting translocation intermediates which span both mitochondrial membranes by import in the presence of the DHFR ligand, methotrexate (MTX) (see Section 6.2.3, *Figure 2B*). The translocation arrest in the presence of MTX is explained by the effect of the MTX on the stability of the native structure of the DHFR domain (14, 15). To allow translocation across the mitochondrial membranes the preprotein must be completely unfolded. In the case of the DHFR domain this unfolding can be prevented by binding of the ligand MTX.

MTX, or (+)amethopterin (Sigma, A-6770; caution, toxic) is insoluble in H_2O but can be dissolved in 100 mM MOPS pH 7.2 at a concentration of 10 mM and is stable at –20 °C. For use in the import experiment, dilute this stock solution 100-fold into H_2O to make a 100 μM solution, which should be used only once and then discarded because MTX is unstable at this dilution. Dilute the MTX further into the import assay to give a final concentration of 0.2–2 μM. Pre-incubate the DHFR fusion protein (as a soluble protein in chemical amounts or as a preprotein synthesized in reticulocyte lysate) with the MTX at 0 °C for 5 min prior to addition of the mitochondria. The translocation intermediate is accumulated at 25 °C and is characterized by processing of the intermediate by the matrix localized processing enzyme MPP and by sensitivity to externally added proteases. The intermediate can be 'chased' by dilution of the MTX. For this purpose, the mitochondria are re-isolated by centrifugation (10 min at 16000–17500 *g*) and resuspended in buffer in the absence of MTX. If necessary the dilution procedure is repeated. The translocation intermediate then will be imported upon further incubation at 25 °C, leading to complete resistance to externally added protease. A detailed characterization of this system can be obtained from ref. 15. To exclude the possibility that the resistance to proteolysis is due to the pressing of the DHFR domain against the outer surface of the mitochondria, incubate the samples with apyrase (as described in Section 3.3.2) to inhibit the mtHsp70 and prevent its effect on the mobility of the polypeptide in transit. In the subsequent incubation with proteinase K, the intermediate will now again be degraded if the DHFR is accessible from the outside.

3.3.4 Translocation in distinct stages: import of the ADP/ATP carrier

The ADP/ATP carrier (AAC) is the most abundant protein of the mitochondrial membranes. Its import pathway into the mitochondrial inner membrane of *N. crassa* has been characterized in five distinct stages (42). The same stages have been reproduced with mitochondria from *S. cerevisiae*. After synthesis in reticulocyte lysate ('Stage 1' in the nomenclature suggested in ref. 42) the AAC can be arrested in the following locations:

(a) *Arrest at the import receptors of the outer membrane (Stage 2)*. Binding to the outer surface of mitochondria is mainly mediated by the outer membrane protein, Tom70. The import intermediate is accumulated after uncoupling of the mitochondria (by the addition of valinomycin) and depletion of ATP (by apyrase). Incubate the reticulocyte lysate containing the radiolabelled AAC with 5 U/ml apyrase (see Section 3.3.2) for 25 min at 25 °C. Separately incubate the mitochondria with the same concentration of apyrase at 0 °C. Add the pre-incubated mitochondria to the import assay containing the ATP-depleted reticulocyte lysate plus 0.5 μM valinomycin, 8 μM antimycin A, and 20 μM oligomycin (added from 100-fold concentrated stock solutions in ethanol, stored at –20 °C, see Section 3.2.2) and incubate the mixture for 5–10 min at 25 °C. If desired, the translocation intermediate can be chased to Stage 3 by addition of 2 mM ATP.

(b) *Accumulation of the AAC in the intermembrane space (Stage 3).* In this stage (the 'GIP intermediate') the AAC has traversed the outer membrane and is exposed to the intermembrane space (43). The intermediate is generated by import in the presence of antimycin A, oligomycin, and valinomycin (see Stage 2) but in the additional presence of ATP. The accumulated AAC is resistant to low amounts of externally added proteases (20 μg/ml proteinase K in the case of mitochondria from *S. cerevisiae*). Any AAC molecules which only proceeded to Stage 2 during the import reaction will be digested under these conditions. Without protease treatment a mixed population accumulates. At very high concentrations of proteinase K (250 μg/ml) the Stage 3 intermediate is degraded. After opening of the intermembrane space by swelling of the mitochondria, the intermediate becomes susceptible low concentrations of proteases (43). An additional characteristic of AAC at Stage 3 is that it is easily extracted by carbonate (see Section 6.1, *Protocol 13*) whereas AAC at its functional location is an integral protein resistant to carbonate extraction.

(c) *Import into the inner membrane (Stage 5).* If the AAC is imported under standard conditions (in the presence of ATP and where the mitochondria establish a membrane potential), it assembles in the inner membrane and forms dimers (and possibly tetramers). At this stage the AAC has reached its functional location and has become resistant to extraction by carbonate. The assembly can be demonstrated by blue native electrophoresis (see Section 6.2.2, *Protocol 18*). An indirect test can be performed, taking advantage of the observation that only the functionally assembled AAC passes through small hydroxylapatite columns while all non-assembled species of the AAC are retained. The procedure described in *Protocol 9* has been adapted from ref. 44. The atractyloside is added to stabilize the AAC dimers. Carboxyatractyloside is more efficient than atractyloside but is not commercially available.

An alternative to TCA precipitation (*Protocol 9*) for the analysis of AAC by SDS–PAGE and fluorography is to use a specific anti-AAC antibody for immunoprecipitation (see Section 6.2.1). An assembly test for the bovine mitochondrial phosphate carrier has been published by Zara *et al.* (45).

Protocol 9. Test for assembly of the ADP/ATP carrier

Equipment and reagents

- Hydroxylapatite (BioGel HTP; Bio-Rad 130-0420)
- Low Triton buffer: 0.5% Triton X-100, 50 mM NaCl, 10 mM MOPS–KOH pH 7.2
- 1 ml micropipettor tips: close the end of each with a small plug of glass wool for use as small chromatography columns
- Mitochondria containing imported radio-labelled AAC (*Protocol 7*)
- Control mitochondria containing imported radiolabelled AAC to Stage 3, prepared as described in Section 3.3.4(b); these serve as a control for non-assembled AAC
- High Triton buffer: 2.5% Triton X-100, 110 mM NaCl, 20 mM MOPS–KOH pH 7.2
- Atractyloside (Sigma, A-6882): prepare a 1 mM stock solution in ethanol
- High Triton buffer containing 10 μM atractyloside (added from the stock solution)
- 0.2% sodium deoxycholate (Sigma, D-6750)
- 70% (w/v) TCA
- Acetone (at −20 °C)
- SDS–PAGE sample buffer (*Protocol 7*)

Method

Perform the steps below in a cold room.

1. Suspend the hydroxylapatite in low Triton buffer.
2. Add a volume of the suspension containing 40 mg hydroxylapatite to several pipette tips closed by glass wool.
3. Test the flow rate of the columns by addition of low Triton buffer. Discard those tips which have a low flow rate or which lose hydroxylapatite.
4. Dissolve samples of the experimental and control mitochondria (50 μg mitochondrial protein) in 100 μl high Triton buffer containing atractyloside.
5. Incubate the samples for 10 min on ice.
6. Remove insoluble material by centrifugation for 5 min at 16 000–17 500 *g*.
7. Apply each supernatant to a hydroxylapatite pipette tip column. Collect the flow-through fraction.
8. Elute the columns with 400 μl low Triton buffer. Combine the eluate and flow-through fractions from each column.
9. Add 0.1 vol. 0.2% deoxycholate to each sample to help precipitate the proteins.
10. Precipitate the proteins by adding 70% TCA to 10% final concentration. Mix carefully and incubate on ice for 15 min.

Protocol 9. *Continued*

11. Pellet the precipitated proteins at 22 000 *g* for 20 min. Remove and discard the supernatants.
12. Spin the tubes again to remove residual supernatant.
13. Add 500 μl cold acetone to each pellet.
14. Pellet the proteins at 22 000 *g* for 20 min. Allow the pellets to air dry at 25 °C.
15. Dissolve the pellets in SDS–PAGE sample buffer (*Protocol 7,* step15) and analyse the proteins by SDS–PAGE and fluorography (ref. 73).

3.4 Localization of imported proteins

To localize a protein inside mitochondria, the mitochondrial compartments are opened to see in which compartment the protein of interest becomes accessible to added proteases. Proteins which are exposed at the outer surface of mitochondria are accessible already to externally added proteases. To open the intermembrane space, the mitochondria are transformed into mitoplasts by swelling. For this purpose, the solution containing the mitochondria (e.g. in 0.6 M sorbitol or 250 mM sucrose) is diluted tenfold with water. Under these conditions, the outer membrane is opened while the inner membrane is kept intact. Incubate the sample for 10–15 min at 0 °C and then re-establish the original osmotic conditions by addition of a concentrated solution of sorbitol or sucrose. Isolate the mitoplasts by centrifugation at 16 000–17 500 *g* for 12 min but note that when resuspending the mitoplasts they have to be treated carefully because they are more fragile than mitochondria. For opening of the matrix space, the mitochondria are sonicated under continuous cooling in ice. Long pulses should be avoided to allow sufficient cooling in between.

The procedures described above are mainly applicable to mitochondria from *S. cerevisiae*. Mitochondria from the fungus *N. crassa* have been fractionated using increasing concentrations of digitonin. Exact protocols for this method can be obtained from Wienhues *et al.* (37). The method is not recommended for mitochondria from *S. cerevisiae*. All fractionations require analysis of compartment-specific marker proteins. To distinguish between proteins of the mitochondrial outer and inner membranes sucrose gradients can be used, and these are described in *Protocol 10*. The protocol has been developed for the analysis of mitochondrial proteins from the yeast *S. cerevisiae*. The vesicles are prepared by sonication. The separation of inner and outer membrane vesicles which is achieved by this method is sufficient for analytical purposes. The density of mitochondrial membrane vesicles corresponds to 32–43% sucrose. For preparative scale separation, step gradients are required as described in refs 46 and 47. Detailed protocols and tables concerning centrifugation techniques can be obtained from ref. 48.

Protocol 10. Sucrose gradient separation of mitochondrial inner and outer membranes

Equipment and reagents

- Ultracentrifuge with a swing-out rotor for 6 ml tubes, and a fixed angle rotor
- Two-chamber gradient mixer (chamber volume 3 ml)
- Buffer A: 2.5 mM EDTA,[a] 100 mM KCl,[b] 20 mM MOPS–KOH pH 7.2[c]
- Sucrose solutions: 20% and 55% (w/v) in buffer A
- Mitochondria after import of radiolabelled protein (*Protocol 7*)
- 10% sucrose in buffer A
- Pepstatin A: 1 mg/ml in ethanol
- SEM buffer, 100 mM PMSF, and SDS–PAGE sample buffer (see *Protocol 7*)
- Sonifier (Branson Sonifier 250, with tip)
- Dounce homogenizer with pestle
- 70% TCA and acetone (*Protocol 9*)

Method

1. Prepare 55–20% sucrose gradients in 6 ml tubes. Add the 55% sucrose solution to the exit chamber of the two-chamber gradient mixer. Cool the gradients on ice.
2. Mix samples of mitochondria containing 250 μg mitochondrial protein in 600 μl buffer A, with 2 μl of 1 mg/ml pepstatin A, and 2 μl of 100 mM PMSF. Sonicate the samples six times for 5 sec at intervals of 15 sec with continuous cooling in ice.
3. Remove any intact mitochondria by centrifugation for 10 min at 16 000 *g*, 2 °C. Discard the pellet.
4. Pellet the membrane vesicles by centrifugation of the supernatant for 30 min at 100 000 *g* in a fixed angle rotor.
5. Remove the supernatant and resuspend the pellet carefully in 200 μl 10% sucrose in buffer A using a Dounce homogenizer.
6. Layer each vesicle suspension on a 55–20% sucrose gradient. Centrifuge for 16 h at 100 000 *g*, 2 °C in the swing-out rotor.
7. Fractionate the gradients into 500 μl fractions.
8. Precipitate the proteins from each fraction using TCA (7.5–10% final concentration), and wash the pellets with acetone (*Protocol 9*, steps 10–14).
9. Analyse the proteins by SDS–PAGE and immunoblotting (labelling marker proteins of outer and inner membrane by specific antibodies).

[a] EDTA should be included to release membrane-bound ribosomes.
[b] The KCl concentration may be varied if necessary.
[c] Hepes may be used instead of MOPS.

4. Import of proteins into microsomes (endoplasmic reticulum)

The machinery which mediates the translocation of proteins into the endoplasmic reticulum is possibly the best studied protein translocation system of the eukaryotic cell. Basic methods and techniques have been published in refs 26 and 49. Microsomes from dog pancreas are now commercially available from Boehringer Mannheim (Cat. No. 1 108 484) and are supplied together with a detailed protocol for import of preproteins from reticulocyte and wheat germ lysates. Dog pancreas microsomes can be used to import precursors of endoplasmic reticulum proteins, secretory proteins, lysosomal proteins, and plasma membrane proteins from different organisms.

4.1 Isolation of microsomes

For experiments on a larger scale microsomes can be prepared using a sucrose step gradient according to *Protocol 11*. What is isolated is the rough microsome fraction. The procedure has been developed essentially by Blobel and Dobberstein (50) and modified by Watts *et al.* (51). A shorter protocol for isolation of dog pancreas microsomes has been published by Walter and Blobel (52).

It should be noted that the pancreas is an organ which secretes digestive enzymes. Therefore, it is essential to keep all samples cooled and to add the protease inhibitors, PMSF and EDTA. Magnesium ions are included in some buffers to stabilize the ribosomes of the rough microsomes. The pancreas is obtained from dogs as described by Walter and Blobel (52). It is recommended to start the preparation of microsomes immediately after sacrificing the dog, but it is also possible to start with a frozen organ.

Protocol 11. Isolation of microsomes from dog pancreas

Equipment and reagents

- Dog pancreas, 100 g frozen[a]
- Cooled centrifuge (Beckman) with rotor (JA-20)
- Ultracentrifuge (Beckman), rotor (Ti70)

The following equipment should be autoclaved (at 121 °C for 20 min):

- Teflon Potter 30 ml homogenizer (motor driven)
- 2 ml Dounce homogenizers (type S) with glass pestles
- Tissue press (diameter of sieve holes 1 mm)
- Centrifuge tubes (25 ml), beakers (500 ml), Erlenmeyer flasks, tweezers, razor blade (to cut the pancreas tissue)

Adjust the pH of the following stock solutions at room temperature. Autoclave them (121 °C for 20 min) except those indicated (*) which should be filter sterilized by passage through a 45 μm Millipore filter.

- 1 M Hepes–KOH pH 7.5
- 2.5 M sucrose (RNase-free): 427.9 g dissolved in 500 ml final volume
- 2 M potassium acetate
- 2 M KCl
- 0.5 M magnesium acetate
- 1 M $MgCl_2$
- 0.2 M $CaCl_2$
- 0.2 M EDTA–KOH pH 7

- 0.2 EGTA–KOH pH 7 (Titriplex VI; 7.607 g in a final volume of 100 ml, soluble only after adjusting the pH) (*)
- 1 M triethanolamine (TEA)–HCl pH 7.5 (*)

The following are also required:

- PMSF (Sigma, P-7626; caution, very toxic): 40 mg/ml in propan-2-ol
- 2-Mercaptoethanol
- 0.5 M DTT
- 2% (w/v) SDS
- Distilled H_2O (autoclaved, 121 °C for 20 min)
- Micrococcal nuclease (Boehringer Mannheim; 8000 U/ml)[b]

Prepare the following solutions using the sterilized stock solutions and autoclaved H_2O in sterilized Erlenmeyer flasks and store them at 0–4 °C:

- TEA buffer: 50 mM TEA–HCl pH 7.5, 50 mM KCl, 5 mM $MgCl_2$, 0.1% (v/v) 2-mercaptoethanol
- Homogenization buffer (prepare 1000 ml): 0.25 M sucrose dissolved in TEA buffer
- 1.3 M, 1.5 M, 1.75 M, and 2.1 M sucrose solutions in TEA buffer (100 ml of each)
- 0.25 M sucrose, 20 mM Hepes–KOH pH 7.5, 50 mM KCl, 1 mM DTT (prepare 200 ml)
- 0.2 M and 0.5 M sucrose (100 ml each) in 20 mM Hepes–KOH pH 7.5, 50 mM KCl, 2 mM magnesium acetate, 1 mM DTT

Method

Carry out the preparation in a cold room.

1. Place 100 g frozen pancreas into a 500 ml beaker, add 100 ml homogenization buffer, and PMSF to a final concentration of 40 μg/ml. Allow the tissue to thaw at 0 °C. Meanwhile prepare the sucrose gradients (step 2).
2. Set up ten sucrose step gradients in centrifuge tubes by layering successively 4 ml 2.1 M sucrose in TEA buffer, 4 ml 1.75 M sucrose in TEA buffer, and 4 ml 1.5 M sucrose in TEA buffer. Keep the gradients in the cold room.
3. Use a razor blade to cut up the tissue. Press the pieces of the thawed tissue through the tissue press and collect the tissue in a beaker. Discard the fat and connective tissue which collect in the sieve holes.
4. Add homogenization buffer to the tissue to give a total volume of about 100 ml.
5. Homogenize the tissue using a motor driven Potter homogenizer with five up-and-down strokes of the Teflon pestle at 1500 r.p.m. applying medium strength.
6. Spin the suspension for 10 min at 12 000 *g*, 4 °C to obtain a post-mitochondrial supernatant (JA-20 rotor). Pool the supernatants and dilute them to a total volume of 200 ml by addition of homogenization buffer.
7. Discard the pellets and layer about 13 ml of supernatant onto each sucrose gradient.
8. Centrifuge the gradients for 15–20 h at 115 000 *g* (45 000 r.p.m. in a Ti70 rotor), 4 °C to isolate the rough microsomes.
9. The rough microsomes should be concentrated at the interphase between the 2.1 M and 1.75 M sucrose layers. However, they are often also contained within the 1.75 M sucrose layer, sometimes visible as

Protocol 11. *Continued*

a sharp yellow/white band. Aspirate and discard the red/brown upper half of the gradient. Collect the rough microsome fraction using a Pasteur pipette. The total volume of the collected microsome fractions may approach 40 ml. Smooth microsomes may contaminate this preparation although they are mainly concentrated at the interphase of the 1.75 M and 1.5 M sucrose layers.

10. Add an equal volume of homogenization buffer to the pooled microsome fraction.

11. Set up four centrifuge tubes, each containing a cushion of 2 ml 1.3 M sucrose in TEA buffer and 40 μg/ml PMSF.

12. Layer 12 ml microsomal suspension on each sucrose cushion and centrifuge for 1 h at 115 000 *g*, 4 °C.

13. Recover the microsomes (which stick to the tube walls). Discard the dense suspension which is found in the lower part of the tubes after the run.[c]

14. Resuspend the microsomes from each tube in 2 ml 0.25 M sucrose, 20 mM Hepes–KOH, 50 mM KCl, 2 mM DTT (i.e. with *no* magnesium ions). Successively rinse the tubes with an additional 2 ml of the same buffer and add this to the microsomes.

15. Homogenize the combined suspension (10 ml total) with three strokes in a small Dounce homogenizer (type S).

16. Remove 50 μl of the homogenate, dilute it 50-fold in 2% SDS, and determine the A_{280}. Adjust the concentration of the isolated microsomes to $A_{280} = 50$ with 0.25 M sucrose, 20 mM Hepes–KOH, 50 mM KCl, 2 mM DTT.

17. Add $CaCl_2$ (final concentration 1 mM), then add micrococcal nuclease (final concentration 40 U/ml). Incubate 10 min at exactly 20 °C to digest the endogenous mRNA of the sample.

18. Stop the reaction by adding EGTA–KOH pH 7 (final concentration 2 mM). Mix and then cool on ice.

19. Add EDTA–KOH pH 7 (final concentration 5 mM) to dissociate the ribosomes from the microsomal membranes. Incubate for 5 min at 0 °C.

20. Re-isolate the microsomes by layering 5 ml on a 5 ml cushion 0.5 M sucrose, 20 mM Hepes–KOH pH 7.5, 50 mM KCl, 2 mM magnesium acetate, 1 mM DTT.

21. Spin for 1 h at 115 000 *g*, 4 °C.

22. Resuspend the microsomal pellet in 5 ml 0.2 M sucrose, 20 mM Hepes–KOH pH 7.5, 20 mM KCl, 2 mM magnesium acetate, 1 mM

DTT, using the Dounce homogenizer (type S). Adjust the microsomes to a concentration of $A_{280} = 30–40$ using the same buffer. Yield: about 20 ml.

23. Freeze the microsomes in 50 μl aliquots of in liquid nitrogen and store them at –70 °C.

[a] The pancreas of a single dog weighs 15–60 g.
[b] Alternatively staphylococcal nuclease (16 000 U/ml; Boehringer Mannheim) may be used. Both enzymes require calcium ions for activity.
[c] The tubes together with the microsomes may be frozen at –70 °C without loss of activity.

Prepared by the method described in *Protocol 11*, microsomes are stable at –70 °C for at least one year. Before use, they should be thawed quickly, for example in a water-bath at room temperature. They tolerate at least three cycles of freezing and thawing but the number of freeze–thaw cycles should be kept limited. Protocols for the isolation of translocation-competent microsomes from yeast have been published elsewhere (9, 53, 54).

4.2 Import of preproteins into microsomes

In contrast to mitochondrial preproteins, the transport of secretory proteins into the mammalian endoplasmic reticulum in most cases occurs co-translationally. However, in yeast most preproteins are translocated post-translationally in both organelles. For reconstitution of co-translational protein import into the endoplasmic reticulum in a cell-free system, isolated microsomes are added to wheat germ lysate or to reticulocyte lysate during translation of the preprotein mRNA. *Protocol 12* describes the procedure. The import reaction is carried out in a final volume of 20 μl. The sample is supplemented by SDS–PAGE sample buffer and analysed by SDS–PAGE and fluorography. Normally the reticulocyte or wheat germ lysate is obtained commercially and will be optimized for efficient mRNA translation. If the lysate is prepared in the laboratory, it should be optimized with respect to magnesium acetate and KCl concentrations. For details see Sections 2.2 and 2.3.

Protocol 12. Co-translational protein import into isolated dog pancreas microsomes

Reagents

- Reticulocyte lysate or wheat germ lysate plus associated solutions and buffers for translation (Amersham, Boehringer Mannheim, Promega)
- [^{35}S]methionine: 10–15 mCi/ml (Amersham SJ 1515, or NEN)
- Microsomes isolated from dog pancreas ($A_{280} = 30–40$); see *Protocol 11*
- mRNA encoding the protein of interest: about 100 ng/μl in H_2O
- 1 mg/ml proteinase K, 100 mM PMSF, and SDS–PAGE sample buffer (*Protocol 7*)
- 20% (v/v) Triton X-100

Protocol 12. *Continued*

Method

1. In 1.5 reaction tubes on ice, mix all the components of the translation system (including 2.5 μl [^{35}S]methionine), except the microsomes and the mRNA, in a volume of 18 μl.
2. Add 1 μl microsomes to each tube.[a] Vortex for 2 sec.
3. Add 1 μl mRNA (50–150 ng) to each tube.
4. Incubate at 30 °C for up to 60 min.
5. Divide the samples into three parts. Treat each part as follows:
   (a) No further additions; the import sample.
   (b) Add 1 mg/ml proteinase K (to 50–175 μg/ml final concentration); to assess the protection of imported protein.
   (c) Add 20% Triton X-100 (to 1% final concentration) plus 1 mg/ml proteinase K (to 50–175 μg/ml final concentration); to show imported protein becomes sensitive to proteinases after membrane dissolution.
6. Incubate samples (a), (b), and (c) at 0 °C for 10 min.
7. Stop the action of proteinase K by adding 0.1 M PMSF to 2 mM final concentration. Mix each sample immediately after addition of the PMSF and incubate for 5 min at 0 °C.
8. Add SDS–PAGE sample buffer and process the samples for SDS–PAGE and fluorography as described in *Protocol 7*, step 15.[b]

[a] The optimum volume of microsomes may vary between 0.5–3 μl.
[b] Re-isolation of the microsomes after the import reaction is not required.

To demonstrate post-translational protein transport, translate the mRNA in the absence of microsomes and stop the reaction by adding cycloheximide (100 μg/ml final concentration) and RNase A (80 μg/ml final concentration) from 25-fold concentrated stock solutions and then incubating for 5 min at 37 °C. If required, the ribosomes can be removed from the translation mixture prior to addition of the microsomes by centrifugation for 15 min at 400 000 *g*. Then add the microsomes and incubate the samples for up to 30 min at 37 °C.

Of special interest in the import of proteins into microsomes is the creation of truncated mRNA transcripts. This technique provides a means to obtain nascent polypeptide chains which remain bound to the ribosome and can be used to accumulate translocation intermediates for chemical cross-linking (55, 56).

5. Import of proteins into other organelles

For methods to study protein import and folding in other cellular systems, consult the references in *Table 1*.

Table 1. Methods for studying protein import and folding in organelles

Membrane system	References
Bacterial plasma membrane	26
Chloroplasts	26, 57
Coated pits and coated vesicles	58–60
Endocytosis	5, 51, 59–61
Endoplasmic reticulum	26, 58
Golgi cisternae	58
Lysosomes	58, 59, 61
Nuclear envelope assembly	58
Mammalian nuclei	58, 61
Yeast nuclei	62
Peroxisomes	26
Plant nuclei	57
Plasma membrane	60, 61
Secretion	5, 58–61
Vacuole	5

6. Analysis of protein import into organelles

6.1 Monitoring the association of proteins with membranes

In characterizing the interactions of proteins with membranes, two simple methods are commonly used to obtain initial information:

(a) By *sonication of the membranes*, one can determine whether or not a protein is associated with the membrane. Moreover, sonication at increasing salt concentrations (0–500 mM KCl) shows whether a protein is only loosely associated or tightly bound. In general, ionic interactions are salt-sensitive, hydrophobic interactions are salt-resistant. The efficiency of sonication must be controlled by testing the association of marker proteins of known topology.

(b) By *carbonate extraction*, one can test if an endogenous or newly imported protein is an integral membrane protein which is inserted in the lipid bilayer or if the protein is associated in a different way (63). Only integral membrane proteins are resistant to extraction by carbonate at pH 11.5. As with other methods, it is necessary to compare the data for a given protein with the behaviour of marker proteins to determine the efficiency of extraction.

Protocol 13 provides a method for carbonate extraction of membrane proteins. To achieve efficient extraction it is essential to use a fresh carbonate solution which is kept on ice, to use low amounts of membranes, and to suspend the membranes carefully. Sometimes it is necessary to extract the membranes twice. After centrifugation, the supernatants are combined and tested for extracted proteins after precipitation by TCA. In parallel the carbonate-resistant membrane pellet should also be treated with TCA. A sample of membranes not subjected to carbonate extraction serves to determine the total protein content of the membranes.

Protocol 13. Carbonate extraction of membrane proteins

Equipment and reagents

- Membrane samples for analysis (*Protocols 7, 8,* or *12*)
- 100 mM Na_2CO_3 in H_2O: make no further additions and do not adjust the pH; cool the solution in ice
- Ice-bath at 0 °C
- Ultracentrifuge, rotor, and tubes for centrifugation up to 250 000 *g*
- 70% and 10% (w/v) TCA
- Cooled centrifuge, rotor, and tubes for centrifugation up to 25 000 *g*
- Acetone (at –20 °C)
- SDS–PAGE sample buffer (*Protocol 7*)

Method

1. Set up two samples of membrane suspensions (50 μg protein) and pellet the membranes by centrifugation. Discard the supernatants.
2. Leave one sample on ice (until step 7) for the determination of total protein (control membranes).
3. Resuspend the other pellet in 400 μl ice-cold 100 mM Na_2CO_3 and extract the membranes for 30 min at 0 °C.
4. Pellet the membranes for 45 min at 250 000 *g*, 2 °C.
5. Remove the supernatant and transfer it to a separate tube. Retain the pellet.
6. Add 70% TCA to the supernatant (to 10% final concentration) and mix carefully. Resuspend the extracted membrane pellet (step 5) in an equal volume of 10% TCA.
7. Suspend the control membrane pellet (from step 2) in an equal volume of 10% TCA.
8. Incubate all three samples for 15 min on ice.
9. Pellet the TCA precipitated proteins in each sample for 20 min at 22 000 *g*. Discard the supernatants. Spin the tubes again for 5 min at 22 000 *g* and discard residual supernatants.
10. Add 500 ml acetone (at –20 °C) to each of the three TCA pellets and vortex.
11. Pellet the proteins at 22 000 *g*, discard supernatants. Allow the pellets to air dry at 25 °C.
12. Dissolve the TCA pellets in SDS–PAGE sample buffer (*Protocol 7*, step 15).
13. Analyse the proteins by SDS–PAGE and Western blotting (ref. 73).

If precipitation by TCA is incomplete, the efficiency can be improved by adding 0.02% (w/v) sodium deoxycholate from a tenfold concentrated stock solution (see *Protocol 9*) or a carrier protein (e.g. 20 mg/ml BSA) prior to the TCA. This may be necessary if the protein concentration of the sample is very low (below some μg/ml). In this case, a prolonged incubation with the TCA can be of additional advantage (up to 1–2 h).

6.2 Analysis of protein complexes involved in organelle import

Many different methods can be applied to investigate the interactions of imported proteins with components of the translocation machinery, with molecular chaperones, or with associated proteins in the functional location. The protocols in this section represent a selection of some very useful techniques to analyse protein complexes in intracellular protein transport. In choosing the method, the following should be considered:

- If only very small amounts of a protein complex are available (not directly detectable in SDS–PAGE), co-immunoprecipitation (Section 6.2.1) is probably the method of choice to isolate the complex. In this case not only an antiserum of sufficient specificity is necessary but also a sensitive assay to detect the small amounts of proteins which will be precipitated. Possible methods are immunostaining of Western blots or the use of radiolabelled protein for import.
- If larger amounts of proteins are to be isolated, for example to obtain sufficient material for sequencing (see Chapter 1), immune precipitations can be performed with antibodies covalently coupled to Protein A–Sepharose (Section 6.2.1). This allows the scaling up of the co-immunoprecipitations, using several millilitres of antiserum.
- If the protein complex of interest is of sufficient stability and the task is not to purify the complex but merely to investigate its formation and its size, molecular size exclusion chromatography can be used.
- An interesting method to detect other proteins in the vicinity of an imported protein *in situ* (in intact membranes or intact organelles) is chemical cross-linking (Section 6.2.3). It should be noted that cross-linking does not prove binding of proteins but only the proximity of the reaction partners.
- Additional methods to be considered in the analysis of protein complexes are affinity chromatography with different types of tagged proteins (such as histidine- or c-myc-tagged constructs) and the yeast two-hybrid system.

6.2.1 Co-immunoprecipitations

Co-immunoprecipitations are very often used to analyse protein complexes. If a specific antiserum against a protein is available, it can be used to test whether other proteins are associated with the imported protein. To test for complexes of membrane proteins, first determine under which conditions the protein (the antigen which is recognized by the antiserum) can be extracted from the membranes. Different detergents can be tested under different salt concentrations (for example 0, 50, 100, 200, and 400 mM KCl). Proteins which stay in the supernatant after centrifugation for 1 h at 100 000 *g* are defined as soluble. To preserve the protein complexes, the conditions of lysis should be as mild as possible.

The co-immunoprecipitations are then performed in a second series of experiments, again in the presence of different detergents and salt concentrations. Before immunoprecipitation, the samples of the lysed membranes should be centrifuged to remove all non-soluble material. Sometimes a spin of 10 min at 25 000 *g* is sufficient, in other cases centrifugation at 100 000 *g* is necessary. The specificity of the precipitations is determined by controls using pre-immune antisera. For ease of use the antibodies are commonly pre-bound to Protein A–Sepharose (binds rabbit IgG and some subtypes of mouse IgG) or Protein G–Sepharose (binds mouse IgGs of all subtypes). If the subtype of a mouse monoclonal antibody is not known, determine the appropriate ligand–Sepharose to use before starting co-immunoprecipitations.

After incubation with the antigen for 30–60 min the Sepharose beads are re-isolated by centrifugation and washed several times. In some cases it is of advantage not to pellet the Sepharose beads in the first washing step but to allow the beads to sediment without centrifugation. Using this procedure, non-specific aggregates remain in the supernatant and can be separated from the Sepharose.

Immunoprecipitations are often facilitated if the antibodies are covalently coupled to Protein A–Sepharose using a cross-linking reagent (64). By eluting at pH 2.5 followed by quick equilibration at pH 7.5 the material can be used about three or four times. Without covalent coupling, the antibodies are released from the Protein A–Sepharose together with the bound antigen when heating the sample. The antibody heavy chains which run at about 50 kDa may distort the protein pattern in subsequent SDS–PAGE. To avoid this problem, it is often sufficient to run the gels under non-reducing conditions, i.e. without mercaptoethanol in the sample buffer; the IgGs are kept intact and stay in the upper part of the gel. However, if more than 15 μl of antiserum are necessary to precipitate the desired amount of protein, covalent coupling of the antibodies to the Protein A–Sepharose provides the only way to obtain precipitations which can be analysed using polyacrylamide gels.

Very small amounts of co-precipitated proteins can easily be detected when starting with membranes from radiolabelled cells. An excellent protocol for ^{35}S-labelling of yeast cells has been published by Kolodziej and Young (65) allowing labelling up to 10^6 c.p.m./μg protein. Alternatively, radiolabelled proteins of interest can be synthesized in reticulocyte lysate and imported prior to lysis of the membranes.

It is often important to exclude the possibility that the precipitated complexes have formed after lysis of the membranes. Here a displacement assay can help. This can be performed by immunoprecipitating the complex from radiolabelled material and subsequently adding some of the same (but non-labelled) proteins. If complexes form after lysis, the non-labelled proteins will displace the labelled proteins in the immune precipitation. This assay can also be performed starting with non-labelled material and subsequently adding labelled proteins.

Protocols 16 and *17* give examples of co-immunoprecipitation in the presence of digitonin on both an analytical and a preparative scale. The protocols were developed for the precipitation of the mitochondrial protein import receptor complex. This complex is very labile and digitonin is the only detergent which was found to keep it stable. As a very mild detergent, digitonin is also of use for proteins of other membranes. Immunoprecipitations in the presence of other detergents can be performed by the same methods. Before the co-immunoprecipitations are described (*Protocols 16* and *17*), *Protocol 14* describes the coupling of antibodies to Protein A–Sepharose, and *Protocol 15* gives a method for recrystallization of digitonin. In work with the mitochondrial outer membrane protein complex of import receptors, the purification of the digitonin by recrystallization was found to be essential.

Protocol 14. Covalent coupling of antibodies to Protein A–Sepharose

Equipment and reagents

- Protein A–Sepharose (Pharmacia)
- 100 mM potassium phosphate buffer pH 7.5
- Rabbit antiserum for the protein of interest (1 ml)
- Rotary mixer (for slow end-over-end mixing)
- 100 mM sodium borate buffer pH 9
- Dimethylpimelimidate (DMP; Sigma, D-8388)
- 1 M Tris–HCl pH 7.5
- TBS: 10 mM Tris–HCl pH 7.5, 0.9% (w/v) NaCl

Method

Perform all steps at room temperature.

1. To 300 μl (wet volume) Protein A–Sepharose, add 2 ml 100 ml potassium phosphate buffer pH 7.5, and 1 ml antiserum.
2. Mix gently for 1 h.
3. Allow the beads to settle and remove the supernatant. Wash the beads twice, by resuspension and settling, in about 20 ml 100 mM sodium borate buffer pH 9.
4. Resuspend the beads in 7 ml borate buffer and add 35 mg solid DMP.
5. Mix the beads gently for 30 min.
6. Wash the beads with about 20 ml 1 M Tris–HCl pH 7.5.

Protocol 14. *Continued*

7. Mix the beads gently for 2 h at room temperature with about 20 ml 1 M Tris–HCl pH 7.5.
8. Wash the beads twice with about 20 ml TBS, and store them in TBS at 4°C.

After cross-linking, the Protein A–Sepharose will still release some antibodies in subsequent immunoprecipitations. Therefore, in some cases the best results are obtained when the material is washed once following the elution protocol. The bound antigen can also be released by low pH (see *Protocol 17*) or by heating to 95 °C, in either the presence or the absence of SDS.

Protocol 15. Recrystallization of digitonin from ethanol

Equipment and reagents

- Digitonin (Merck, 3043)
- Ethanol (absolute)
- Vacuum desiccator

Method

1. Boil 1 g digitonin in 20 ml absolute ethanol until it is completely dissolved.
2. Cool the solution in the freezer at –20 °C for 30 min. Slow decrease of the temperature improves the purity of the digitonin, but is usually not necessary.
3. Collect the digitonin crystals by centrifugation at 4 °C, 8000–18000 *g* for 10 min.
4. Dry the digitonin for several hours under vacuum.
5. Store the digitonin at –4 °C and keep it dry.

Protocol 16 has been developed to precipitate the receptor complex from the yeast *S. cerevisiae* and the fungus *N. crassa* (23). The mitochondria are lysed in the presence of 0.5% digitonin which has been purified by recrystallization in ethanol (*Protocol 15*). The antiserum which is used in this procedure was raised against Tom40 (previously termed Isp42 or Mom38), the central component of the import receptor complex. The antibodies are prebound but not cross-linked to Protein A–Sepharose. Covalently coupled antibodies can also be used (*Protocol 14*). About 100 μg mitochondrial protein are used per sample in the co-immunoprecipitation (corresponding to one lane in the final SDS–PAGE).

Protocol 16. Co-immunoprecipitation on an analytical scale of the mitochondrial import receptor complex from radiolabelled mitochondria

Equipment and reagents

- Radiolabelled mitochondria: at least 100 μg protein, 20 000 c.p.m./μg, isolated from [^{35}S]methionine labelled cells
- Recrystallized digitonin (*Protocol 15*)
- Digitonin buffer: 250 mM sucrose, 10 mM MOPS pH 7.2, 1 mM EDTA, 3% (w/v) BSA, 130 mM NaCl, 10% (v/v) glycerol, 0.5% (w/v) recrystallized digitonin, 1 mM PMSF—prepare this just before use
- 0.1 M PMSF and SDS–PAGE sample buffer (*Protocol 7*)
- Vortex mixer
- Protein A–Sepharose (Pharmacia): make a 1:3 (v/v) suspension by adding 1 vol. beads to 3 vol. H_2O
- Rotary mixer (to prevent sedimentation of the Protein A–Sepharose beads; *Protocol 14*)
- Sucrose/MOPS buffer: 250 mM sucrose, 10 mM MOPS pH 7.2, 1 mM EDTA, 3% (w/v) BSA, 130 mM NaCl, 10% (v/v) glycerol
- Antiserum directed against Tom40 (a polyclonal rabbit antiserum raised against the isolated Tom40 protein)
- Wash buffer: 250 mM sucrose, 10 mM MOPS pH 7.2, 1 mM EDTA, 130 mM NaCl, 10% (v/v) glycerol, 0.5% (w/v) recrystallized digitonin
- 10 mM Tris–HCl pH 7.5

A. *Lysis of mitochondria*

1. Lyse a sample of the ^{35}S-labelled mitochondria (100 μg mitochondrial protein, at least 20 000 c.p.m./μg) in 0.1 ml freshly prepared digitonin buffer, by pipetting the sample up and down.
2. Vortex for 10 sec at medium force.
3. Incubate at 0 °C for 10 min.
4. Remove insoluble material by centrifugation at 18 000 *g* for 10 min at 2 °C.
5. Retain the supernatant (lysed mitochondria) on ice until part C, step 1.

B. *Pre-binding of antibodies to Protein A–Sepharose*

1. Add 30 μl of a 1:3 (v/v) suspension of Protein A–Sepharose beads to 470 μl sucrose/MOPS buffer. Incubate for 10 min at 4 °C with continuous rotation.
2. Add Tom40 antiserum (10–15 μl) to the suspension and incubate for 1 h at 4 °C with gentle rotation.
3. Wash the Protein A–Sepharose (with the pre-bound antibodies) with digitonin buffer. To do this, allow the beads to settle, remove the supernatant, and resuspend the beads in 1 ml digitonin buffer. Repeat this step.

C. *Co-immunoprecipitation*

1. Suspend the Protein A–Sepharose/antibody beads from part B, step 3 in 400 μl digitonin buffer and add the lysed mitochondria (from part A, step 5).

Protocol 16. *Continued*

2. Incubate for 1 h at 4°C with constant rotation.
3. Wash the Protein A–Sepharose beads three times with 1 ml lots of wash buffer.
4. Wash the beads once with 1 ml 10 mM Tris–HCl pH 7.5.
5. Add 15 μl SDS–PAGE sample buffer to the beads, mix, and heat for 3 min at 95°C.
6. Analyse the proteins by SDS–PAGE and fluorography (*Protocol 7*).

Co-immunoprecipitation on a preparative scale is described in *Protocol 17*. It is similar to *Protocol 16*, with the following differences:

- Radiolabelled mitochondria are not required.
- Bound proteins are not released by denaturation at 95°C but by lowering the pH to 2.5.
- The precipitated proteins are obtained in chemical amounts, often sufficient for microsequencing.

Protocol 17. Precipitation on a preparative scale of the mitochondrial import receptor complex by covalently coupled antibodies

Equipment and reagents

- Mitochondria (which may contain newly imported preproteins), as described in *Protocols 7, 8,* or *12*
- Digitonin buffer, 10 mM Tris–HCl pH 7.5, and wash buffer (see *Protocol 16*)
- Antibodies covalently bound to Protein A–Sepharose prepared as in *Protocol 14*, starting with 2 ml antiserum against Tom40 (*Protocol 16*)
- Rotary mixer (*Protocol 14*)
- Elution buffer: 100 mM glycine–HCl pH 2.5 (store at 4°C)
- 70% (w/v) TCA
- 10 mM potassium phosphate buffer pH 7.2
- SDS–PAGE sample buffer (*Protocol 7*)

Method

1. Lyse the isolated mitochondria (10 mg mitochondrial protein per sample) according to *Protocol 16*, part A, steps 1–5.
2. Add about 600 mg Protein A–Sepharose with covalently coupled antibodies.
3. Incubate for 1 h at 4°C with constant rotation.
4. Wash the Protein A–Sepharose three times with 1 ml lots of wash buffer. Wash once with 1 ml 10 mM Tris–HCl pH 7.5.
5. Wash the beads with 2 ml cold elution buffer. Pellet the Protein A–Sepharose and retain the supernatants.
6. Repeat step 5 and pool the supernatants.

7. Precipitate the eluted proteins with TCA (10% final concentration).
8. Wash and resuspend the Protein A–Sepharose beads quickly (immediately after step 6) in 10 mM phosphate buffer pH 7.2, and store them at 4°C for reuse.
9. Analyse the TCA precipitated proteins by SDS–PAGE (*Protocol 7*, step 15) followed by Coomassie staining or Western blotting and immunostaining (ref. 73).

An excellent collection of protocols covering all methods of work with antibodies has been provided by Harlow and Lane (64).

6.2.2 Blue native polyacrylamide gel electrophoresis

Blue native electrophoresis is not a very common method for the analysis of protein complexes. However it has proved to be a very powerful tool in the analysis of respiratory chain complexes and the protein import complexes of chloroplasts and mitochondria, both for preparative and analytical purposes (66–68). The system, as developed by H. Schägger (66), has some crucial advantages:

- In contrast to other systems of native electrophoresis, blue native electrophoresis allows separation of both negatively and positively charged proteins.
- Not only soluble but also membrane proteins and complexes of membrane proteins can be separated. An example is shown in *Figure 2*.
- The blue native electrophoresis system allows separation over a wide range of molecular weights (10–1000 kDa) at high resolution.

The major difference of the blue native electrophoresis in comparison to common SDS–PAGE is the use of the dye Coomassie Blue G-250 instead of SDS. Besides staining the proteins during the electrophoresis, the Coomassie dye has three additional functions:

- It keeps membrane proteins soluble.
- It is a negatively charged dye which can bind to all membrane proteins and mediate their migration to the anode at pH 7.5.
- The negative charges of the Coomassie dye repel protein complexes and thus prevent aggregation.

Protocol 18 describes the method. Technically the following points are important:

- The polyacrylamide gel is prepared as a gradient gel (6–13% acrylamide or even up to 20%).
- The gel is run in a conventional gel chamber in a cold room or (for better results) in a water-cooled chamber. The run takes about 5 h for gels of 14 cm length. Gels of a higher concentration of acrylamide may require up to 8 h.

Marker proteins must be run to determine the molecular sizes of the proteins of interest. *Protocol 18* (equipment and reagents) lists a satisfactory range. Catalase (Sigma, C-100) and β-lactoglobulin B (Sigma, L-8005) are also good markers. The different marker proteins should first be tested in separate lanes on a gel because some of the marker proteins run in part as additional higher molecular weight complexes.

Protocol 18. Blue native polyacrylamide gel electrophoresis

Equipment and reagents

- Electrophoresis chamber: the best is a system with direct cooling in a water-bath with gels 1 mm thick and 14 cm long (e.g. the SE600 gel chamber, Hoefer Scientific Instruments)
- Two-chamber gradient mixer with 10 ml chambers
- Acrylamide (Serva, 10675)
- *Bis*acrylamide (Serva, 29125)
- AB-mix: acrylamide 49.5% (w/v), *bis*acrylamide 3% (relative to the total amount of acrylamide). Thus 100 ml of AB-mix will contain 48 g acrylamide and 1.5 g *bis*acrylamide.
- ϵ-aminocaproic acid (6-aminocaproic acid; Sigma, A-2504)
- Bis-Tris (Sigma, B-9754)
- 3 × gel buffer: 200 mM ϵ-aminocaproic acid, 150 mM bis-Tris pH 7 at 4°C[a]
- 99% glycerol (Serva)
- 10% (w/v) ammonium persulfate (APS), freshly made
- *N,N,N′,N′*-tetraethylmethylenediamine (TEMED; Serva)
- Propan-2-ol
- Samples of isolated mitochondria (or other organelle membranes)[b]
- Lysis buffer. This may contain different detergents (digitonin, laurylmaltoside, Triton X-100, CHAPS, Mega 9, octyl glycoside, Brij 35, Tween 20) or salt concentrations. It should be selected to extract the desired proteins and preserve the protein complexes of interest. It additionally may contain EDTA, glycerol, or protease inhibitors.
- Ultracentrifuge (e.g. a Beckman TLX centrifuge with TLA-45 rotor)
- Coomassie G-250 (Serva)
- Sample buffer A: 5% (w/v) Coomassie Blue G-250, 500 mM ϵ-aminocaproic acid, 100 mM bis-Tris pH 7. Dissolve the solid Coomassie dye and the solid ϵ-aminocaproic acid in the bis-Tris buffer.
- Marker protein mixture: 10–20 μg each of albumin, β-amylase, apoferritin, and thyroglobulin (Sigma, MW-GF-1000)
- Anode buffer: 50 mM bis-Tris pH 7 (with HCl) at 4°C
- Tricine (Sigma, T-7911)
- Cathode buffer A: 50 mM Tricine, 15 mM bis-Tris, 0.02% (w/v) Coomassie Blue G-250 pH 7 at 4°C
- Cathode buffer B: 50 mM Tricine, 15 mM bis-Tris pH 7 at 4°C (correction of pH is usually not required)

A. *Preparation of the gel*

1. For a 6–13% acrylamide separation gel make up the 6% and 13% acrylamide solutions according to the table below. For 6–16.5% or 6–20% gels make up the 6%, 16.5%, and the 20% solutions, respectively.

	6%	13%	16.5%	20%
• AB-mix	1.07 ml	2.35 ml	3.05 ml	3.75 ml
• 3 × gel buffer	3 ml	3 ml	3 ml	3 ml
• 99% glycerol	–	1.8 g	1.8 g	1.8 g
• H_2O to a final volume of 9 ml				
• 10% APS	38 μl	30 μl	30 μl	30 μl
• TEMED	3.8 μl	3 μl	3 μl	3 μl

2. Mix the solutions and pour 9 ml of each solution into the appropriate chamber of the gradient mixer (the higher concentration in the exit chamber).
3. Pour the acrylamide gradient between the gel plates and overlay it with propan-2-ol. Allow polymerization to occur at room temperature.
4. Prepare the stacking gel solution (containing 4% acrylamide) as below:
   - AB-mix 0.3 ml
   - 3 × gel buffer 1.25 ml
   - H_2O to a final volume of 3.75 ml
   - APS 10% 30 μl
   - TEMED 3 μl
5. Remove the propan-2-ol from the separation gel and replace it with the stacking gel solution. Insert the well-forming comb and allow the gel to set.[c]

B. *Preparation of the samples*

1. Lyse the samples of isolated membranes in lysis buffer (up to 50 μg protein/lane, 1 μg protein/μl lysis buffer) under the conditions which are required to extract the desired proteins and to keep their complexes intact.[d]
2. Remove insoluble material by centrifugation for 30 min at 190000 *g*.
3. Mix 45 μl of the supernatants with 5 μl sample buffer.
4. Prepare the marker protein mixture (albumin, β-amylase, apoferritin, and thyroglobulin, 10–20 μg each) in sample buffer for loading on to the gel (centrifugation is not necessary).

C. *Electrophoresis*

1. Load the membrane protein samples (part B, step 3) and marker proteins (part B, step 4) in separate lanes of the gel.
2. Assemble the gel (part A, step 5) in the electrophoresis unit and start the electrophoresis with anode buffer and cathode buffer A (containing Coomassie dye) in their respective chambers.
3. Run the gel at 100 V until the Coomassie dye has reached the separation gel, then set the electrical limits of the electrophoresis at 500 V and 15 mA.
4. After the dye has travelled about one-third of the way down the gel, stop the electrophoresis. Discard cathode buffer A and replace it with cathode buffer B (without Coomassie dye).
5. Continue the electrophoresis until the dye is within 1 cm of the end of the gel. Turn off the power.

Protocol 18. *Continued*

6. Remove the gel from the apparatus and analyse the separated proteins as described in the following text.[e]

[a] Schägger *et al.* (66) routinely use a concentration of 1.5 M ε-aminocaproic acid. The precise concentration depends on the stability of the protein complexes and may need to be determined in a pilot experiment.
[b] Samples of proteins dissolved in the presence of 2% SDS can be used to demonstrate the protein pattern of the monomeric subunits.
[c] After polymerization the gel may be stored at 4°C for two to three days.
[d] Determine the best lysis buffer by lysis under different conditions and subsequent centrifugation for 45 min at 100 000 *g*. Analyse pellet and supernatant for the protein of interest. Proteins which remain in the supernatant at 100 000 *g* are defined as 'soluble'.
[e] The anode buffer can be used two or three times if the gels are stopped early enough (5).

After electrophoresis the gel can be analysed as a blue gel (after staining with regular Coomassie dye, destaining, and drying) or it can be used for Western blotting. For blotting, the gel should be soaked for 5 min in a blot buffer containing 0.02% SDS. Nitrocellulose membranes or PVDF (polyvinylidene difluoride) membranes can be used. The latter have the advantage that they can also be used for Coomassie staining.

6.2.3 Chemical cross-linking of proteins by Sulfo-MBS and DSS

Chemical cross-linking can be done in a very simple but efficient way. *Protocols 19* and *20* were originally developed for the cross-linking of membrane proteins of the endoplasmic reticulum (69) but have also been applied successfully to mitochondrial proteins (70, 71). The sucrose/EDTA/MOPS buffers of the protocols can be substituted by other buffers which are appropriate for the

Figure 2. Detection of protein complexes by immunoblotting after blue native electrophoresis. (A) The blue native gel was prepared as described in *Protocol 18*. Shown are complexes containing Tim23, a component of the mitochondrial inner membrane protein transport machinery. Tim23, an integral membrane protein of 23 kDa, is a component of a protein complex of 90 kDa (the TIM complex) which was extracted from the membrane by digitonin. After accumulation of the hybrid protein $b_2\Delta$-DHFR spanning both mitochondrial membranes, the TIM complex is connected to TOM proteins of the outer membrane, yielding a complex of 600 kDa. See panel (B) for a diagrammatic representation of the protein–membrane complexes formed and the reagents present in each lane. Ref. 67 contains additional information. (B) Scheme showing the three different import stages of the preprotein $b_2\Delta$-DHFR analysed for complex formation with Tim23 in the gel of panel (A). (1) Import of $b_2\Delta$-DHFR into the matrix: no binding to Tim23 (which is part of the TIM complex). (2) Import of $b_2\Delta$-DHFR in the absence of a membrane potential: $b_2\Delta$-DHFR partially binds to the TOM complex of the outer membrane but does not insert into the TIM complex. (3) Import of $b_2\Delta$-DHFR in the presence of the ligand methotrexate (MTX): MTX stabilizes the structure of the DHFR domain of the hybrid protein and thus prevents complete translocation across the mitochondrial membranes. $b_2\Delta$-DHFR accumulates as a translocation intermediate spanning both the TOM and the TIM machinery. See Section 3.3.3 for technical details.

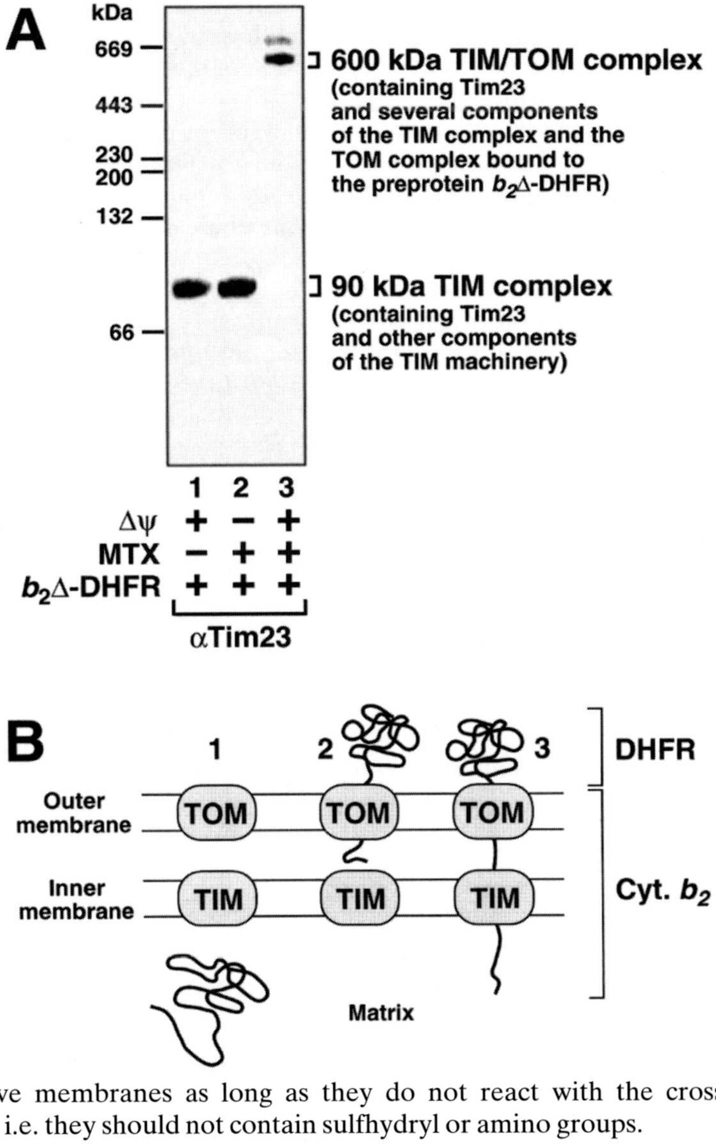

respective membranes as long as they do not react with the cross-linking reagent, i.e. they should not contain sulfhydryl or amino groups.

The protein to be cross-linked is synthesized in reticulocyte lysate in the presence of [^{35}S]methionine and imported into the membrane (see *Protocols 7*, *8*, and *12*). A prerequisite with this approach is a sufficiently high specific radioactivity of the preprotein. Therefore, efficient mRNA translation is essential. Usually at least 75 000 c.p.m. of [^{35}S]methionine labelled protein/μl of reticulocyte lysate is required and higher specific radioactivities are of advantage. The assay of incorporation of radioactivity into translation products is described in *Protocol 1*.

The efficiency of cross-linking usually reaches only a few per cent. Higher efficiencies require purified components in higher concentration or more specific approaches such as direct incorporation of the cross-linker into one of the desired reaction partners.

The products of the reaction can be analysed by SDS–PAGE. Immunoprecipitations are performed to determine the reaction partner of the radioactive protein. With higher amounts of protein and an appropriate assay, cross-linking of non-labelled proteins can also be carried out following the same protocols.

In every case the optimal concentration of the cross-linking reagent has to be determined by titration, increasing and decreasing the standard concentration in twofold steps as appropriate. In these titrations keep the amount of protein constant or vary it in similar small steps. In general it is often difficult to find suitable conditions for a cross-linking reaction by arbitrary titrations of concentrations across orders of magnitudes. It is much easier to start with the standard conditions of *Protocols 19* and *20* or use data from the literature. Several simpler and more sophisticated methods of chemical cross-linking of components involved in intracellular protein transport have been published (26).

Cross-linking reagents are labile chemicals which should be stored at the appropriate temperature and handled carefully. To keep the chemicals dry, they should be stored in a closed tube (together with a drying agent if provided by the manufacturer). After taking a cross-linking reagent out of the freezer, the tube should be allowed to reach room temperature before opening it, to avoid the condensation of moisture inside.

For cross-linking with Sulfo-MBS (*m*-maleimidobenzoyl-*N*-hydroxysulfosuccinimide ester) as in *Protocol 19* the following characteristics of the reagent are important:

- Sulfo-MBS is soluble in water.
- It is a hetero-bifunctional cross-linker which cross-links thiol groups to amino groups.

Protocol 19. Chemical cross-linking of mitochondrial proteins by Sulfo-MBS

Reagents

- Mitochondria (or other organelles) after import of radioactively labelled preproteins (*Protocols 7*, *8*, or *12*, or see Section 3.3 for translocation intermediates)
- S_{500}EM buffer: 0.5 M sucrose, 1 mM EDTA, MOPS–KOH pH 7.2
- SEM buffer: 250 mM sucrose, 1 mM EDTA, MOPS–KOH pH 7.2
- Sulfo-MBS (Pierce; keep reagent dry)
- 2-Mercaptoethanol
- 1 M glycine–HCl pH 7.2
- 2% SDS, 60 mM Tris–HCl pH 6
- Immunoprecipitation buffer: 1% Triton X-100, 0.3 M NaCl, 10 mM Tris–HCl pH 7.5

Method

1. Follow the procedures in *Protocols 7, 8,* or *12* to import the protein into the mitochondria (30 μg mitochondrial protein per sample) from the reticulocyte lysate (10–50 μl) for accumulation as a translocation intermediate.[a]
2. Separate the mitochondria from the BSA buffer used in the import reaction and from the free [^{35}S]methionine of the reticulocyte lysate by centrifugation through a sucrose cushion. To do this layer the import mixture on top of a 500 μl cushion of S_{500}EM buffer and centrifuge for 10 min at 20 000 *g*. Remove the supernatant.
3. Wash the mitochondria once with 0.5 ml cold SEM buffer. It may be sufficient to rinse the tube walls and the pellet of mitochondria without resuspension of the mitochondria, and then centrifuge again for 6 min at 16 000 *g*. Discard the supernatant of this centrifugation.
4. Resuspend the washed mitochondria in 200 μl cold SEM buffer and then add an additional 700 μl.
5. Dissolve Sulfo-MBS in cold SEM buffer at a concentration of 0.5 mg/ml and immediately add 100 μl of the solution to the suspension of mitochondria. Mix and incubate the suspension at 0 °C for 20 min.[b]
6. To stop the reaction, add 1 μl of 2-mercaptoethanol and 200 μl of 1 M glycine–HCl pH 7.2. Mix carefully and incubate at 0 °C for 20 min.
7. Pellet the mitochondria for 10 min at 16 000 *g* and 2 °C.
8. Analyse products of the reaction by SDS–PAGE (see ref. 73) and/or immunoprecipitation with specific antibodies. For immunoprecipitation dissolve the pelleted mitochondria in 20 μl of 2% SDS, 60 mM Tris–HCl pH 6.8.[c] Heat the sample at 95 °C for 3 min. Cool the sample on ice, and add 500 μl immunoprecipitation buffer. Remove insoluble material by centrifugation at 10 000 *g* for 5 min. Immunoprecipitate the cross-linking products of interest using specific antibodies pre-bound to Protein A–Sepharose as described in *Protocol 16*.

[a] In the import reaction (step 1), a high amount of reticulocyte lysate is required to obtain a sufficiently high amount of labelled protein for detection after the reaction. The amount of mitochondrial protein (or protein of the corresponding membrane) in the sample is important relative to the amount and the concentration of the cross-linking reagent in step 5, and thereby for the efficiency of the cross-linking reaction.

[b] Final concentration of Sulfo-MBS during cross-linking is 50 μg/ml (0.12 mM).

[c] Higher concentrations of SDS interfere with the immunoprecipitation.

Protocol 20 describes cross-linking with disuccinimidyl suberate (DSS). For cross-linking with DSS the following characteristics are important:

- DSS is insoluble in water. Therefore, DSS is first dissolved in DMSO and only subsequently diluted into the reaction buffer. The DSS starts precipitating as soon as it is diluted into aqueous solution. Only part of the DSS molecules reach the membranes for the cross-linking reaction. However, cross-linking is often very efficient.

- DSS is a homo-bifunctional cross-linker which cross-links neighbouring amino groups.
- DSS easily destroys membranes. Therefore the cross-linking products are usually not recovered by centrifugation of the membranes but by TCA precipitation.

Protocol 20. Chemical cross-linking of mitochondrial proteins by DSS

Reagents

- Sample of mitochondria (or other organelle) for analysis (see *Protocol 19*)
- DSS (Pierce; keep dry)
- Dimethyl sulfoxide (DMSO; analytical grade)
- S_{500}EM buffer, SEM buffer, and 2% SDS, 60 mM Tris–HCl pH 6 (see *Protocol 19*)
- 1 M Tris–HCl pH 7.2
- 70% (w/v) TCA
- Acetone (at –20 °C)

Method

1. Prepare the mitochondria as in *Protocol 19*, steps 1–4.
2. Dissolve DSS in DMSO at a concentration of 20 mg/ml. Dilute 100 μl of this stock solution in 900 μl SEM buffer at 25 °C and mix rapidly. Immediately add 100 μl of the mixture to the suspension of mitochondria and mix quickly.[a]
3. Incubate the mixture at 0 °C for 20 min.
4. Add 200 μl 1 M Tris–HCl pH 7.2. Mix carefully and incubate at 0 °C for 20 min.
5. Precipitate the proteins using TCA (*Protocol 9*, steps 10–14).
6. Redissolve the TCA precipitate in 2% SDS, 60 mM Tris–HCl pH 6, and analyse the products of the reaction as suggested in *Protocol 19*.

[a] The SEM buffer which is used for the dilution of the DSS stock solution should be pre-incubated at 25 °C to improve the solubility of the DSS in the aqueous solution. The final concentration of DSS during cross-linking is 0.2 mg/ml (0.54 mM).

6.3 Analysis of protein folding after import into organelles

Protein transport across membranes in most cases requires an unfolding reaction. Subsequently the proteins have to refold and adopt their final three-dimensional structure. Therefore the characterization of protein folding reactions has become an important task in studying intracellular protein traffic.

The folding reactions of proteins *in vivo* are often mediated by chaperone proteins and protein translocation is often driven by distinct members of this group. The interaction of preproteins or translocation intermediates with chaperone proteins can be monitored using the methods described in the preceding sections. Particularly co-immunoprecipitations are used to test for transient complex formation (see *Protocol 16*).

To characterize the folding state of proteins during translocation, in many cases fusion proteins containing a DHFR moiety are used, taking advantage of the properties of this enzyme as described in Section 2.6. Organelles can be lysed after different times of import and tested for the resistance of the DHFR domain by treatment with proteinase K. Resistance against proteolysis is taken as an indicator of the native folding state of the DHFR. This test can be combined with co-immunoprecipitations using antisera raised against the components which interact with the preprotein during the import process. Thereby the folding pathway of a preprotein can be traced through the different stages of the import into the organelle. Examples have been published in refs 18, 19, 21, 22, and 72.

Hybrid proteins containing a DHFR domain can also be used for studies on protein folding *in vitro*. The hybrid protein is dissolved in 8 M urea or in 6 M guanidinium as described in *Protocol 2* and then diluted into a refolding buffer at 20 °C. The folding reaction can take place at 30 °C or at 0 °C. Aliquots are removed after different times of incubation (0–60 min) and are rapidly mixed with a solution containing proteinase K at 0 °C. To allow quick proteolysis, up to 200 μg/ml proteinase K can be used, but concentrations as low as 10 μg/ml may be sufficient. The proteinase K is inactivated by addition of PMSF (see *Protocol 7*) and the proteins are precipitated using TCA (see *Protocol 9*) for analysis by SDS–PAGE and fluorography. The folding pathway of mouse DHFR has not yet been resolved in detail. However, in may be of interest to compare folding reactions of a substrate protein *in vitro* and *in vivo*, and this option is particularly offered by the DHFR.

Defects of protein folding systems often cause aggregation of the substrate protein. The aggregation of many proteins can easily be tested by lysis of the organelle and subsequent centrifugation. In some cases a centrifugation at 16 000 *g* for 15 min is sufficient to pellet the aggregated material, in other cases 100 000 *g* may be required. Assays of this type are often combined with import into organelles from mutant strains and serve to demonstrate the effect of the depletion or inactivation of a component which is involved in keeping the substrate protein in a soluble state.

Acknowledgements

I thank Richard Zimmermann (University of Saarland, Homburg) for advice in writing the section on protein import into the endoplasmic reticulum and the isolation of microsomes, Michaela Stoltz (University of Freiburg) for help concerning the isolation of mitochondria from rat liver, and Ulf Bömer (Schering AG, Berlin) for helpful comments on the manuscript. Work of the author's laboratory was supported by the Deutsche Forschungsgemeinschaft and the Sonderforschungsbereich 388.

References

1. Schatz, G. and Dobberstein, B. (1996). *Science*, **271**, 1519.
2. Schekman, R. (1985). *Annu. Rev. Cell Biol.*, **1**, 115.
3. Kübrich, M., Dietmeier, K., and Pfanner, N. (1995). *Curr. Genet.*, **27**, 393.
4. Kunau, W.-H., Beyer, A., Franken, T., Götte, K., Marioch, M., Saidowsky, J., *et al.* (1993). *Biochimie*, **75**, 209.
5. Guthrie, C. and Fink, G. R. (ed.) (1991). In *Methods in enzymology*, Vol. 194. Academic Press, London.
6. Johnston, J. J. (ed.) (1994). *Molecular genetics of yeast: a practical approach*. IRL Press, Oxford.
7. Jackson, R. J. and Hunt, T. (1983). In *Methods in enzymology* (ed. S. Fleischer and B. Fleischer), Vol. 96, p. 50. Academic Press, London.
8. Erickson, A. H. and Blobel, G. (1983). In *Methods in enzymology* (ed. S. Fleischer and B. Fleischer), Vol. 96, p. 38. Academic Press, London.
9. Garcia, P. D., Hansen, W., and Walter, P. (1991). In *Methods in enzymology* (ed. C. Guthrie and G. R. Fink), Vol. 194, p. 675. Academic Press, London.
10. Hönlinger, A., Keil, P., Nelson, R. J., Craig, E. A., and Pfanner, N. (1995). *Biol. Chem. Hoppe-Seyler*, **376**, 515.
11. Shatzman, A. R. (1998). In *Protein expression: a practical approach*, p. 170. IRL Press, Oxford.
12. Stone, D. and Phillips, A. W. (1977). *FEBS Lett.*, **74**, 85.
13. Nunberg, J. H., Kaufman, R. J., Chang, A. C. Y., Cohen, S. N., and Schimke, R. T. (1980). *Cell*, **19**, 355.
14. Eilers, M. and Schatz, G. (1986). *Nature*, **322**, 228.
15. Rassow, J., Guiard, B., Wienhues, U., Herzog, V., Hartl, F.-U., and Neupert, W. (1989). *J. Cell Biol.*, **109**, 1421.
16. Wienhues, U., Becker, K., Schleyer, M., Guiard, B., Tropschug, M., Horwich, A. L., *et al.* (1991). *J. Cell Biol.*, **115**, 1601.
17. Häusler, T., Stierhof, Y. D., Wirtz, E., and Clayton, C. (1996). *J. Cell Biol.*, **132**, 311.
18. Ostermann, J., Horwich, A. L., Neupert, W., and Hartl, F.-U. (1989). *Nature*, **341**, 125.
19. Manning-Krieg, U. C., Scherer, P. E., and Schatz, G. (1991). *EMBO J.*, **10**, 3273.
20. Mayhew, M., da Silva, A. C. R., Martin, J., Erdjument-Bromage, H., Tempst, P., and Hartl, F.-U. (1996). *Nature*, **379**, 420.
21. Rospert, S., Looser, R., Dubaquié, Y., Matouschek, A., Glick, B. S., and Schatz, G. (1996). *EMBO J.*, **15**, 764.
22. Rassow, J., Mohrs, K., Koidl, S., Barthelmess, I. B., Pfanner, N., and Tropschug, M. (1995). *Mol. Cell. Biol.*, **15**, 2654.
23. Matouschek, A., Rospert, S., Schmid, K., Glick, B. S., and Schatz, G. (1995). *Proc. Natl. Acad. Sci. USA*, **92**, 6319.

24. Stammers, D. K., Champness, J. N., Beddell, C. R., Dann, J. G., Eliopoulos, E., Geddes, A. J., *et al.* (1987). *FEBS Lett.*, **218**, 178.
25. Attardi, G. M. and Chomyn, A. (ed.) (1995). *Methods in enzymology*, Vol. 260. Academic Press, London.
26. Tartakoff, A. M. (ed.) (1991). *Methods in cell biology*, Vol. 34. Academic Press, London.
27. Zara, V., Palmieri, F., Mahlke, K., and Pfanner, N. (1992). *J. Biol. Chem.*, **267**, 12077.
28. Pon, L. and Schatz, G. (1991). In *The molecular and cellular biology of the yeast Saccharomyces cerevisiae* (ed. J. R. Broach, J. R. Pringle, and E. W. Jones), Vol. 1, p. 333. Cold Spring Harbor Laboratory Press, New York.
29. Campbell, I. and Duffus, J. H. (ed.) (1988). *Yeast: a practical approach.* IRL Press, Oxford.
30. Daum, G., Böhni, P. C., and Schatz, G. (1982). *J. Biol. Chem.*, **257**, 13028.
31. Glick, B. S. and Pon, L. A. (1995). In *Methods in enzymology* (ed. G. M. Attardi and A. Chomyn), Vol. 260, p. 213. Academic Press, London.
32. Stoltz, M., Rysavy, P., Kalousek, F., and Brandsch, R. (1995). *J. Biol. Chem.*, **270**, 8016.
33. Darley-Usmar, V. M., Rickwood, D., and Wilson, M. T. (ed.) (1987). *Mitochondria: a practical approach.* IRL Press, Oxford.
34. Zara, V., Dietmeier, K., Palmisano, A., Vozza, A., Rassow, J., Palmieri, F., *et al.* (1996). *Mol. Cell. Biol.*, **16**, 6524.
35. Glick, B. S. (1995). In *Methods in enzymology* (ed. G. M. Attardi and A. Chomyn), Vol. 260, p. 224. Academic Press, London.
36. Becker, K., Guiard, B., Rassow, J., Söllner, T., and Pfanner, N. (1992). *J. Biol. Chem.*, **267**, 5637.
37. Wienhues, U., Koll, H., Becker, K., Guiard, B., and Hartl, F. U. (1992). In *Protein targeting: a practical approach* (ed. A. I. Magee and T. Wileman), p. 135. IRL Press, Oxford.
38. Söllner, T., Rassow, J., and Pfanner, N. (1991). In *Methods in cell biology* (ed. A. M. Tartakoff), Vol. 34, p. 345. Academic Press, London.
39. Cyr, D. M., Ungermann, C., and Neupert, W. (1995). In *Methods in enzymology* (ed. G. M. Attardi and A. Chomyn), Vol. 260, p. 241. Academic Press, London.
40. Alconada, A., Gärtner, F., Hönlinger, A., Kübrich, M., and Pfanner, N. (1995). In *Methods in enzymology* (ed. G. M. Attardi and A. Chomyn), Vol. 260, p. 263. Academic Press, London.
41. Pfanner, N., Tropschug, M., and Neupert, W. (1987). *Cell*, **49**, 815.
42. Pfanner, N. and Neupert, W. (1987). *J. Biol. Chem.*, **262**, 7528.
43. Rassow, J. and Pfanner, N. (1991). *FEBS Lett.*, **293**, 85.
44. Schleyer, M. and Neupert, W. (1984). *J. Biol. Chem.*, **259**, 3487.
45. Zara, V., Rassow, J., Wachter, E., Tropschug, M., Palmieri, F., Neupert, W., *et al.* (1991). *Eur. J. Biochem.*, **198**, 405.
46. Pon, L., Moll, T., Vestweber, D., Marshallsay, B., and Schatz, G. (1989). *J. Cell Biol.*, **109**, 2603.
47. Söllner, T., Griffith, G., Pfaller, R., Pfanner, N., and Neupert, W. (1989). *Cell*, **59**, 1061.
48. Rickwood, D. (ed.) (1992). *Preparative centrifugation: a practical approach.* IRL Press, Oxford.

49. Fleischer, S. and Fleischer, B. (ed.) (1983). *Methods in enzymology*, Vol. 96. Academic Press, London.
50. Blobel, G. and Dobberstein, B. (1975). *J. Cell Biol.*, **67**, 852.
51. Watts, C., Wickner, W., and Zimmermann, R. (1983). *Proc. Natl. Acad. Sci. USA*, **80**, 2809.
52. Walter, P. and Blobel, G. (1983). In *Methods in enzymology* (ed. S. Fleischer and B. Fleischer), Vol. 96, p. 84. Academic Press, London.
53. Rothblatt, J. A. and Meyer, D. I. (1986). *Cell*, **44**, 619.
54. Lyman, S. K. and Schekman, R. (1995). *J. Cell Biol.*, **131**, 1163.
55. Gilmore, R., Collins, P., Johnson, J., Kellaris, K., and Rapiejko, P. (1991). In *Methods in cell biology* (ed. A. M. Tartakoff), Vol. 34, p. 224. Academic Press, London.
56. Görlich, D., Kurzchalia, T. V., Wiemann, M., and Rapoport, T. A. (1991). In *Methods in cell biology* (ed. A. M. Tartakoff), Vol. 34, p. 241. Academic Press, London.
57. Galbraith, D. W., Bourque, D. P., and Bohnert, H. J. (ed.) (1995). *Methods in cell biology*, Vol. 50. Academic Press, London.
58. Rothman, J. E. (ed.) (1992). *Methods in enzymology*, Vol. 219. Academic Press, London.
59. Tartakoff, A. M. (ed.) (1989). *Methods in cell biology*, Vol. 31. Academic Press, London.
60. Tartakoff, A. M. (ed.) (1989). *Methods in cell biology*, Vol. 32. Academic Press, London.
61. Magee, A. I. and Wileman, T. (ed.) (1992). *Protein targeting: a practical approach*. IRL Press, Oxford.
62. Schlenstedt, G., Hurt, E., Doye, V., and Silver, P. A. (1993). *J. Cell Biol.*, **123**, 785.
63. Fujiki, Y., Hubbard, A. L., Fowler, S., and Lazarow, P. B. (1982). *J. Cell Biol.*, **93**, 97.
64. Harlow, E. and Lane, D. (ed.) (1988). *Antibodies: a laboratory manual*. Cold Spring Harbor Laboratory, New York.
65. Kolodziej, P. A. and Young, R. A. (1991). In *Methods in enzymology* (ed. C. Guthrie and G. R. Fink), Vol. 194, p. 508. Academic Press, London.
66. Schägger, H. and von Jagow, G. (1991). *Anal. Biochem.*, **199**, 223.
67. Dekker, P. J. T., Martin, F., Maarse, A. C., Bömer, U., Müller, H., Guiard, B., *et al.* (1997). *EMBO J.*, **16**, 5408.
68. Caliebe, A., Grimm, R., Kaiser, G., Lübeck, J., Soll, J., and Heins, L. (1997). *EMBO J.*, **16**, 7342.
69. Görlich, D., Prehn, S., Hartmann, E., Herz, J., Otto, A., Kraft, R., *et al.* (1990). *J. Cell Biol.*, **111**, 2283.
70. Söllner, T., Rassow, J., Wiedmann, M., Schlossmann, J., Keil, P., Neupert, W., *et al.* (1992). *Nature*, **355**, 84.
71. Blom, J., Kübrich, M., Rassow, J., Voos, W., Dekker, P. J. T., Maarse, A. C., *et al.* (1993). *Mol. Cell. Biol.*, **13**, 7364.
72. Hammond, C. and Helenius, A. (1994). *Science*, **266**, 456.
73. Hames, B. D. (ed.) (1998). *Gel electrophoresis of proteins: a practical approach*, 3rd edn. Oxford University Press.

3

Analysis of protein phosphorylation

S. IVAR WALAAS and ANNE CARINE ØSTVOLD

1. Introduction

Extensive studies have demonstrated that extracellular signals, e.g. hormones, neurotransmitters, growth factors, as well as intracellular signals and effectors, e.g. cyclic nucleotides, Ca^{2+} ions, and phospholipid derivatives, mediate many of their effects by regulating the activities of protein phosphorylation systems (1–3). The role of reversible protein phosphorylation in regulating and integrating the activity of eukaryotic cells has thus become widely appreciated during the last decades (4). Cellular protein phosphorylation systems consist of a minimum of three components:

(a) The *phosphorylated proteins* themselves.
(b) A large number of *protein kinases* which catalyse the incorporation of phosphate from the terminal phosphate of ATP into specific serine, threonine, tyrosine, and possibly histidine (5) residues in the substrate phosphoproteins.
(c) A distinct set of *phosphoprotein phosphatases* responsible for dephosphorylation of the various phosphorylated residues on the substrate proteins (6–8).

Studies on protein phosphorylation systems in intact tissues and cells should ideally include an examination of the levels, the cellular and intracellular localizations, and the biochemical characteristics and state of activity of each of these components. This chapter will describe a number of well-established methods that may be employed to investigate protein phosphorylation systems. Most of the protocols described have been used in our laboratory, although a limited number of protocols will essentially be based on published methods. General methods in protein biochemistry will not be described in detail. For a more complete coverage of specific aspects, the reader is also directed to recent compilations of methodological articles (9–13).

2. Investigating protein phosphorylation systems

Protein phosphorylation is a highly dynamic phenomenon, and investigators should have a clear understanding of possible pitfalls inherent in the different experimental approaches. The method most frequently employed to measure protein phosphorylation in intact tissue and cells will include pre-incubation of the tissue preparation with [^{32}P]orthophosphate to label intracellular ATP pools, followed by stimulation of the tissue. Experimental manipulation of cellular signalling during such incubations may be achieved by addition of compounds which activate synthesis or decrease degradation of intracellular messengers. Examples include forskolin, which activates the catalytic domain of adenylyl cyclase (14), and phosphodiesterase inhibitors such as isobutylmethylxanthine, which inhibit the breakdown of cyclic AMP (15). Direct modulation of some phosphorylation systems can also be achieved, e.g. by thiosubstituted analogues of cyclic nucleotides which directly activate or inhibit the relevant cyclic nucleotide-dependent protein kinases (16), and tumour-promoting phorbol esters, which mimic the action of diacylglycerols on the protein kinase C family (17). Finally, modulation of protein phosphatases may be achieved by using membrane-permeant phosphatase inhibitors (see Section 7). Typical problems in such systems include unintended disturbances in plasma membrane ion and phosphate transport, changes in mitochondrial function, and unknown changes in the equilibration of various cellular ATP pools (18, 19).

Another approach, which avoids many of these problems, but gives less complete information, includes incubation of cell-free preparations with [γ-^{32}P]ATP under conditions where specific protein kinases or phosphatases are modulated, in the absence or presence of added substrate proteins, and intracellular effectors or inhibitors (Section 4).

3. Phosphorylation of proteins in intact preparations

3.1 General considerations

The experimental preparations most frequently employed present considerable differences in levels of complexity, ease of manipulation, and types of question that can be answered. This section briefly reviews the basic properties of these systems and provides protocols for using some of them.

3.2 Intact animals

Intact animals have been used only infrequently for investigations of protein phosphorylation. However, some studies clearly indicate that analysis of protein phosphorylation is possible in such systems, for instance intact rodent brain (see discussion presented by Rodnight *et al.* in ref. 20).

3.3 Tissue slices

Tissue slices of different stages of anatomical intactness, as well as other dissected preparations (e.g. the superior cervical ganglion, the pituitary stalk-posterior pituitary preparation), have been used extensively for the study of cellular signalling. Problems with such preparations, particularly with slices, include their less than complete structural preservation, their sensitivities to hypoxia, and their limited survival times (21, 22). Other major problems concern differences in the efficiency of phosphate uptake and ATP synthesis in the various tissue preparations. 'Difficult' preparations include isolated skeletal muscle, e.g. hemidiaphragm preparations, which transport and incorporate [^{32}P]orthophosphate extremely slowly, while other tissues, e.g. isolated adipocytes, may label the ATP pool and phosphoproteins more rapidly (19, 23). Such differences must be examined for each preparation and optimal incubation conditions determined for efficient labelling of proteins.

One should also be aware of the need for rapid inactivation when terminating these incubations, employing freezing, boiling, and/or mechanical disruption in the presence of inactivating agents such as the strong ionic detergent, SDS.

Protocol 1 describes the standard procedure for phosphorylation of proteins in tissue slices.

Protocol 1. Protein phosphorylation in tissue slices

Equipment and reagents

- Krebs Ringer bicarbonate medium (KRB): 125 mM NaCl, 3.5 mM KCl, 1.5 mM KH_2PO_4, 1.5 mM $MgSO_4$, 25 mM $NaHCO_3$, 10 mM glucose pH 7.4—gas this with 95% O_2/5% CO_2 before use
- Tissue slicer (e.g. from Sorvall)
- Tissue for analysis, freshly obtained from the biological source
- Incubation flasks: 50 ml Erlenmeyer flasks[a]
- Incubator, gassed with 95% O_2/5% CO_2
- KRB with Ca^{2+}: KRB plus 1 mM $CaCl_2$
- Phosphate-free KRB: KRB with Ca^{2+} but without the KH_2PO_4
- [^{32}P]orthophosphate (e.g. in aqueous solution, 10 mCi/ml; Amersham)
- Liquid N_2 and storage vials[a]
- Forceps, scissors for tissue dissection
- Stock solution (10 × concentrated) of drug to be examined, dissolved in KRB

Method

1. Dissect the organs of interest and place them in 5–10 ml ice-cold KRB.
2. Slice the tissue (preferably in a cold room) with a tissue slicer set at 0.3–0.4 mm. Add the slices to flasks containing 10 ml KRB and swirl lightly. Aspirate the buffer, add 10 ml KRB, and pre-incubate the slices for 30 min at 30°C with continuous carboxygenation.
3. Aspirate the medium, add 10 ml KRB with Ca^{2+}, and continue pre-incubation for 30–90 min.

Protocol 1. *Continued*

4. Aspirate the medium and add a small volume of phosphate-free KRB with Ca^{2+}, containing carrier-free [^{32}P]orthophosphate (0.5–1 mCi/ml), to cover the slices. Purge each flask with 95% O_2/5% CO_2 and pre-incubate for 30–120 min with gentle shaking.
5. Initiate the incubation by the addition of 0.1 vol. KRB containing the drug of interest.
6. Terminate the reaction by rapid aspiration of the medium (NB: dispose of this safely). Remove the slices with forceps and freeze them in scintillation vials by adding liquid N_2. Process the tissue by one of the protocols in Section 5, depending on the specific variable to be investigated.

[a] Glass or plastic scintillation vials (20 ml vol.) may also be used.

3.4 Isolated cells

3.4.1 General considerations

Preparations containing isolated cells have become powerful tools in the analysis of protein phosphorylation systems at the single cell level (11, 24–26). Such preparations avoid the cellular and morphological complexity of intact tissue, and allow one to examine homogeneous populations of cells, all presumably with identical properties and responses to physiological stimuli. Isolated brain nerve terminals (synaptosomes) may also be used for similar protein phosphorylation studies (27).

The major criteria used for selection of cell type include the cell viability, responsiveness to relevant extracellular physiological stimuli, and ability of the cells to perform the process under investigation. One should also note developments in molecular biological methods which allow targeted manipulations of gene expression in isolated cells, and examination of aspects of protein function regulated by phosphorylation with a high degree of precision (28).

3.4.2 Labelling and stimulation of isolated cells

Ideally, cells should be pre-labelled with [^{32}P]orthophosphate to label the cellular ATP pool without inducing a high level of background labelling of phospholipids, nucleic acids, and metabolites, and initial studies should therefore try to identify suitable pre-incubation conditions (26). *Protocol 2* describes a procedure for protein phosphorylation in isolated cells and nerve terminals.

Protocol 2. Protein phosphorylation in isolated cells and nerve terminals

Equipment and reagents

- Appropriate cells grown in suitable medium
- Phosphate-free medium (Gibco), Hepes-containing medium for suspension cultures, specific medium for monolayer cultures
- [^{32}P]orthophosphate (*Protocol 1*)
- Rubber policeman (for adherent cells)
- Incubator for suspension culture (e.g. Innova 4000 shaking incubator, New Brunswick Scientific)
- Incubator for monolayer culture (e.g. Forma Scientific CO_2 incubator)
- TCA, acetone, or 10% SDS (depending on the analysis method, see step 4)

Method

1. For cells growing in suspension, centrifuge the culture at 750 *g* for 5 min and discard the medium. Resuspend cells in phosphate-free medium containing [^{32}P]orthophosphate (0.5–2 mCi/ml). For cells grown as monolayers, aspirate the medium and add phosphate-free medium containing [^{32}P]orthophosphate (0.2–0.5 mCi/ml) using a rubber policeman to scrape the cells off the dishes.
2. Pre-incubate the cells at 25–37 °C for 30–120 min.[a]
3. Remove the medium, add non-radioactive medium, and divide the suspensions into aliquots. Incubate the cells at the optimal temperature for various predetermined time periods.[a]
4. At the appropriate times, terminate the incubation. For precipitation of proteins, add 100% (w/v) TCA to 10%, or ice-cold acetone to 50–90% (v/v) final concentration. For solubilization of proteins, add 10% (w/v) SDS to 1% final concentration and heat the sample at 100 °C for 2 min. For immunoprecipitations of specific proteins, add appropriate lysis buffers (see *Protocols 14–16*).[b]
5. Analyse the samples by one of the methods in Section 5.

[a] The optimal temperature and times for the pre-incubation and incubation phases must be determined in pilot experiments.
[b] Select a method compatible with the analysis to be performed (Section 5).

4. Phosphorylation of proteins in cell-free preparations

4.1 General considerations

Cell-free incubations can be performed under defined conditions and in the absence of diffusion barriers, which allows enzymes, substrates, and cofactors to gain access to each other under optimal conditions (29, 30). Such experiments may, therefore, demonstrate the capacity of the tissue preparation to phosphorylate and/or dephosphorylate unknown or identified proteins. Use of radiolabelled ATP makes this approach particularly suitable for partial characterization of those protein phosphorylation systems where phosphoserine and phosphothreonine represent the major amino acid species labelled. One should note that phosphotyrosine usually constitutes a minute fraction of total protein-bound phosphoamino acid, and requires slightly modified incubation conditions and analytical approaches (see *Protocol 1*, and ref. 31).

The limitations inherent in these preparations should also be kept in mind. The cellular architecture is lost, the biochemical composition of the medium surrounding the protein phosphorylation systems becomes unphysiological, and the spatial and functional relations between enzymes, substrates, and regulators become distorted. Such artefacts make the physiological interpretations of the results difficult. In particular, physiological time courses and proofs of phosphorylation of specific proteins in intact cells cannot be obtained from these experiments.

4.2 Labelling and stimulation of cell-free preparations

Investigation of cell-free protein phosphorylation systems necessitates selection of suitable and appropriate tissue fractions as well as incubation conditions (e.g. buffers, pH, ionic compositions, and activating agents such as Ca^{2+} or cyclic nucleotides). *Protocol 3* describes a method for the phosphorylation of proteins *in vitro*. Specific protein phosphorylation systems can be modulated by the addition of specific effectors or inhibitors, e.g. cyclic AMP analogues, Ca^{2+} and calmodulin, phospholipid derivatives and phorbol esters, or peptide inhibitors specific for distinct protein kinases (see Section 6).

Protocol 3. Phosphorylation of proteins *in vitro*

Equipment and reagents

- Appropriate tissue
- Tissue grinders: Potter-Elvehjem, Waring Blender, or Ultra Turrax models
- Homogenization buffer: 10 mM Tris–HCl pH 7.4, 1 mM EDTA, 1 mM DTT, 50 U/ml aprotinin, and 0.1 mM PMSF[a]
- Centrifuge and tubes for preparing tissue fractions (e.g. Sorvall HiSpeed and Beckman Ultracentrifuges)
- Reagents for protein determination
- 10 × phosphorylation buffer: 0.2 M Hepes pH 7.4, 0.1 M $MgCl_2$, 10 mM EDTA, 10 mM EGTA, 10 mM DTT[b]
- Vessels for incubating samples: preferably 1.5 ml microcentrifuge tubes or 10 × 75 mm glass test-tubes
- [γ-^{32}P]ATP: 10 Ci/ml, > 2500 Ci/mmol (Amersham)
- 3 × SDS sample buffer: 9% (w/v) SDS, 0.186 M Tris–HCl pH 6.8, 15% (v/v) glycerol, 6% (v/v) 2-mercaptoethanol, 0.1 mg/ml bromophenol blue

Method

1. Dissect and homogenize the tissue in 10 vol. ice-cold homogenization buffer (i.e. 1 g tissue/10 ml buffer).

2. Prepare subcellular fractions by differential centrifugation of the homogenate using standard methods. Resuspend the particulate fractions by rehomogenization in homogenization buffer. Measure the protein concentration of each fraction.
3. Prepare samples for *in vitro* phosphorylation by adding aliquots (10–100 μg protein) of the tissue fractions to microcentrifuge tubes containing 10 μl ice-cold phosphorylation buffer and store them on ice. Add effectors or H_2O to 90 μl total volume.
4. Pre-incubate the samples for 60 sec at 30°C. Initiate the reactions by addition of 10 μl ATP containing 1 μCi [γ-^{32}P]ATP to each tube (final concentration 1–10 μM, final volume 100 μl).
5. Terminate the reactions after 10–120 sec by the addition of 50 μl 3 × SDS sample buffer. Mix well and analyse the samples as in Section 5.

[a] Additional protease inhibitors, such as leupeptin, antipain, pepstatin A, and chymostatin, may also be added if necessary.
[b] Add 0.1 M $MnCl_2$ and 0.5 mM sodium orthovanadate for labelling of phosphotyrosine residues.

5. Analysis of phosphorylated proteins

5.1 General considerations

Following the incubations described in *Protocols 1–3*, the phosphorylated proteins derived from the different preparations must be separated, characterized, and, if possible, identified. A large variety of protein chemistry methods is available for these purposes, detailed descriptions of which are outside the scope of this chapter. We briefly review some standard methods which we have found to be particularly helpful. For further information see refs 9–13, 32–34.

5.2 Quantification of phosphoproteins

To compare phosphorylation reliably in different samples, aliquots which contain equal amounts of the protein under study should be compared. Methods which allow quantification of the specific protein (e.g. immunoblotting) or, alternatively, the total amount of protein represent the preferred approaches. In samples labelled with ^{32}P, the total amount of TCA and alcohol insoluble ^{32}P in the sample (most of which is incorporated into protein) may also be used as a reference value. *Protocol 4* describes the determination of ^{32}P incorporation in cells and tissues.

Protocol 4. Determination of total ^{32}P incorporation in intact cells and tissue

Equipment and reagents

- Tissue or cell samples for analysis (*Protocols 1* or *2*)
- 1% (w/v) SDS
- Sonicator (Vibra Cell, Sonics and Materials Inc.)
- Tissue grinders (Potter-Elvehjem type, optional)
- GF-C filter paper (Whatman)
- 10% (w/v) TCA
- 95% ethanol
- Scintillation counter, vials, and cocktail

Method

1. Take 0.1–1 ml of tissue or cell sample for analysis, add an equal volume of hot 1% (w/v) SDS, and disrupt the tissue by sonication or homogenization.
2. Take a 20 cm × 30 cm piece of GF-C paper and rule it into 1 cm × 1 cm squares using a pencil.
3. Measure the TCA and ethanol insoluble radioactivity by spotting 5–10 μl aliquots of the homogenized samples onto the GF-C filter paper; air dry.
4. Soak the paper in 10% (w/v) TCA for 15 min. Discard the TCA and repeat the soaking three times, each for 5 min. Dispose of radioactive solutions according to regulations.
5. Rinse the paper in 95% ethanol for 5 min, air dry the paper, cut out the squares, and measure the ^{32}P by scintillation counting.[a]

[a] Cherenkov counting in aqueous medium may also be used.

5.3 Phosphoprotein separation

Electrophoresis of proteins in polyacrylamide gels remains unsurpassed as a general method for protein separation in terms of simplicity, resolution, and reliability (34). Performing electrophoresis in the presence of the strong ionic detergent SDS has the added advantage that subunits of polymeric proteins are dissociated and proteins become denatured and inactivated. Although many electrophoresis buffers have been used, buffers containing SDS, Tris–HCl, and glycine (35) appear to be most suitable for separation of proteins in the range of 15–300 kDa, while buffers containing SDS, Tris, and Tricine (36) appear to be preferable for separation of proteins and peptides in the range of 2–50 kDa. For one-dimensional SDS–PAGE protein separations, see ref. 34 for example.

The number of proteins that can be separated by one-dimensional electrophoresis remains limited. Two-dimensional electrophoresis methods allow several hundred individual proteins to be detected on a single slab gel, particularly when using either isoelectric focusing or non-equilibrium pH gradient gel electrophoresis in the first dimension (37, 38). Both these methods rely heavily on urea to solubilize proteins prior to separation in the first dimension, and many particulate proteins remain difficult to handle (39). Using a less sensitive method (developed for membrane proteins), and including SDS in the buffers for both dimensions (40), is an alternative way of solubilizing and separating several difficult phosphoproteins (41). For further information, the reader is directed to specific publications dealing with these problems (42, 43).

5.4 Protein isolation by immunomethods

The complexity of protein expression in many tissue preparations, which frequently makes electrophoresis insufficient for the study of individual proteins, can often be overcome by partial purification of the protein(s) in question. When a suitable antibody is available, several immunoadsorption methods can be used to isolate the protein from highly complex mixtures. For these purposes, proteins are solubilized in suitable lysis buffers, with compositions depending on the nature of the protein under study and frequently including chelator of divalent cations, and inhibitors of phosphatases and proteases. For soluble proteins, a simple hypotonic medium will suffice, while preparations derived from cytoskeletal or membrane components may necessitate mixtures of chaotropes, strong detergents, and salts (44). Some of these components may then have to be removed to allow antibody–antigen interactions to occur. A well-tested approach utilizes solubilization of proteins in media containing hot 1% (w/v) SDS, followed by addition of a fivefold excess of non-ionic detergents (e.g. Triton X-100) before the immunoreaction takes place (45). Boiling of the sample in SDS has the added advantage of rapidly inactivating protein kinases, phosphatases, and proteases. For further information about these methods, the reader is referred to specialist literature (44).

5.5 Analysis of multisite phosphorylation

Many proteins can be phosphorylated on multiple amino acid residues. In the case of ^{32}P-labelled proteins, a classical approach for distinguishing between these phosphorylation sites has been to use phosphopeptide mapping. Reproducible cleavage of proteins in polyacrylamide gel pieces or of proteins adsorbed to polyvinylidene difluoride (PVDF) membranes can be obtained with proteases (trypsin, thermolysin, or *S. aureus* V8 protease), or by chemical treatment (e.g. with cyanogen bromide which cleaves proteins at methionine residues) (46–48). The phosphopeptides obtained may then be examined and characterized by methods such as one-dimensional electrophoresis, isoelectric focusing, two-dimensional thin-layer electrophoresis chromatography, HPLC, or microsequencing. One-dimensional peptide mapping by SDS–PAGE is described in *Protocol 5*.

Protocol 5. One-dimensional peptide mapping of proteins separated by SDS–PAGE

Equipment and reagents

- Protein samples (*Protocols 1–3*) analysed by SDS–PAGE and autoradiography
- Reswelling buffer: 62.5 mM Tris–HCl pH 6.8, 0.1% SDS, 5% (v/v) glycerol, 2% (v/v) 2-mercaptoethanol
- SDS–PAGE gel and buffers (see ref. 34): the gel should have a stacking gel 4 cm deep and 15% polyacrylamide separating gel
- *Staphylococcus aureus* V8 protease (Sigma)
- Gel drier
- X-ray film and facilities for autoradiography

Method

1. Identify the ^{32}P-labelled proteins of interest in the SDS–PAGE gel from the autoradiograph of the gel. Cut out the protein bands and incubate the gel pieces in reswelling buffer for 30 min at room temperature.
2. Insert the pieces into the wells of a second SDS–PAGE gel. Mount the gel in the electrophoresis apparatus.
3. Overlay the gel pieces with 50 μl reswelling solution containing 0.1–5 μg/lane of V8 protease.
4. Digest the proteins during electrophoresis at low voltage (50–60 V) through the stacking gel. Separate the phosphopeptides by electrophoresis through the separating gel at voltages used for standard SDS–PAGE.
5. Dry the gel and autoradiograph it to locate the phosphopeptides.

Figure 1 shows a comparison between the neuronal protein synapsin I, phosphorylated by cyclic AMP-dependent protein kinase and Ca^{2+}-calmodulin-dependent kinases (CaM kinases) *in vitro* (*Protocol 3*), and synapsin I phosphorylated *in situ* (*Protocol 2*) following incubation of isolated nerve terminals with [^{32}P]orthophosphate (27). In this experiment, a method for one-dimensional peptide mapping based on incomplete proteolysis was used (49, 50). Phosphorylation of synapsin I *in vitro* with activated cAMP-dependent protein kinase generated one, while incubation with activated CaM kinases generated two V8 protease-derived phosphopeptides. This accurately reflects the presence of one phosphorylation site (site 1) for cAMP-dependent protein kinase which is shared with CaM kinase I, and multiple, but closely spaced, sites for phosphorylation by CaM kinase II (site 2) (50). Moreover, these sites were differentially labelled *in situ* following increases in cAMP (induced by forskolin) or Ca^{2+} (induced by nerve terminal depolarization).

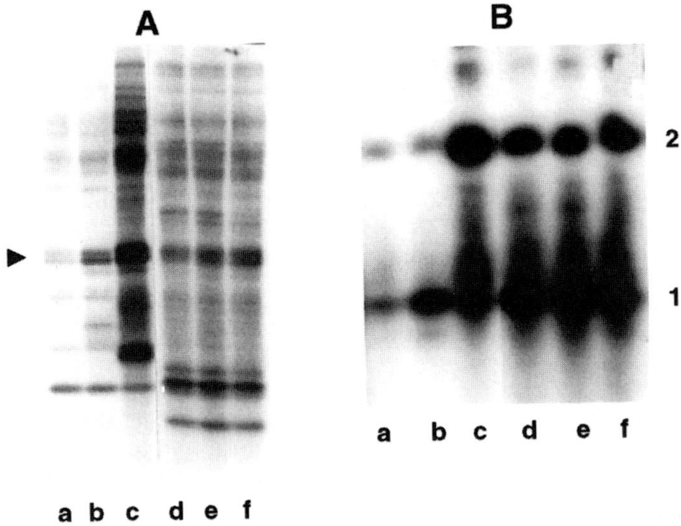

Figure 1. Autoradiogram showing one-dimensional peptide map of the ^{32}P-labelled nerve terminal protein synapsin I from isolated rat cerebrocortical nerve terminals and separated by SDS–PAGE. (A) Lanes a–c, lysed nerve terminal preparation incubated *in vitro* (*Protocol 3*) in the absence (a) or presence of 1 μM 8-Br-cAMP (b) or 0.5 mM Ca^{2+} (c). Lanes d–f, intact nerve terminal preparation pre-labelled with [^{32}P]orthophosphate and incubated *in situ* in the absence (d) or presence of 10 μM forskolin (e) or 60 mM KCl (f) (*Protocol 2*). Arrowhead indicates the position of synapsin I. (B) Peptide map of phosphorylated synapsin I. Gel pieces containing synapsin I from the experiment shown in (A) were treated with the V8 protease from *S. aureus* as described in *Protocol 5*. Position 1, phosphopeptide containing phosphorylation site 1 which is labelled by cyclic AMP-dependent protein kinase and CaM kinase I. Position 2, phosphopeptide containing phosphorylation site 2 which is labelled by CaM kinase II.

When one-dimensional mapping of peptides derived from incomplete proteolysis is insufficient to detect all phosphorylation sites in a protein, complete digestion of the phosphoprotein and (usually) two-dimensional separation (48) must be performed as described in *Protocol 6*.

Protocol 6. Two-dimensional enzymatic phosphopeptide mapping

Equipment and reagents

- Protein samples (*Protocols 1–3*) analysed by SDS–PAGE and autoradiography
- 50% (v/v) methanol
- 0.1 M NH_4HCO_3
- Proteolytic enzyme appropriate for analysis of the phosphoprotein of interest
- Vacuum centrifuge (e.g. SpeedVac, Savant)
- Solvent A: 10% (v/v) acetic acid, 1% (v/v) pyridine pH 3.5 in water
- Solvent A containing 0.05% (w/v) phenol red
- Thin-layer silica plates (Merck)
- Fume-hood
- Thin-layer electrophoresis unit (e.g. Multiphor II, Pharmacia)

Protocol 6. *Continued*

- Chromatography tank for ascending TLC
- Solvent B: acetic acid:pyridine:butan-1-ol: H_2O (3:15:10:12)
- X-ray film and facilities for autoradiography

Method

1. Cut out the gel pieces containing the proteins of interest from the SDS–PAGE gel, wash the pieces in 50% (v/v) methanol for at least 6 h, and dry them by vacuum centrifugation.
2. Reswell the gel pieces in 0.1 M NH_4HCO_3 containing 0.1–1 mg/ml of the proteolytic enzyme.[a]
3. Incubate at 30°C for 8–12 h.
4. Centrifuge the samples at 10 000 *g* for 1 min, collect the supernatant from each and evaporate it by vacuum centrifugation.
5. Resuspend the dried phosphopeptides in 20 μl solvent A containing phenol red, and centrifuge the sample at 10 000 *g* for 2 min.
6. Apply 5–10 μl of the sample 4 cm from the edge in the middle of a thin-layer silica plate.
7. Subject the plate to electrophoresis in solvent A for approx. 2 h at 400 V until the phenol red dye has migrated 6 cm along the edge of the plate.
8. Dry the plate in a fume-hood. Subject it to ascending chromatography in the second direction in solvent B. Dry the plate, and localize the phosphopeptides by autoradiography.

[a] The concentration of protease required will depend on the protease used and may need to be determined in pilot studies.

Figure 2 shows a two-dimensional phosphopeptide map of the nuclear phosphoprotein HMG I, a protein that can be phosphorylated by a variety of protein kinases *in vitro* (51, 52). The protein was ^{32}P-labelled either (panel A) by pre-labelling of metaphase arrested HeLa cells with [^{32}P]orthophosphate (*Protocol 2*), or (panel B) by incubating recombinant HMG I *in vitro* with purified p34$^{cdc2/cyclinB}$ kinase (cdc2 kinase) (*Protocol 3*). Following complete digestion with thermolysin and phosphopeptide separation by two-dimensional thin-layer electrophoresis and chromatography as described in *Protocol 6*, several phosphorylation sites, some of which overlapped, were found. Since the primary sequence of HMG I (53) indicates only three putative phosphorylation sites for the cdc2 kinase, some incomplete thermolytic digestion might have occurred. This demonstrates that the ability to achieve complete digestion with site-specific proteolytic enzymes may depend on the nature of both the substrate and protease employed.

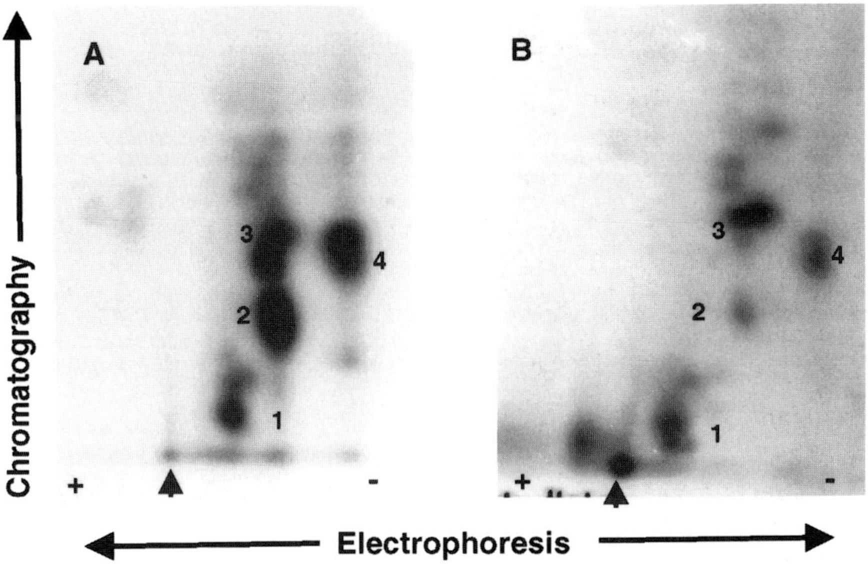

Figure 2. Autoradiogram showing two-dimensional phosphopeptide mapping of the nuclear phosphoprotein HMG I, labelled in HeLa cells arrested in metaphase of the cell cycle (A), or recombinant HMG I phosphorylated with purified p34cdc2/cyclinB kinase (Cdkl) *in vitro* (B). HMG I was isolated from the cells by acid extraction (53) and SDS–PAGE. The gel pieces were subjected to digestion with thermolysin and the resulting phosphopeptides were separated as in *Protocol 6*. Arrowheads indicate the application point, numbers indicate the major phosphopeptides present in both maps.

5.6 Phosphoamino acid analysis

The importance of phosphoamino acid analysis, which sometimes may suffice for identification of the regulated phosphorylation site (54), should also be stressed. Identification of phosphorylated amino acids is most commonly performed on proteins separated by SDS–PAGE and subjected to acid hydrolysis, followed by one- or two-dimensional separation of the amino acids and autoradiography. Such acid hydrolysis may be performed directly on phosphoproteins in gel pieces, on proteolytic phosphopeptides eluted from gel pieces, or on phosphoproteins blotted onto PVDF membranes (54–56).

Phosphotyrosine (P-Tyr) residues in phosphoproteins frequently represent a minor fraction of total phosphoamino acids. To remove most of the labelled phosphoserine (P-Ser) residues, the polyacrylamide gels or PVDF membranes may be incubated in hot alkali (11, 57). Alternatively, P-Tyr can be separately identified by immunoblotting with P-Tyr antibodies (58), many of which are commercially available. *Protocol 7* describes the analysis of phosphoamino acids by acid hydrolysis and two-dimensional separation.

Protocol 7. Detection of phosphoamino acids

Equipment and reagents

- ^{32}P-labelled phosphopeptides eluted from an SDS–PAGE gel (*Protocol 6*, steps 1–4) or [^{32}P]phosphoprotein blotted onto PVDF membrane (ref. 34)
- 6 M HCl
- Thin-layer electrophoresis and chromatography facilities, vacuum centrifuge (*Protocol 6*)
- Solvent C: 8% (v/v) acetic acid, 2% (v/v) formic acid pH 1.9
- Solvent C containing 0.01% (w/v) phenol red
- Phosphotyrosine (P-Tyr), phosphoserine (P-Ser), and phosphothreonine (P-Thr) standards (from Sigma): make a 1 mg/ml stock solution of each in water
- Plastic TLC plates coated with cellulose (Merck) or thin -layer silica plates (*Protocol 6*)
- Solvent D: isobutyric acid, 0.5 M NH_4OH (5:3, v/v)
- 0.5% (w/v) ninhydrin in butan-1-ol
- Autoradiography facilities and X-ray film

Method

1. Resuspend the dried [^{32}P]phosphopeptides in 0.2 ml 6 M HCl and hydrolyse them at 110 °C for 90 min. Alternatively, add pieces of PVDF membrane containing the ^{32}P-labelled protein directly to 6 M HCl and hydrolyse as above. Collect the hydrolysate and evaporate by vacuum centrifugation.
2. Dissolve the amino acids in 5 µl solvent C containing 0.01% phenol red, add 5 µl each of the stock P-Tyr, P-Ser, and P-Thr solutions as internal markers. Centrifuge the samples at 5000 *g* for 2 min, and spot 5–10 µl (containing at least 100 c.p.m.) in the corner of a TLC plate 3 cm from each edge.
3. Perform electrophoresis at 800 V in solvent C in one dimension until the phenol red is 2 cm from the edge of the plate, and then chromatography in solvent D in the second dimension in a fume-hood.
4. Dry the plate in a fume-hood, visualize the amino acid spots by ninhydrin staining. To do this submerge the TLC plate briefly in a glass tray containing the ninhydrin solution, or apply as aerosol in a fume-hood from a distance of 30–50 cm. Dry the plate for 5 min, then place in an oven at 110 °C for 5 min.
5. Locate ^{32}P-labelled amino acids by autoradiography.

5.7 Analysis of the state of phosphorylation

Understanding the mechanisms involved in physiological regulation of phosphorylation systems *in situ* requires that the state of phosphorylation of each phosphorylated amino acid residue in the protein in intact cells can be determined. Most investigators will rely on *in situ* labelling with [^{32}P]orthophosphate for these experiments (*Protocol 2*), followed by quantitative analysis of specific phosphorylation sites. This approach may, however, under certain conditions give misleading results (18), and should preferably be compared to results obtained with other methods.

Chemical detection of phosphate linked to specific sites in proteins may be done by non-radioactive methods based on mass spectrometry (ref. 59), but these methods necessitate highly specialized equipment not usually available to standard laboratories. At present, the most convenient approach employs phosphoprotein-specific antibodies directed against distinct phosphorylation sites in the protein under study. Such antibodies can be prepared against synthetic phosphopeptides with sequences derived from the protein sequence, and an increasing number of such antibodies are becoming available commercially. Use of such antibodies in immunoblots allows quantitative determination of the state of phosphorylation of specific sites in identified proteins (60, 61). *Protocol 8* gives a procedure for the detection of phosphoproteins by immunoblotting. Procedures for immunoblotting are given in refs 34, 35, and 44.

Protocol 8. Analysis of the state of phosphorylation of proteins in intact cells by immunoblotting

Equipment and reagents

- Tissue frozen in liquid nitrogen (*Protocol 1*) or cells processed in hot SDS (*Protocol 2*, step 4)
- Lysis buffer: 1% (w/v) SDS
- Sonicator (Vibra Cell, Sonics and Materials Inc.)
- Reagents for protein determination by a standard method
- SDS–PAGE gels, buffers, and electrophoresis equipment (ref. 34)
- Electroblotting reagents and equipment (Chapter 5, *Protocol 14*)
- Antibody recognizing phosphorylated forms of the protein under investigation
- Antibody recognizing the holoprotein irrespective of the phosphorylation state
- Equipment and reagents for visualizing antibodies; this will depend on the nature of the antibodies used

Method

1. Add lysis buffer to samples containing intact tissue in liquid N_2 (approx. 1 ml buffer/100 mg tissue).
2. Rapidly disrupt the tissue by sonication,[a] transfer the solution to a microcentrifuge tube, cap the tube, punch a small hole in the cap, and heat the tube in a boiling water-bath for 1 min.
3. Assay the sample for protein content using a standard method. Add lysis buffer so that all the samples have the same protein concentration (preferably in the range of 0.5–1 mg/ml).
4. Separate the proteins in the samples by SDS–PAGE and electrotransfer them from the gel onto nitrocellulose or PVDF as described in Chapter 5, *Protocol 14*, or in refs 34, 35, and 44.

Protocol 8. *Continued*

5. Determine the amount of phosphorylated protein by probing the immunoblot with the phospho-specific antibody. Estimate the total amount of the protein under study by probing a second immunoblot with antibody recognizing both phosphorylated and dephosphorylated forms of the protein.

[a] Samples may also be homogenized using Potter-Elvehjem tissue grinders.

6. Analysis of protein kinases

6.1 General considerations

Although the physiologically relevant result of protein phosphorylation in intact cells is a change in the state of phosphorylation of a protein, estimation of protein kinase and phosphatase activities may also be of interest. Since methods for studying protein kinases differ somewhat from those used in phosphoprotein analysis, a brief discussion is presented here. More general information on protein kinases is given by Hardie and Hanks (2, 3).

6.2 Analysis of protein kinase activity *in vitro*

The basic approach to protein kinase analysis consists of measuring the ability of an extract containing the activity of interest to catalyse phosphorylation of a suitable exogenous substrate under 'initial rate' conditions. The extracts must, therefore, be prepared in a buffer which avoids denaturation, aggregation, and degradation of proteins, and with suitable amounts of activating factors added. Since many protein kinases are themselves regulated by phosphorylation, addition of phosphatase inhibitors may also be necessary. Finally, non-specific inhibitory compounds may be removed by dilutions or by simple chromatographical methods.

Protein kinase activities can be examined both in solution and in 'solid phase' assays, where extracted proteins have been adsorbed to filter papers or bound to immunoadsorbents or other kinds of binding matrices, which then are used as the source of kinase activities (62). A number of protein kinases can also be separated by SDS–PAGE, followed by renaturation and assay in the gel or on PVDF membranes (see *Protocol 16* and ref. 63).

The substrates to be added for the phosphorylation reaction include a phosphate donor and a phosphate acceptor. All protein kinases can employ [γ-^{32}P]ATP as phosphate donor, with P-Ser/Thr-specific protein kinases preferring Mg^{2+} and P-Tyr-specific protein kinases preferring Mn^{2+} as the cation. The enzymes display apparent K_m values for ATP below 100 μM (usually around 10 μM), and ATP levels near saturating concentrations (five- to tenfold over K_m) will usually give the best results. The assays are performed with added exogenous substrates and under initial rate conditions, i.e. with activity linear with respect to the amount of enzyme added and the incubation time, with substrate concentrations in excess of apparent K_m values, and with not more than 5–10% of substrate being consumed (64).

Several types of protein substrates can be used as phosphate acceptors. Ideally the protein should represent a physiological monospecific substrate, but multikinase substrates like histones, myelin basic protein, or casein are more frequently employed. Enzymes are then distinguished by exploiting other enzymatic properties, most frequently by comparing activity in the absence and presence of specific activators or inhibitors (e.g. cyclic nucleotides, Ca^{2+}-calmodulin, phospholipid derivatives, specific inhibitor peptides). Alternatively, synthetic peptides with 'consensus sequences', which represent major specificity determinants for many protein kinases (65), can tentatively be used as 'monospecific' substrates. Such peptides have the added advantages that they can be designed to suit specific needs, for example by addition of charged residues to simplify product purification (66).

Following transfer of ^{32}P to the protein substrate, the phosphorylated product is separated from unreacted [γ-^{32}P]ATP and quantified. This is usually achieved by exploiting differences between these components in solubility, size, and/or charge. Separation by means of SDS–PAGE, where the labelled ATP, inorganic phosphate, and phospholipids migrate in front of the electrophoretic dye while proteins are retarded, generally gives the lowest background, and is also suitable when more than one protein is present in the substrate preparation. However, this time-consuming method is not suitable when small peptides are employed (36). When single proteins are used as substrates, precipitation with TCA is a suitable and rapid alternative method. Most (although not all) proteins will be denatured and precipitated in the presence of 5–25% (w/v) TCA, while free phosphate and ATP remain in solution. An acceptable signal-to-noise ratio is usually achieved by adding BSA (usually at 0.1 mg/ml final concentration) as a carrier during precipitation, and washing the pellet three to five times before counting the radioactivity. In contrast, passing the reaction mixture through anionic exchangers will allow basic peptide substrates to elute in the flow-through, while ATP is quantitatively retained (65). Although this method is somewhat more time-consuming when large numbers of samples are analysed, the ease of collection and quantification of the phosphorylated peptide substrate makes this the preferred method when an initial purification of substrates is desired, followed by further characterization of product, e.g. by HPLC, determination of phosphoamino acids, or amino acid sequencing (67). Finally, substrates may be isolated by binding to phosphocellulose paper. This cation exchanger will bind peptides or proteins with basic residues (e.g. arginine), while ATP and free phosphate remain in solution (68). In the case of a more acidic peptide substrate (such as casein kinase 2), the necessary basic residues may be added using standard peptide synthesis chemistry (66). Since it is possible to analyse a large number of samples at the same time, this rapid and simple method is very efficient for screening of multiple samples, and is described in *Protocol 9*.

Protocol 9. Assay of protein kinase activity by substrate binding to phosphocellulose

Equipment and reagents

- Phosphocellulose paper (P81, Whatman)
- 10 × phosphorylation buffer (*Protocol 3*)
- Tissue or cell extracts for assay (*Protocol 3*, steps 1–2); assay for protein content, store on ice
- Peptide or protein substrates appropriate for kinase to be analysed, with basic residues (Arg or Lys) in excess of acidic or phosphorylated residues
- 1 mM ATP containing [γ-^{32}P]ATP (approx. 10 μCi/ml)
- Acetic acid (glacial)
- 75 mM phosphoric acid
- Facilities for scintillation or Cherenkov counting (*Protocol 4*)

Method

1. Mark out a piece of P81 paper with a pencil into 2 × 2 cm squares.
2. In a microcentrifuge tube or a 10 × 75 mm glass tube, mix the following:

• 10 × phosphorylation buffer	10 μl
• extract to be assayed (0.1–10 μg protein)	10 μl
• peptide or protein substrate	10 μl
• activating factors (10 × stock solutions)	10 μl
• H_2O	to 90 μl volume

3. Pre-incubate the tubes for 60 sec at 30 °C.
4. Initiate the reactions by adding 10 μl 1 mM [γ-^{32}P]ATP and incubate for 1–10 min in a shaking water-bath.
5. Terminate the reactions by adding 10 μl cold glacial acetic acid to each tube.
6. Spot aliquots (e.g. 10 μl) onto the marked out P81 paper grid and air dry it.
7. Immerse the P81 paper in a beaker containing 75 mM phosphoric acid and wash five times, each for 5 min with gentle stirring. Dispose of radioactive solution according to local safety regulations.
8. Rinse the P81 paper in water and air dry. Cut out the squares and measure their ^{32}P content by Cherenkov or scintillation counting. Also spot 10 μl [γ-^{32}P]ATP, used as substrate, to determine its specific radioactivity (c.p.m./mol ATP, usually in the range of 10^4–10^5 c.p.m./nmol ATP).

9. Determine the protein kinase activity as follows:

   (a) Calculate the amount (mol) of product formed:

   $$\frac{(\text{c.p.m.}_{\text{formed}} - \text{c.p.m.}_{\text{blank}})}{\text{specific radioactivity of ATP}}$$

   (b) Correct for the volume spotted on the P81 paper as a fraction of the total sample (e.g. if 10 μl was spotted from 110 μl total sample, then multiply by 110 ÷ 10 = 11).

   (c) Divide by the amount of protein in the extract (mg protein) and incubation time (min):

   $$= \frac{\text{amount product formed}}{\text{incubation time} \times \text{amount protein present}} \text{ (mol/min/mg)}$$

6.3 Analysis of specific protein kinases *in vitro*

6.3.1 Cyclic AMP-dependent protein kinase (PKA)

This enzyme consists of two regulatory (inhibitory) subunits which bind two catalytic subunits. Added cyclic AMP binds to the regulatory subunits and releases the catalytic units, which now are free to phosphorylate suitable substrates. Even under 'basal' conditions, many tissues contain some free catalytic subunit. The holoenzyme may be activated by the addition of 1 μM 8-bromo-cyclic AMP, for example. The free catalytic subunit of this enzyme may be inhibited by addition of the heat stable inhibitor peptide which binds strongly to the catalytic domain of the enzyme and completely prevents any substrate binding and catalysis from taking place (65). The assay should, therefore, ideally be performed under three sets of conditions:

(a) Without any addition (which will give background activity plus that caused by free catalytic subunit).

(b) In the presence of a cyclic AMP analogue (to give both background and total activity of the enzyme).

(c) In the presence of the heat stable inhibitor peptide (to give background activity).

Histones are frequently used substrates, with histone H2B being particularly useful. A number of synthetic peptides containing the sequence Arg-X-X-Ser(Thr), which constitutes the preferred motif for this kinase (65) may also be used. The most popular of these is the octapeptide termed 'Kemptide' (commercially available) which is derived from the phosphorylation site of pyruvate kinase (65). *Protocol 10* describes the assay of PKA.

Protocol 10. Assay of PKA

Equipment and reagents

- Tissue or cells for study (*Protocols 3* and *9*)
- Homogenization buffer (*Protocol 3*) containing 5 mM zinc acetate, 0.1% Triton X-100 (v/v); also make up the same buffer with 5 mM EDTA
- 10 × phosphorylation buffer, reagents for protein determination (*Protocol 3*)
- 1 mM Kemptide substrate peptide (Sigma)
- 0.1 mM 8-bromo-cyclic AMP (Sigma)
- 1 mM protein kinase inhibitor peptide (Sigma)
- Acetic acid, phosphoric acid, P81 phosphocellulose paper, scintillation or Cherenkov counting equipment, 1 mM [γ-^{32}P]ATP (*Protocol 9*)

Method

1. Homogenize the cells or tissue in 10 vol. cold homogenization buffer containing 5 mM zinc acetate and 0.1% (v/v) Triton X-100. Centrifuge the homogenate at 10 000 *g* for 10 min and then resuspend the pellet in 1 ml ice-cold homogenization medium containing 5 mM zinc acetate, 5 mM EDTA, and 0.1% (v/v) Triton X-100.
2. In a microcentrifuge tube or a 10 × 75 mm glass tube, mix the following:

• 10 × phosphorylation buffer	10 μl
• extract to be assayed (0.1–10 μg protein)	10 μl
• 1 mM Kemptide substrate	10 μl
• 0.1 mM 8-bromo-cyclic AMP or H_2O	10 μl
• 1 mM protein kinase inhibitor or H_2O	10 μl
• H_2O	to 90 μl volume

3. Pre-incubate the mixtures for 60 sec at 30°C. Initiate the reactions by addition of 10 μl [γ-^{32}P]ATP. Incubate the reactions at 30°C.
4. Terminate the reactions after 10–120 sec by adding acetic acid and spotting 10 μl aliquots onto P81 phosphocellulose paper (*Protocol 9*, step 5).
5. Process the P81 phosphocellulose paper as in *Protocol 9*, steps 5–7.
6. Calculate the enzyme activities (nmol/min/mg protein, as in *Protocol 9*). Let activity in the presence of cyclic AMP represent A, let activity in the absence of cyclic AMP represent B, and let activity in the presence of peptide inhibitor represent C. Then the total cyclic AMP-dependent protein kinase is given by (A – C); free catalytic subunit in the sample is given by (B – C); and holoenzyme present in the sample is given by (A – B). The activity ratio (free catalytic subunit as fraction of total cyclic AMP-dependent protein kinase) equals (B – C)/(A – C).

6.3.2 Protein kinase C

This enzyme (diacylglycerol-activated, Ca^{2+}-phospholipid-dependent protein kinase or PKC) is a complicated family made up of monomeric enzymes, most of which have regulatory domains with binding sites for unsaturated diacylglycerols (where tumour-promoting phorbol esters act as artificial ligands), and distinct binding domains for Ca^{2+} and acidic phospholipids (69). The conventional type of PKC may, therefore, be activated by addition of 1.5 mM $CaCl_2$, 50 μg/ml phosphatidylserine, and 10 nM phorbol-12,13-myristate acetate or 1 μg/ml diolein. It should be noted that addition of detergents may interfere with the binding of both diacylglycerol and phospholipids to the enzyme, and could, therefore, be detrimental to enzyme assays *in vitro*.

The enzyme has a pseudosubstrate domain (residues 19–36) which binds tightly to the catalytic domain, and a synthetic PKC_{19-36} peptide can be used as a specific inhibitor (65). The enzyme activity is conveniently assayed using histone H1 as a substrate, followed by the phosphocellulose paper method (*Protocol 9*) to isolate the product. Alternatively, synthetic peptides containing the Arg-X-X-Ser(Thr)-X-Arg motif, several of which are commercially available, such as residues 4–14 of myelin basic protein, can also be used with the phosphocellulose paper method. Each sample is assayed by one set of tubes receiving $CaCl_2$ and phospholipid cofactors, while another set of tubes receives EGTA-containing phosphorylation buffer only. *Protocol 11* describes the procedure.

Protocol 11. Assay of protein kinase C

Equipment and reagents

- L-α-phosphatidyl-L-serine (Sigma)
- 1,2-dioleoyl-*sn*-glycerol (diolein, Sigma)[a]
- Chloroform
- Compressed dry nitrogen
- Sonicator (Vibra Cell, Sonics and Materials Inc.)
- Resuspension buffer: 20 mM Tris–HCl pH 7.4
- 10 × phosphorylation buffer (*Protocol 3*)
- Tissue or cell extract for analysis prepared as in *Protocol 3*, steps 1–2, using homogenization buffer containing 10 μM leupeptin and 0.1 mM EGTA; assay the extract for protein content
- 15 mM $CaCl_2$
- Peptide substrate: 1 mM stock solution myelin basic protein (4–14) (Research Biochemicals International)[b]
- Acetic acid, 75 mM phosphoric acid, phosphocellulose paper, and facilities for scintillation or Cherenkov counting, 1 mM [γ-^{32}P]ATP (*Protocol 9*)

Method

1. Make up phospholipid stock solution as follows:
   (a) Mix 10 mg phosphatidylserine and 0.2 mg diolein, both dissolved in chloroform (Sigma).
   (b) Dry the mixture under nitrogen in a fume-hood.

Protocol 11. *Continued*

(c) Resuspend the residue in resuspension buffer to form a 10 × stock solution containing 0.5 mg/ml phosphatidylserine and 10 μg/ml diolein, employing brief sonication on ice.

(d) Store the 10 × phospholipid stock solution in the dark, frozen in aliquots at –20 °C.

2. In microcentrifuge tubes or 10 × 75 mm test-tubes, mix the following:

• 10 × phosphorylation buffer	10 μl
• extract to be assayed (0.1–10 μg protein)	10 μl
• 10 × phospholipid stock solution (or H_2O)[d]	10 μl
• 15 mM $CaCl_2$ (or H_2O)[d]	10 μl
• 1 mM peptide substrate[b]	10 μl
• H_2O	to 90 μl volume

3. Pre-incubate the tubes for 60 sec at 30 °C. Initiate the reactions by adding 10 μl [γ-^{32}P]ATP.

4. Terminate the reactions after 10–120 sec by adding acetic acid and analyse the phosphorylated products by spotting aliquots on P81 phosphocellulose paper (*Protocol 9,* steps 5–8).

5. Calculate the PKC activity by subtracting total activity from activity seen in the absence of Ca^{2+} and phospholipid, employing the formula in *Protocol 9,* step 9.

[a] Phorbol esters, such as phorbol-12,13-dibutyrate may be used instead of diolein since they are more stable to metabolism. However, they must be handled with caution due to their actions as tumour promoters. Use a 1 μM stock solution in 10% (v/v) DMSO and a final concentration of 10–100 nM.

[b] Alternatively histone type III S (1 mg/ml stock solution, Sigma) may be used.

[c] If the tissue or cell extract contains other kinases, peptide inhibitors of PKA and/or calmodulin-dependent kinase II may have to be added, to final concentrations of 2 μM and 5 μM, respectively.

[d] For the control activity tube substitute H_2O for the phospholipid and $CaCl_2$.

6.3.3 Ca^{2+}-calmodulin-dependent protein kinase II

CaM kinase II which is particularly enriched in neural tissues, is composed of several subunits with binding sites for CaM and several activating and inhibitory autophosphorylation sites (70). Activation by CaM preferentially autophosphorylates a site (Thr-286 in the α-subunit) which creates an autonomous enzyme no longer dependent on CaM (71, 72). CaM kinase II has been characterized using exogenous protein substrates such as synapsin I or microtubule-associated protein-2 (73, 74). However, the enzyme is most conveniently assayed using synthetic peptides, such as 'syntide-2' or 'autocamtide-2' (75, 76), both of which are commercially available, in the presence of $CaCl_2$

and calmodulin. A synthetic peptide, comprising residues 281–302 of the α-subunit of CaM kinase II (with Ala substituted in position 286, commercially available) can be used as an inhibitor (77). *Protocol 12* describes the assay of CaM kinase II.

Protocol 12. Assay of CaM kinase II

Equipment and reagents

- Tissue or cell extract for assay (*Protocol 3*, steps 1–2); assay the extract for protein and store it on ice
- 10 × phosphorylation buffer (*Protocol 3*)
- 1 mM peptide substrate (Research Biochemicals International), dissolved in H_2O
- 15 mM $CaCl_2$
- 0.03 mM calmodulin (Calbiochem-Novabiochem or Boehringer), dissolved in H_2O
- 1 mM [γ-^{32}P]ATP (*Protocol 9*)
- Acetic acid, 75 mM phosphoric acid, phosphocellulose paper, and facilities for scintillation or Cherenkov counting (*Protocol 9*)

Method

1. In 10 × 75 mm glass test-tubes or microcentrifuge tubes, mix:

• 10 × phosphorylation buffer	10 μl
• extract to be assayed (0.1–10 μg protein)	10 μl
• 1 mM peptide substrate	10 μl
• 15 mM $CaCl_2$ (or H_2O)[a]	10 μl
• 0.03 mM calmodulin (or H_2O)[a]	10 μl
• H_2O	to 90 μl volume

2. Pre-incubate the mixture for 60 sec at 30 °C.
3. Initiate the reactions by addition of 10 μl [γ-^{32}P]ATP.
4. Terminate the reactions after 10–120 sec by adding acetic acid and spot 10 μl aliquots on P81 phosphocellulose paper (*Protocol 9*, steps 5–8).
5. Calculate the enzyme activity by subtracting background (assayed in EGTA) from total activity (assayed with $CaCl_2$ and calmodulin). Calculate the activity by employing the formula in *Protocol 9*, step 9.

[a] For the control activity tube, substitute H_2O for the $CaCl_2$ and calmodulin.

6.3.4 Casein kinase 2

This multifunctional and widely distributed enzyme is characterized by its ability to use GTP as well as ATP as substrate, its inhibition by low amounts of heparin, and its activation by relatively high salt concentrations (78). Casein kinase 2, which appears to phosphorylate a variety of proteins *in situ*, preferentially phosphorylates peptides with acidic residues carboxy terminal to the phosphorylated serine/threonine, such as containing the motif Ser-X-X-Glu (79). For such peptides, basic residues must be added at the N-terminus before the substrate will bind to phosphocellulose paper, as in Arg-Arg-Arg-Asp-Asp-Asp-Ser-Asp-Asp-Asp (which is commercially available). Tubes without substrate are used as blanks. *Protocol 13* describes the assay of casein kinase 2.

Protocol 13. Assay of casein kinase 2

Equipment and reagents

- Tissue or cell extract for assay (*Protocol 3*, steps 1–2); assay the extract for protein and store it on ice
- 10 × casein kinase 2 buffer: 0.5 M Tris–HCl pH 7.6, 0.1 M $MgCl_2$, 2 M NaCl[a]
- 5 mM casein kinase 2 peptide substrate (Boehringer)
- [γ-^{32}P]ATP (*Protocol 3*)
- Acetic acid, 75 mM phosphoric acid, phosphocellulose paper, and facilities for scintillation or Cherenkov counting (*Protocol 9*)

Method

1. In microcentrifuge tubes or 10 × 75 mm glass tubes, mix the following:

• extract to be analysed (0.1–10 μg protein)	10 μl
• 10 × casein kinase 2 buffer	10 μl
• 5 mM casein kinase 2 peptide substrate	10 μl
• H_2O	to 90 μl volume

2. Pre-incubate the mixtures for 60 sec at 30°C. Initiate the reactions by addition of 10 μl [γ-^{32}P]ATP.
3. Terminate the reactions after 1–5 min by adding acetic acid and analyse the phosphorylation of the substrate by spotting 10 μl aliquots on P81 phosphocellulose paper (*Protocol 9*, steps 5–6).
4. Wash the phosphocellulose paper, count the samples, and calculate the enzyme activity as in *Protocol 9*, steps 7–9.

[a] KCl may be used instead of NaCl.

6.3.5 MAP kinases

MAPKs (mitogen-activated protein kinases) or ERKs (extracellular-regulated kinases) are serine/threonine-specific, proline-directed protein kinases which are regulated by a number of extracellular stimuli. MAPK is activated by sequential phosphorylation on both tyrosine and threonine as a result of a protein kinase cascade (80, 81). The MAPK cascade connects cell surface receptors to specific transcription factors and other regulatory proteins, thus allowing extracellular signals to regulate expression of specific genes (82, 83). MAPK activity can be detected by immunoprecipitation of kinase activity from cell lysates, followed by kinase assay using specific MAPK substrates (84). Assay of MAPK is described in *Protocol 14*.

Protocol 14. MAP kinase assay

Equipment and reagents

- Phosphate-buffered saline (PBS): 137 mM NaCl, 2.6 mM KCl, 8.1 mM Na_2HPO_4, 17.6 mM KH_2PO_4 pH 7.35
- PBS-Tween: PBS containing 0.05% (v/v) Tween 20
- Protein A–Sepharose (Pharmacia) in a slurry (0.1 g/ml PBS plus 0.05% sodium azide)
- Anti-MAPK antibodies (anti-p42^{mapk} and anti-p44^{mapk}; Upstate Biotechnology)
- Cells or tissue for assay
- Lysis buffer: 50 mM Tris–HCl pH 7.4, 100 mM NaCl, 1% (v/v) Triton X-100, 1 mg/ml benzamidine, 1 mM sodium orthovanadate, 50 mM NaF, 10 mM sodium pyrophosphate, 1 mM PMSF, 10 μg/ml each of pepstatin A, aprotinin, leupeptin, and soybean trypsin inhibitor
- Kinase buffer: 25 mM Tris–HCl pH 7.4, 10 mM $MgCl_2$, 1 mM DTT
- Myelin basic protein substrate (Sigma)
- [γ-^{32}P]ATP, acetic acid, 75 mM phosphoric acid, phosphocellulose paper, and facilities for scintillation or Cherenkov counting (*Protocol 9*)

Method

1. Wash 50 μl Protein A–Sepharose beads twice in PBS-Tween and incubate the beads with 200 μl of antibodies to MAPK (diluted in PBS-Tween), for 1 h at 37 °C using gentle agitation. Following incubation, remove the antibody solution and wash the beads twice in PBS Tween.
2. Wash the cells twice in ice-cold PBS and lyse them on ice in lysis buffer. If using tissue, homogenize it in ice-cold lysis buffer. Centrifuge the lysate at 4 °C for 20 min, 10 000 *g*, and add 250 μl of the cleared cell lysate to the pelleted Protein A–Sepharose beads. Incubate for 12 h at 4 °C with gentle agitation.
3. Wash the beads twice in 0.5 ml lysis buffer and twice in 0.5 ml kinase buffer. Resuspend the pelleted beads in 20 μl kinase buffer containing 1 mg/ml myelin basic protein and 40 μM ATP containing 5 μCi [γ-^{32}P]ATP. Run blanks without myelin basic protein substrate. To avoid possible contamination with endogenous kinases other than MAPK, include inhibitors of PKA, PKC, and cyclin-dependent kinases in the incubation mixture. (An inhibitor cocktail is available from Upstate Biotechnology.)
4. Incubate the mixture at 30 °C for 10 min. Terminate the reactions by adding acetic acid and analyse the phosphorylation of the substrate by spotting samples onto phosphocellulose paper as described in *Protocol 9*, steps 5–8.
5. Calculate the enzyme activity following the procedure in *Protocol 9*, step 9.

As an alternative to the phosphocellulose-based assay described in *Protocol 14*, the samples may be analysed by SDS–PAGE. After the incubation (step 4) add SDS–PAGE sample buffer and carry out SDS–PAGE (and autoradiography) as described in ref. 34. The myelin basic protein substrate band can be excised from the gel and counted in a scintillation counter. Alternatively, the number of c.p.m. in the substrate can be estimated by drying the gel and employing a PhosphoImager instrument.

MAPK can also be analysed by separating the tissue or cell extracts on substrate-containing SDS–PAGE gels, followed by renaturation of the separated proteins in the gels, and incubation of the gels with [γ-^{32}P]ATP (85, 86). A procedure for this is provided in *Protocol 15.*

Protocol 15. Detection of MAPK activity in renatured SDS gels

Equipment and reagents

- Tissue or cell extract (*Protocol 3*, steps 1–2) or immunoprecipitated MAPK (*Protocol 14*) for assay
- Myelin basic protein (Sigma)
- SDS–PAGE gel (10% polyacrylamide with 0.1 mg/ml myelin basic protein in the separating gel), facilities for slab gel electrophoresis
- Buffer A: 50 mM Hepes pH 7.4, 5 mM 2-mercaptoethanol
- Buffer A containing 20% (v/v) propan-2-ol
- Buffer A containing 6 M guanidine–HCl
- Buffer A containing 0.04% (v/v) Tween 20
- Buffer B: 25 mM Hepes pH 7.4, 10 mM $MgCl_2$, 90 μM sodium orthovanadate, 5 mM 2-mercaptoethanol
- [γ-^{32}P]ATP (*Protocol 9*)
- 5% (w/v) TCA, 10 mM sodium pyrophosphate
- X-ray film and facilities for autoradiography

Method

1. Separate the proteins of the cell extracts on the SDS–PAGE gel (10–20 μg total protein per lane). If the sample for assay is immunoprecipitated MAPK (*Protocol 14*) similarly run it on the SDS–PAGE gel.
2. Wash the gel twice in buffer A with 20% (v/v) propan-2-ol to remove the SDS. Re-equilibrate the gel with buffer A for 1 h at room temperature.
3. Incubate the gel for 1 h with two changes in buffer A containing 6 M guanidine–HCl to denature the proteins.
4. Renature the proteins in two changes of buffer A with 0.04% (v/v) Tween 20 for 16 h at 4°C.
5. Pre-incubate the gel for 30 min in buffer B. Detect MAPK activity by incubation at 30°C for 1 h in 10 ml buffer B plus 250 μCi [γ-^{32}P]ATP (50 μM final concentration).
6. Terminate the reaction by immersing the gel in 5% TCA containing 10 mM sodium pyrophosphate. Wash the gel several times in the same buffer before drying. Autoradiograph the gel to detect MAPK activity.

6.3.6 Cyclin-dependent kinases

Cyclins and cyclin-dependent kinases (Cdks) are universal regulators of cell cycle progression in eukaryotic cells (87, 88). At present, the genes of eight related Cdks have been cloned from mammalian genomes. The catalytic subunits of Cdk require association with a cyclin regulatory subunit for kinase activity, with different cyclins being involved in the G1/S transition (G1 cyclins) and the G2/M transition (mitotic cyclins) (89, 90). Other regulatory mechanisms include the accumulation or activation of inhibitory proteins, and positive and negative phosphorylation at three conserved kinase sites (91–93).

Because of the existence of so many related Cdks in cell and tissue lysates, the assay of an individual Cdk necessitates the removal of other Cdks by pre-clearing of the lysate. For immunodepletion of interfering activities, lysates are initially pre-cleared by immunoprecipitation with normal serum or with excess of an antiserum specific for the Cdk in question. This is crucial when investigating individual members of this family of enzymes (e.g. deplete the lysate of cdk2 activity when cdk1 activity is to be investigated). *Protocol 16* describes the procedure for the assay of Cdk.

Protocol 16. Cdk assay

Equipment and reagents

- SDS–PAGE sample buffer, SDS–PAGE gel, buffers, and electrophoresis equipment (ref. 34)
- Cells for assay growing as monolayers or in suspension as appropriate for the individual cell lines
- PBS (*Protocol 14*)
- Lysis buffer: 50 mM Tris–HCl pH 7.5, 120 mM NaCl, 0.5% (v/v) Nonidet P-40, 2 mM EGTA, 1 mM sodium orthovanadate, 50 mM NaF, 100 μM benzamidine, 1 mM PMSF, 10 μg/ml each of leupeptin, aprotinin, and soybean trypsin inhibitor
- TBS-Tween: 20 mM Tris–HCl pH 7.35, 137 mM NaCl, plus 0.05% (v/v)Tween 20
- Cdk kinase buffer: 20 mM Tris–HCl pH 7.4, 10 mM $MgCl_2$, 1 mM DTT
- Cdk1/Cdk2 substrate (Histone H1; Boehringer) or Cdk4/Cdk6 substrate (GST-RB; Upstate Biotechnology) as appropriate
- Protein A–Sepharose (*Protocol 14*)
- Normal rabbit serum (antibody that does not recognize the antigen being studied)
- Appropriate anti-cyclin kinase antibody (Upstate Biotechnology)

Method

1. Collect 5×10^6 cells by centrifugation (750 *g* for 5 min) and wash them once in 5 ml ice-cold PBS. Resuspend the cells in 200 μl lysis buffer and lyse them for 20 min at 4°C followed by centrifugation at 10 000 *g* for 5 min. Collect the supernatant (lysate).
2. Pre-clear the lysate by adding 10 μl normal rabbit serum to 200 μl lysate and incubate for 1 h on ice. Add 50 μl Protein A–Sepharose beads and incubate for 30 min at 4 °C using agitation. Spin 2 min at 10 000 *g* and collect the supernatant. Repeat the incubation with a fresh portion of Protein A–Sepharose beads, and carefully collect the supernatant.
3. Prepare Protein A–Sepharose–antibody conjugate beads (*Protocol 14*, step 1).

Protocol 16. *Continued*

4. Add 200 μl pre-cleared lysate (step 2) to the conjugate beads and incubate for 3–6 h at 4°C using gentle agitation.
5. Wash the conjugate beads twice in 500 μl TBS-Tween and once in 500 μl kinase buffer.
6. Resuspend the conjugate beads in 40 μl kinase buffer containing 10 μCi [γ-^{32}P]ATP (20 μM final ATP concentration) and 0.5 mg/ml of the appropriate protein substrate. After 25 min incubation at 25°C, terminate the reaction by addition of SDS–PAGE sample buffer and analyse the sample by SDS–PAGE and autoradiography (ref. 34).
7. Calculate the activity as described (*Protocols 9* and *15*).

6.4 Analysis of protein kinase activity in intact cells

Pharmacological approaches, using membrane permeable inhibitors, have been employed to identify tentatively changes in a number of protein kinase activities *in situ* (94). However, many of these compounds are not very specific, and clear conclusions are often difficult to draw. Interested readers are directed to an extensive discussion of the usefulness of synthetic modulators of protein kinase activities in ref. 95.

Direct measurement of the state of activity of a given protein kinase in intact tissue is also of considerable interest to the investigator. At the present time, methods for determination of *in situ* activity of cyclic AMP-dependent protein kinase and Ca^{2+}-calmodulin-dependent protein kinase II appear valuable. Activation of protein kinases *in situ* may also be examined by studying the state of phosphorylation of specific substrates in the intact cells, for example, by analysing the activity of protein kinase C through MARCKS phosphorylation (96) or the activity of neuronal CaM kinase II through the phosphorylation of sites 2/3 on synapsin I (61). However, it should be remembered that such phosphoproteins may be substrates for as yet unidentified kinases, and that they also are under regulation by phosphatases.

6.4.1 Cyclic AMP-dependent protein kinase (protein kinase A)

The amount of free, dissociated catalytic subunit present in the tissue indicates the state of activity of protein kinase A (PKA) *in situ* (97, 98). The state of activity is usually presented as an 'activity ratio', i.e. the amount of activated enzyme as a fraction of the total PKA present in the cell, and can be estimated as described in *Protocol 10*. It should be noted that this approach is valid only if the enzyme can be extracted from the tissue in a medium which prevents further dissociation and reassociation of the holoenzyme from occurring, for example by using 0.5 M NaCl and/or phosphodiesterase inhibitors (98, 99).

6.4.2 Ca^{2+}-calmodulin-dependent protein kinase II

Since CaM kinase II becomes autonomous following Ca^{2+}-calmodulin-induced activation and autophosphorylation (71, 72), comparison of CaM kinase II activity *in vitro* in the absence and presence of Ca^{2+}-calmodulin can be used to estimate changes in the amount of activated enzyme present *in situ* (100–103). Alternatively, it is possible to follow the state of phosphorylation of Thr-286 in the α-subunit, either by labelling *in situ* with ^{32}P followed by phosphoamino acid analysis or phosphopeptide fingerprinting (100), or by using phospho-specific antibodies specifically recognizing this epitope (103).

6.5 Purification of protein kinases

Historically, purification of protein kinases has been necessary for characterization of these enzymes, and for examining their involvement in many physiological processes (for example, see ref. 104). While all protein kinases appear to contain highly homologous catalytic domains (2), other parts of these proteins are distinct, and form the basis for their biochemical purification. A large literature on protein purification exists in which the various steps involved are extensively discussed (105–107), and the reader is referred to these for a more detailed treatment. More recently, molecular biological methods have been employed to prepare some of these enzymes, and many purified protein kinases are now commercially available. This has obviated the necessity to employ difficult and time-consuming protein purification procedures. For further discussions see ref. 108.

7. Phosphoprotein phosphatases

7.1 General considerations

The characterization of a protein phosphatase involved in the dephosphorylation of a physiologically important phosphoprotein, and its regulation *in situ*, would ideally include demonstration that a protein is dephosphorylated *in situ*, that the tissue expresses an enzyme which can catalyse this dephosphorylation, and that the same reaction occurs both *in situ* and *in vitro*. Although some examples of dual-specificity phosphatases are known, present knowledge indicates that two major classes of phosphoprotein phosphatases exists, i.e. enzymes which selectively dephosphorylate P-Ser/Thr residues (109), and a completely distinct group of enzymes which dephosphorylate P-Tyr residues (110).

7.2 P-Ser/P-Thr phosphoprotein phosphatases

7.2.1 General considerations

Recent advances in phosphoprotein phosphatase research, made possible by molecular biological approaches, indicate the existence of a large number of phosphoprotein phosphatases isoforms (7, 8), but the original classification of these enzyme families put forward by Ingebritsen and Cohen (111) remains the most convenient way of describing these enzymes. Based on substrate specificity and inhibitor sensitivity, two major enzyme groups with broad substrate specificities can be distinguished. The first group, comprising the various forms of phosphoprotein phosphatase 1 (PP1), selectively dephosphorylates the β-subunit of phosphorylase kinase, is sensitive to the proteins inhibitor-1 and inhibitor-2, and has a wide substrate specificity. The second group, comprising PP2A, PP2B, and PP2C, selectively dephosphorylates the α-subunit of phosphorylase kinase, and is not sensitive to the inhibitor proteins. Among the type 2 enzymes, PP2A, which also has a wide substrate specificity, is not dependent on cations for its activity. PP2B, which has a more limited substrate specificity, is activated by Ca^{2+}-calmodulin. PP2C, which also appears to have limited specificity, is Mg^{2+}-dependent. Molecular cloning studies further indicate that these enzymes are encoded by two gene families, one of which, designated PPP, encodes for the PP1, PP2A, and PP2B enzymes, while another, designated MMM, encodes for the Mg^{2+}-dependent phosphoprotein phosphatases and the substrate-specific enzyme, pyruvate dehydrogenase phosphatase (8).

The regulation of these enzymes is complex. In addition to direct or indirect regulation caused by second messengers, such as the cAMP-induced activation of inhibitor-1 leading to inhibition of PP1, and the Ca^{2+}-calmodulin-induced activation of PP2B, the catalytic subunits of many of these enzymes are bound to regulatory proteins which may modulate both activity and subcellular distribution (7, 8).

Here we discuss briefly the examination of phosphatase activities in intact cells and extracts and provide basic protocols. For detailed characterization, including purification protocols, the reader is directed to more comprehensive papers (109, 110, 112).

7.2.2 Protein dephosphorylation in intact cells

The methods described in *Protocols 1, 2*, and *8* for following stimulus-induced increases in the state of phosphorylation of identified proteins in intact cells can also be used for following protein dephosphorylation. For the tentative identification of the enzymes responsible for dephosphorylation *in situ*, several recently described and commercially available membrane permeable protein phosphatase inhibitors may be employed (*Table 1*). Okadaic acid, a marine toxin produced by dinoflagellates, is a very potent inhibitor of PP2A and a less potent inhibitor of PP1. In contrast, calyculin A inhibits both PP1 and PP2A with the same potencies *in vitro* (113). The hepatotoxic microcystins, membrane impermeable cyclic peptides produced by blue-green algae and specifically transported into hepatocytes, are equally potent inhibitors of PP1 and 2A and may be used in studies on liver cells. PP2B appears to be specifically inhibited by the immunosuppressants cyclosporin A and FK 506

Table 1. Protein (Ser/Thr) phosphatase inhibitors[a]

Compound	**Commercial source**	**Relative potency**[b]
Okadaic acid	Sigma	PP2A >> PP1 >> PP2B
Tautomycin	Wako	PP1 > PP2A >> PP2B
Calyculin A	Sigma	PP2A = PP1 >> PP2B
Microcystin LR[c]	Sigma	PP2A = PP1 >> PP2B
Cyclosporin A	Sandoz	PP2B >> PP1, PP2A
FK 506	Fugisawa Pharmaceuticals	PP2B >> PP1, PP2A

[a] The table indicates potency of cell permeable inhibitors of protein phosphatase (PP) types 1, 2A, and 2B.
[b] > indicates 'greater than'.
[c] Microcystin LR enters hepatocytes only, others are cell permeable.

(114) in association with endogenous immunophilin binding proteins (115). The range of concentrations necessary to achieve effects of these agents *in situ* (usually 0.1–2 μM) may vary in different cells and should be determined experimentally. For a more detailed discussion of these inhibitors, see ref. 95.

7.2.3 Protein dephosphorylation in cell-free preparations

Extracting phosphoprotein phosphatases from intact cells for subsequent *in vitro* assays will usually disrupt most of the regulatory influences active *in situ*, and allow assay of free catalytic subunits only (112). To retain the activity state of the enzymes, samples must be prepared in media which neither activate nor deactivate the enzymes. Usually, these media include the metal chelators, EDTA and EGTA, which inactivate protein kinases. This has turned out to be detrimental for analysis of PP2B, since chelator-induced removal of Ca^{2+} inactivates the enzyme, while the absence of a chelator may lead to artefactual activation caused by Ca^{2+} release from intracellular stores. Analysis of changes in Ca^{2+}-independent phosphatase activity (mainly PP1 plus PP2A) appears, however, possible.

Selection of a substrate that can be used with these enzymes is important. Phosphorylase kinase labelled with ^{32}P by PKA has typically been used as a substrate. PP1 and PP2A can also be assayed with phosphorylase *a* labelled by phosphorylase kinase, while PP2B can be assayed with the regulatory subunit (RII) of PKA, or the protein inhibitor-1, both labelled with PKA. Finally, several peptides or non-peptide substrates may be used for type 2 phosphatases (109). *Protocol 17* gives a method for assaying P-Ser/Thr phosphoprotein phosphatases *in vitro*.

To be able to distinguish between different protein phosphatases, various protein phosphatase inhibitors (*Table 1*) may be used in conjunction with specific incubation conditions for *in vitro* assay (*Protocol 17*). While NaF (25–100 mM) and sodium pyrophosphate (2–10 mM) both represent examples of broad spectrum inhibitors which will efficiently inhibit all phosphatase activities, the conditions described in *Table 2* may be used to distinguish between PP1 and PP2A, since okadaic acid is much more potent against PP2A than PP1 (IC_{50} = 0.1–1 nM towards PP2A versus 2–300 nM versus PP1) (116), while calyculin A is equipotent against both (IC_{50} = 1–2 nM) (117). Certain peptide inhibitors may also be used, including the microcystin peptides which are less expensive than okadaic acid, and can be used to inhibit PP1 and PP2A (*Table 1*). In addition, PP2B appears to contain an autoinhibitory sequence, and a synthetic peptide comprising this domain may be used as a peptide inhibitor for PP2B (118).

Protocol 17. Assay of P-Ser/Thr phosphoprotein phosphatase *in vitro*

Equipment and reagents

- Appropriate substrate protein or peptide (see text), phosphorylated *in vitro* to high stoichiometry with [γ-^{32}P]ATP (109)
- Protein kinase suitable for labelling the substrate (see text)
- Materials for anion exchange chromatography, ammonium sulfate precipitation, and dialysis (109)
- Extracts for analysis: prepare tissue or cell extracts containing enzyme to be investigated (*Protocol 3*, steps 1–2); store on ice
- 10 × phosphatase buffer: 0.5 M Tris–HCl pH 7, 10 mg/ml BSA, 10 mM DTT
- Appropriate inhibitor/activator for phosphatase under study (see *Table 2*)
- 40% (w/v) TCA
- 20 mg/ml BSA
- Materials for Cherenkov or scintillation counting (*Protocol 9*)

Method

1. Label substrate with ^{32}P by incubating [γ-^{32}P]ATP (final concentration 1 mM) with the protein substrate and high concentrations of suitable protein kinase (*Protocol 9*) for 60 min.
2. Remove the free [^{32}P]ATP from the labelled substrate by anion exchange chromatography and concentrate the substrate protein by ammonium sulfate precipitation and dialysis (109). Measure the protein concentration.
3. In microcentrifuge tubes, mix the following:

• extract to be assayed	10 μl
• 10 × phosphatase buffer	10 μl
• appropriate inhibitor/activator (*Table 2*)	10 μl
• H_2O	to 90 μl volume

   Blank tubes receive homogenization buffer instead of extract.
4. Pre-incubate the tubes for 1 min at 30°C. Start the incubation by addition of the phosphorylated substrate (10 μl volume, preferably 10 mM final concentration, usually 1000–5000 c.p.m. is adequate).
5. Terminate the reactions by adding 100 μl of 40% TCA (final concen-

tration 20%) and 25 μl of 20 mg/ml BSA (final concentration 4 mg/ml). After 5 min on ice, centrifuge the samples at 10 000 *g* for 5 min. Collect 100 μl aliquots of supernatant and measure the ^{32}P content by Cherenkov or scintillation counting.

6. Determine the specific radioactivity (c.p.m./mol) of the substrate by counting a 10 μl aliquot (step 1).
7. Determine the protein phosphatase activity (mol/min/mg) as follows:
   (a) Calculate the amount of product formed (mol) as follows:
   $$\frac{\text{c.p.m.}_{\text{sample}} - \text{c.p.m.}_{\text{blank}}}{\text{specific radioactivity}}$$
   (b) Correct for the dilution and volume used for counting in step 5 (i.e. divide by 100 and multiply by 225).
   (c) Divide by the amount of protein in the extract (mg protein) and the incubation time (in minutes).

7.3 Tyrosine-specific phosphoprotein phosphatases

On the basis of amino acid sequencing comparisons, phosphotyrosine-specific phosphatases (PTPs) belong to two enzyme families distinct from other protein phosphatases (110). One of these families comprises most of the known PTPs, including dual specificity enzymes which are able to dephosphorylate tyrosine as well as serine and threonine residues, while the second group comprises small, low molecular mass PTPs. Many PTPs are transmembrane proteins and may be involved in intercellular signalling, while others contain targeting domains such as SH2 domains (119, 120). Purified PTPs often have a very high specific activity when analysed *in vitro* with various non-physiological substrates, and little is presently known about the regulatory mechanisms or the physiological substrates for most of these enzymes *in situ* (110).

Table 2. Conditions for distinguishing Ser/Thr phosphatases in extracts[a]

Agent	Commercial source	PP1	PP2A	PP2B	PP2C	Concentration required
Phosphoinhibitor-1	Promega	Inhibits	–	–	–	2 μM
Inhibitor-2	Promega	Inhibits	–	–	–	4 μM
Okadaic acid	Sigma	Inhibits	Inhibits	–	–	0.5 μM
$CaCl_2$	–	–	–	Activates	–	0.5 μM
Calmodulin	Calbiochem	–	–	Activates	–	2 μM
$MgCl_2$	–	–	–	–	Activates	10 mM

[a] The table indicates the change in activity of Ser/Thr phosphoprotein phosphatases found when assayed in the presence of the indicated agents. No change in activity is indicated by –. See also ref. 109.

Myelin basic protein from bovine brain (110) is frequently used as a substrate for assaying these enzymes. Myelin basic protein may be phosphorylated with [^{32}P]ATP and tyrosine-specific protein kinases (104), conveniently employing insulin or PDGF receptors prepared from human placental membranes, or commercially available protein tyrosine kinases, such as *src* or *abl* (110). Analysis of PTP activity then employs the general approach described in *Protocol 17*, with preliminary experiments being used to determine conditions for initial rate activity measurements. Orthovanadate represents a convenient and specific inhibitor of these enzymes. A method for assaying PTPs, based on that described by Tonks (110), is presented in *Protocol 18*.

Protocol 18. Analysis of tyrosine-specific phosphoprotein phosphatase

Equipment and reagents

- Myelin basic protein substrate (Sigma) labelled with [^{32}P]ATP to high stoichiometry (104) using a suitable tyrosine-specific protein kinase (109)
- Materials for anion exchange chromatography on Dowex AG1-X8, ammonium sulfate precipitation and dialysis, 40% (w/v) TCA, 20 mg/ml BSA (*Protocol 17*)
- Materials for determination of ^{32}P content of substrate (*Protocol 4*)
- Extract to be examined (*Protocol 3*)
- 10 × phosphatase buffer: 0.25 M imidazole–HCl pH 7.2, 10 mg/ml BSA (fatty acid- and globulin-free), 10 mM DTT

Method

1. Remove the [^{32}P]ATP from the labelled substrate by anion exchange chromatography on Dowex AG1-X8 and concentrate the protein by ammonium sulfate precipitation and dialysis (see *Protocol 17*). Check that essentially all the ^{32}P is TCA precipitable (*Protocol 4*).
2. Mix 10 μl of cell or tissue extract in microcentrifuge tubes with 10 μl 10 × phosphatase buffer and keep the tubes on ice (*Protocol 17*). Add H_2O to final 90 μl volume. Blanks receive H_2O instead of extract.
3. Pre-incubate the extract for 1 min at 30 °C. Start the incubation by addition of phosphorylated substrate (10 μl, usually 1000–5000 c.p.m. is adequate, *Protocol 17*, step 4).
4. Terminate the reaction by adding 40% TCA (final concentration 20%) and 20 mg/ml BSA (final concentration 4 mg/ml), and proceed as in *Protocol 17*, steps 5–7.

References

1. Walaas, S. I. and Greengard, P. (1991). *Pharmacol. Rev.*, **43**, 299.
2. Hardie, G. and Hanks, S. (ed.) (1995). *The protein kinase factsbook—protein-serine kinases.* Academic Press, London.

3. Hardie, G. and Hanks, S. (ed.) (1995). *The protein kinase factsbook—protein-tyrosine kinases.* Academic Press, London.
4. Krebs, E. G. (1994). *Trends Biochem. Sci.*, **19**, 439.
5. Hunter, T. (1991). In *Methods in enzymology* (ed. T. Hunter and B. M. Sefton), Vol. 200, p. 3. Academic Press, San Diego.
6. Shenolikar, S. and Nairn, A. C. (1993). In *Advances in second messenger and phosphoprotein research* (ed. P. Greengard and G. A. Robison), Vol. 23, p. 1. Raven Press, New York.
7. Cohen, P. (1994). *Bioessays*, **16**, 583.
8. Cohen, P. T., Chen, M. X., and Armstrong, C. G. (1996). *Adv. Pharmacol.*, **36**, 67.
9. Hunter, T. and Sefton, B. M. (ed.) (1991). *Methods in enzymology*, Vol. 200, part A. Academic Press, San Diego.
10. Hunter, T. and Sefton, B. M. (ed.) (1991). *Methods in enzymology*, Vol. 201, part B. Academic Press, San Diego.
11. Woodgett, J. R. (1992). In *Neuronal cell lines: a practical approach* (ed. J. N. Wood), p. 133. IRL Press, Oxford.
12. Hardie, D. G. (ed.) (1993). *Protein phosphorylation: a practical approach.* IRL Press, Oxford.
13. Hemmings, H. C., Jr. (1997). *Neuromethods 30. Regulatory protein modification. Techniques and protocols*. Humana Press, Totowa, New Jersey.
14. Seamon, K. B. and Daly, J. W. (1978). *Adv. Cyclic Nucleotide Protein Phosphorylation Res.*, **20**, 1.
15. Forn, J. and Greengard, P. (1978). *Proc. Natl. Acad. Sci. USA*, **75**, 5195.
16. Gjertsen, B. T., Mellgren, G., Otten, A., Maronde, E., Genieser, H.-G., Jastorff, B., *et al.* (1995). *J. Biol. Chem.*, **270**, 1.
17. Castagna, M., Takai, Y., Kaibuchi, K., Sano, K., Kikkawa, U., and Nishizuka, Y. (1982). *J. Biol. Chem.*, **257**, 7847.
18. Rudolph, S. A., Beam, K. G., and Greengard, P. (1978). In *Membrane transport processes*, Vol. 1 (ed. J. F. Hoffman), p. 107. Raven Press, New York.
19. Hopkirk, T. J. and Denton, R. M. (1986). *Biochim. Biophys. Acta*, **885**, 195.
20. Rodnight, R., Trotta, E. E., and Perrett, C. (1985). *J. Neurosci. Methods*, **13**, 87.
21. Garthwaite, J., Woodhams, P. L., Collins, M. J., and Balazs, R. (1979). *Brain Res.*, **173**, 373.
22. Schurr, A., Teyler, T. J., and Tseng, M. T. (ed.) (1987). *Brain slices: fundamentals, applications and implications*. Karger, Basel.
23. Walaas, O., Horn, R. S., and Walaas, S. I. (1991). *Biochim. Biophys. Acta*, **1094**, 92.
24. Rudolph, S. A. and Krueger, B. K. (1979). *Adv. Cyclic Nucleotide Res.*, **10**, 107.
25. Garrison, J. C. (1983). In *Methods in enzymology* (ed. J. D. Corbin and J. G. Hardman), Vol. 99, p. 20. Academic Press, New York.
26. Garrison, J. C. (1993). In *Protein phosphorylation: a practical approach* (ed. D. G. Hardie), p. 1. IRL Press, Oxford.
27. Sihra, T. S. (1997). In *Neuromethods. Regulatory protein modification: techniques and protocols* (ed. H. C. Hemmings, Jr.), Vol. 30, p. 67. Humana Press, Totowa, New Jersey.
28. Moss, S. J., Blackstone, C. D., and Huganir, R. L. (1993). *Neurochem. Res.*, **18**, 105.
29. Wiegant, V. M., Zwiers, H., Schotman, P., and Gispen, W. H. (1978). *Neurochem. Res.*, **3**, 443.
30. Walaas, S. I., Nairn, A. C., and Greengard, P. (1983). *J. Neurosci.*, **3**, 291.

31. Derkinderen, P. and Giraud, J.-A. (1997). In *Neuromethods. Regulatory protein modification. Techniques and protocols* (ed. H. C. Hemmings, Jr.), Vol. 30, p. 251. Humana Press, Totowa, New Jersey.
32. Walker, J. M. (ed.) (1984). *Methods in molecular biology*, Vol. 1. *Proteins.* Humana Press, Clifton, New Jersey.
33. Darbre, A. (ed.) (1986). *Practical protein chemistry - a handbook.* John Wiley & Sons Ltd., New York.
34. Hames, B. D. (ed.) (1998). *Gel electrophoresis of proteins: a practical approach.* 3rd edn. IRL Press, Oxford.
35. Laemmli, U. K. (1970). *Nature*, **227**, 680.
36. Schägger, H. and von Jagow, G. (1987). *Anal. Biochem.*, **166**, 368.
37. O'Farrell, P. H. (1975). *J. Biol. Chem.*, **250**, 4007.
38. O'Farrell, P. Z., Goodman, H. M., and O'Farrell, P. H. (1977). *Cell*, **12**, 1133.
39. Kelly, P. T., Yip, R. K., Shields, S. M., and Hay, M. (1985). *J. Neurochem.*, **45**, 1620.
40. Imada, M. and Sueoka, N. (1980). *Biochim. Biophys. Acta*, **625**, 179.
41. Wang, J. K., Walaas, S. I., and Greengard, P. (1988). *J. Neurosci.*, **8**, 281.
42. Hochstrasser, D. F., Harrington, M. G., Hochstrasser, A.-C., Miller, M. J., and Merril, C. R. (1988). *Anal. Biochem.*, **173**, 424.
43. Dunbar, B. S., Kimura, H., and Timmons, T. M. (1990). In *Methods in enzymology* (ed. M. P. Deutscher), Vol. 182, p. 441. Academic Press, San Diego.
44. Harlow, E. and Lane, D. (ed.) (1988). *Antibodies. A laboratory manual.* Cold Spring Harbor Laboratory, New York.
45. Goelz, S. E., Nestler, E. J., Chehrazi, B., and Greengard, P. (1981). *Proc. Natl. Acad. Sci. USA*, **78**, 2130.
46. Jahnen, W., Ward, L. D., Reid., G. E., Moritz, R. L., and Simpson, R. J. (1990). *Biochem. Biophys. Res. Commun.*, **166**, 139.
47. Scott, M. G., Crimmins, D. L., McCourt, D. W., Tarrand, J. J., Eyerman, M. C., and Nahm, M. H. (1988). *Biochem. Biophys. Res. Commun.*, **155**, 1353.
48. van der Geer, P., Luo, K., Sefton, B. M., and Hunter, T. (1993). In *Protein phosphorylation: a practical approach* (ed. D. G. Hardie), p. 31. IRL Press, Oxford.
49. Cleveland, D. W., Fischer, S. G., Kirschner, M. W., and Laemmli, U. K. (1977). *J. Biol. Chem.*, **252**, 1102.
50. Huttner, W. B. and Greengard, P. (1979). *Proc. Natl. Acad. Sci. USA*, **76**, 5402.
51. Palvimo, J. and Linnala-Kankkunen, A. (1989). *FEBS Lett.*, **257**, 101.
52. Meijer, L., Østvold, A. C., Walaas, S. I., Lund, T., and Laland, S. G. (1991). *Eur. J. Biochem.*, **196**, 557.
53. Lund, T., Dahl, K., Mørk, E., Holtlund, J., and Laland, S. G. (1987). *Biochem. Biophys. Res. Commun.*, **146**, 725.
54. Halpain, S., Girault, J. A., and Greengard, P. (1990). *Nature*, **343**, 369.
55. Hildebrandt, E. and Fried, V. A. (1989). *Anal. Biochem.*, **177**, 407.
56. Kamps, M. P. (1991). In *Methods in enzymology* (ed. T. Hunter and B. M. Sefton), Vol. 201, p. 21. Academic Press, San Diego.
57. Cooper, J. A. and Hunter, T. (1981). *Mol. Cell. Biol.*, **1**, 165.
58. Kamps, M. P. and Sefton, B. M. (1988). *Oncogene*, **2**, 305.
59. Cohen, P., Gibson, B. W., and Holmes, C. F. (1991). In *Methods in enzymology* (ed. T. Hunter and B. M. Sefton), Vol. 201, p. 153. Academic Press, San Diego.

60. Czernik, A. J., Girault, J.-A., Nairn, A. C., Chen, J., Snyder, G., Kebabian, J., *et al.* (1991). In *Methods in enzymology* (ed. T. Hunter and B. M. Sefton), Vol. 201, p. 264. Academic Press, San Diego.
61. Czernik, A. J., Mathers, J., and Mische, S. M. (1997). In *Neuromethods. Regulatory protein modification. Techniques and protocols* (ed. H. C. Hemmings, Jr.), Vol. 30, p. 219. Humana Press, Totowa, New Jersey.
62. Clover, C. V. C. and Allis, C. D. (1991). In *Methods in enzymology* (ed. T. Hunter and B. M. Sefton), Vol. 200, p. 85. Academic Press, San Diego.
63. Ferrell, J. E., Jr. and Martin, G. S. (1991). In *Methods in enzymology* (ed. T. Hunter and B. M. Sefton), Vol. 200, p. 430. Academic Press, San Diego.
64. Segel, I. H. (ed.) (1975). *Enzyme kinetics: behavior and analysis of rapid equilibrium and steady-state enzyme systems.* Wiley, New York.
65. Pearson, R. B. and Kemp, B. E. (1991). In *Methods in enzymology* (ed. T. Hunter and B. M. Sefton), Vol. 200, p. 62. Academic Press, San Diego.
66. Casnellie, J. E. (1991). In *Methods in enzymology* (ed. T. Hunter and B. M. Sefton), Vol. 200, p. 115. Academic Press, San Diego.
67. Walaas, S. I., Czernik, A. J., Olstad, O. K., Sletten, K., and Walaas, O. (1994). *Biochem. J.*, **304**, 635.
68. Witt, J. J. and Roskoski, R., Jr. (1975). *Anal. Biochem.*, **66**, 253.
69. Nishizuka, Y. (1995). *FASEB J.*, **9**, 484.
70. Hanson, P. I. and Schulman, H. (1992). *Annu. Rev. Biochem.*, **61**, 559.
71. Miller, S. G. and Kennedy, M. B. (1986). *Cell*, **44**, 861.
72. Lai, Y., Nairn, A. C., and Greengard, P. (1986). *Proc. Natl. Acad. Sci. USA*, **83**, 4253.
73. Bennett, M. K., Erondu, N. E., and Kennedy, M. B. (1983). *J. Biol. Chem.*, **258**, 12735.
74. Schulman, H. (1984). *J. Cell Biol.*, **99**, 11.
75. Pearson, R. B., Woodgett, J. R., Cohen, P., and Kemp, B. E. (1985). *J. Biol. Chem.*, **260**, 14471.
76. Hanson, P. I., Kapiloff, M. S., Lou, L. L., Rosenfeld, M. G., and Schulman, H. (1989). *Neuron*, **3**, 59.
77. Hvalby, O., Hemmings, H. C., Jr., Paulsen, O., Czernik, A. J., Nairn, A. C., Godfraind, J.-M., *et al.* (1994). *Proc. Natl. Acad. Sci. USA*, **91**, 4761.
78. Tuazon, P. T. and Traugh, J. A. (1991). In *Advances in second messenger and phosphoprotein research* (ed. P. Greengard and G. A. Robison), Vol. 23, p. 123. Raven Press, New York.
79. Pinna, L. A. (1990). *Biochim. Biophys. Acta*, **1054**, 267.
80. Seger, R. and Krebs, E. G. (1995). *FASEB J.*, **9**, 726.
81. Guan, K. L. (1994). *Cell Signal.*, **6**, 581.
82. Treisman, R. (1996). *Curr. Opin. Cell. Biol.*, **8**, 205.
83. Su, B. and Karin, M. (1996). *Curr. Opin. Immunol.*, **8**, 402.
84. Geahler, R. L., Anostario, M., Jr., Low, P. S., and Harrison, M. L. (1986). *Anal. Biochem.*, **153**, 151.
85. Chao, T.-S. O., Byron, K. L., Lee, K.-M., Villereal, M., and Rosner, M. R. (1992). *J. Biol. Chem.*, **267**, 19876.
86. Kulas, D. T., Freund, G. G., and Mooney, R. A. (1996). *J. Biol. Chem.*, **271**, 755.
87. Nurse, P. (1990). *Nature*, **344**, 503.
88. Reed, S. I. (1992). *Annu. Rev. Cell Biol.*, **8**, 529.

89. Pines, J. and Hunter, T. (1992). *Ciba Found. Symp.*, **170**, 187.
90. Sherr, C. (1993). *Cell*, **73**, 1059.
91. Krek, W. and Nigg, E. (1991). *EMBO J.*, **10**, 305.
92. Solomon, M. J. (1993). *Curr. Opin. Cell Biol.*, **5**, 180.
93. Reed, S. I., Bailly, E., Dulic, V., Hengst, L., Resnitzky, D., and Slingerland, J. (1994). *J. Cell Sci. Suppl.*, **18**, 69.
94. Hidaka, H. and Kobayashi, R. (1993). In *Protein phosphorylation: a practical approach* (ed. D. G. Hardie), p. 87. IRL Press, Oxford.
95. Hemmings, H. C., Jr. (1997). In *Neuromethods. Regulatory protein modification: techniques and protocols* (ed. H. C. Hemmings, Jr.), Vol. 30, p. 121. Humana Press, Totowa, New Jersey.
96. Robinson, P. J., Liu, J.-P., Chen, W., and Wenzel, T. (1993). *Anal. Biochem.*, **210**, 172.
97. Corbin, J. D., Keely, S. L., and Park, C. R. (1975). *J. Biol. Chem.*, **250**, 218.
98. Palmer, W. K., McPherson, J. M., and Walsh, D. A. (1980). *J. Biol. Chem.*, **255**, 2663.
99. Chneiweiss, H., Cordier, J., and Glowinski, J. (1991). *J. Neurochem.*, **57**, 1708.
100. Gorelick, F. S., Wang, J. K. T., Lai, Y., Nairn, A. C., and Greengard, P. (1988). *J. Biol. Chem.*, **263**, 17209.
101. MacNicol, M. and Schulman, H. (1992). *J. Biol. Chem.*, **267**, 12197.
102. Fukunaga, K., Soderling, T. R., and Miyamoto, E. (1992). *J. Biol. Chem.*, **267**, 22527.
103. Suzuki, T., Okumura-Noji, K., Ogura, A., Kudo, Y., and Tanaka, R. (1992). *Proc. Natl. Acad. Sci. USA*, **89**, 109.
104. Kaczmarek, L. K., Jennings, K. R., Strumwasser, F., Nairn, A. C., Walter, U., Wilson, F. D., *et al.* (1980). *Proc. Natl. Acad. Sci. USA*, **83**, 5968.
105. Scopes, R. (1982). *Protein purification. Principles and practice.* Springer–Verlag, New York.
106. Harris, E. L. V. and Angal, S. (ed.) (1990). *Protein purification applications: a practical approach*. IRL Press, Oxford.
107. Harris, E. L. V. and Angal, S. (ed.) (1989). *Protein purification methods: a practical approach.* IRL Press, Oxford University Press, Oxford.
108. Wang, Y. and Roach, P. J. (1993). In *Protein phosphorylation: a practical approach* (ed. D. G. Hardie), p. 121. IRL Press, Oxford.
109. MacKintosh, C. (1993). In *Protein phosphorylation: a practical approach* (ed. D. G. Hardie), p. 197. IRL Press, Oxford.
110. Tonks, N. K. (1993). In *Protein phosphorylation: a practical approach* (ed. D. G. Hardie), p. 231. IRL Press, Oxford.
111. Ingebritsen, T. S. and Cohen, P. (1983). *Science*, **221**, 331.
112. Halpain, S. (1997). In *Neuromethods. Regulatory protein modification: techniques and protocols* (ed. H. C. Hemmings, Jr.). Vol. 30, p. 45. Humana Press, Totowa, New Jersey.
113. Hardie, D. G. (1993). In *Protein phosphorylation: a practical approach* (ed. D. G. Hardie), p. 109. IRL Press, Oxford.
114. Kunz, J. and Hall, M. N. (1993). *Trends Biochem. Sci.*, **18**, 334.
115. Liu, J., Farmer, J. D., Lane, W. S., Friedman, J., Weissman, I., and Schreiber, S. (1991). *Cell*, **66**, 807.
116. Cohen, P., Holmes, C. F. B., and Tsukitani, Y. (1990). *Trends Biochem. Sci.*, **15**, 98.

117. Ishihara, H., Martin, B. L., Brautigan, D. L., Karaki, H., Ozaki, H., Kato, Y., *et al.* (1989). *Biochem. Biophys. Res. Commun.*, **159**, 871.
118. Hashimoto, Y., Perrino, B. A., and Soderling, T. R. (1990). *J. Biol. Chem.*, **265**, 1924.
119. Brady-Kalnay, S. M. and Tonks, N. K. (1994). *J. Biol. Chem.*, **269**, 28472.
120. Mauro, L. J. and Dixon, J. E. (1994). *Trends Biochem. Sci.*, **19**, 151.

4

Protein glycosylation

DAVID A. ASHFORD and FRAN PLATT

1. Introduction

The most common co- and post-translational modification of eukaryotic proteins is the covalent attachment of oligosaccharides. The vast majority of cell surface and secreted proteins are glycosylated, as are the envelope proteins of many viruses. It is now clear that the presence of glycans on glycoproteins affects both the structural and functional properties of the protein, such as protein folding and conformation, stability to denaturation, solubility, and resistance to proteolysis as well as key biological properties such as receptor binding, modulation of enzyme activity, and cellular recognition events (1–5). This chapter is intended as a guide to the analysis of the glycosylation of proteins and to techniques for the manipulation of protein glycosylation. A previous title in the Practical Approach series (6) and other recent books and articles (7–10) have covered glycobiology, glycosylation, and carbohydrate analysis more extensively and they (and the references they contain) can provide more detail for readers who wish to know more.

1.1 Protein glycosylation

Glycans (oligosaccharides) can be attached to proteins by *N*-glycosidic linkages via the amide group of asparagine (Asn) or by *O*-glycosidic linkages to hydroxy amino acids such as serine (Ser), threonine (Thr) and, in plants, hydroxyproline (Hyp). *N*-glycosylation occurs only at the specific peptide sequence or sequon, Asn-X-Ser/Thr-Y (where X and Y can be any amino acid except proline), whereas sites of *O*-glycosylation show no specific amino acid sequence. The main mechanisms of the biosynthesis of *N*-linked oligosaccharides and their attachment to proteins appear to be conserved between animals, plants, and fungi. *O*-glycosylation is more variable and in plants is markedly different to that seen in yeast and animals.

The glycosylation of proteins shows characteristic properties (1). The cell determines the level of expression of enzymes of oligosaccharide processing, resulting in cell- or tissue-specific glycosylation patterns. The protein structure also influences glycosylation, and different proteins from the same cell can contain different glycan structures. These interactions lead to protein-specific and site-specific glycosylation. Considerations of protein glycosylation reach a higher degree of complexity when it is appreciated that a cell can code for the same protein but with different glycans. This variety can be both in the glycans at a single glycosylation site and in the combinations of glycans at multiple sites. These factors mean that a purified glycoprotein exhibits microheterogeneity in its glycans. Incomplete utilization of potential *N*- and *O*-glycosylation sites, leading to variable occupancy of a given site by carbohydrate, adds additional complications. A rationalization of this complexity is to consider a 'glycoprotein' as a population of molecules, each an individual glycosylation variant or glycoform. Glycoforms have the same polypeptide and vary only in their oligosaccharide chains. They also have slightly different chemical and biological properties. The pattern of glycoforms of a glycoprotein shows conserved heterogeneity, i.e. during a constant physiological state it is reproducible and not random. However, glycosylation has been shown to be affected by environmental factors in animal cell culture (11) and can be altered by disease (12).

1.2 Oligosaccharide structures

The structures of all *N*-linked oligosaccharides share a common core composed of three mannose (Man) and two *N*-acetylglucosamine (GlcNAc) residues (*Figure 1*). Yeast *N*-glycans contain only these monosaccharides with from 8–50 mannoses in a single structure, whereas many plant and animal glycans have more complicated compositions. Animal glycans can be divided into four classes (for representative structures see *Figure 1*):

(a) The *oligomannose* (often called high mannose) type; these contain only Man and GlcNAc with between 5–9 Man and 2 GlcNAc residues.

(b) The *complex* type; which commonly contain galactose (Gal), fucose (Fuc), and sialic acid (*N*-acetylneuraminic acid, NeuNAc), in addition to Man and GlcNAc. These oligosaccharides can be highly branched with up to six 'arms' or 'antennae'.

(c) The *hybrid* type; which have compositional and structural characteristics of both oligomannose and complex oligosaccharides.

(d) The *polylactosamine* type; these are sometimes grouped together with the complex type. They have repeating linear sequences of Gal-GlcNAc (lactosamine) in their antennae.

Plant *N*-glycoproteins, in common with those from animals, can carry oligomannose type structures and they can also contain 'xylose/fucose' or 'modified' type glycans (sometimes referred to as 'plant complex' type) that typically contain Man, GlcNAc, xylose (Xyl), and Fuc residues (*Figure 1*). No NeuNAc has been found in plant glycans but terminal Gal residues do occur.

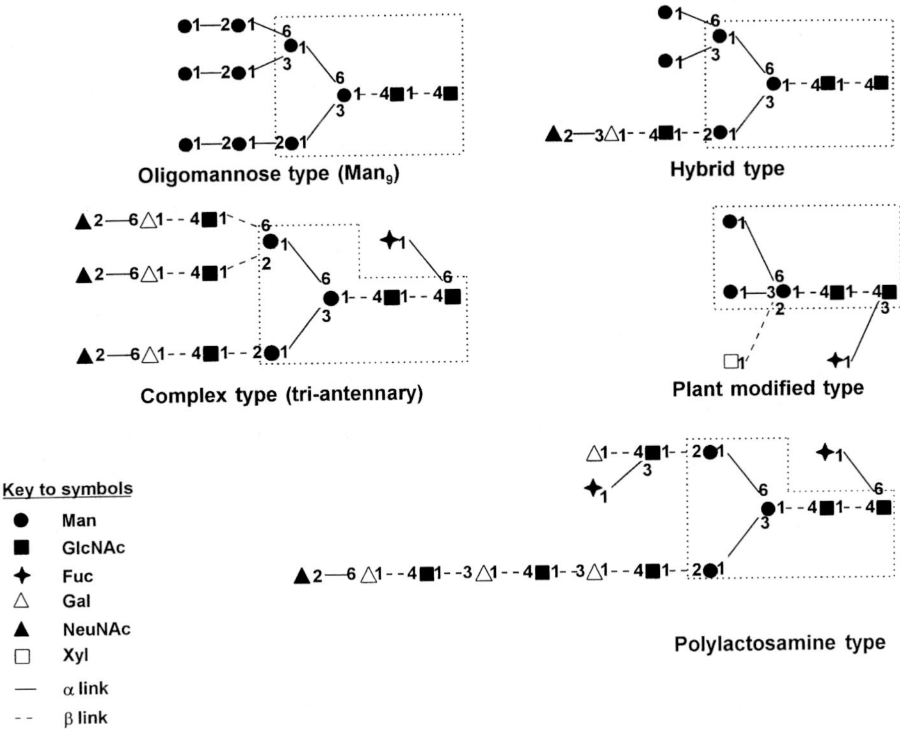

Figure 1. The structure of typical *N*-glycans. The common pentasaccharide core in each structure is indicated by the boxed area. Man, mannose; GlcNAc, *N*-acetylglucosamine; Fuc, fucose; Gal, galactose; NeuNAc, *N*-acetylneuraminic acid; Xyl, xylose. The numbers refer to the positions involved in the glycosidic linkages.

O-glycosylation in animals usually involves the linkage of *N*-acetylgalactosamine (GalNAc) to Thr or Ser, and such *O*-glycans can contain Gal, GlcNAc, Fuc, and NeuNAc, but no Man. Conversely, yeast *O*-glycans are composed wholly of Man. In plant *O*-linked glycoproteins, arabinose is often linked to Hyp, and Gal can be linked to Hyp or Ser. The size of *O*-linked chains can vary from one residue to several hundreds.

1.3 Glycosylation pathways

1.3.1 *N*-glycosylation

N-linked glycoproteins are formed through the covalent attachment of a preformed glycan to an asparagine residue within the sequon Asn-X-Ser/Thr as described above. The details of *N*-glycan biosynthesis have been reviewed extensively elsewhere (13) and are summarized in *Figure 2*. The precursor glycan is synthesized in the endoplasmic reticulum (ER) through the sequential action of glycosyltransferases which build up the oligosaccharide

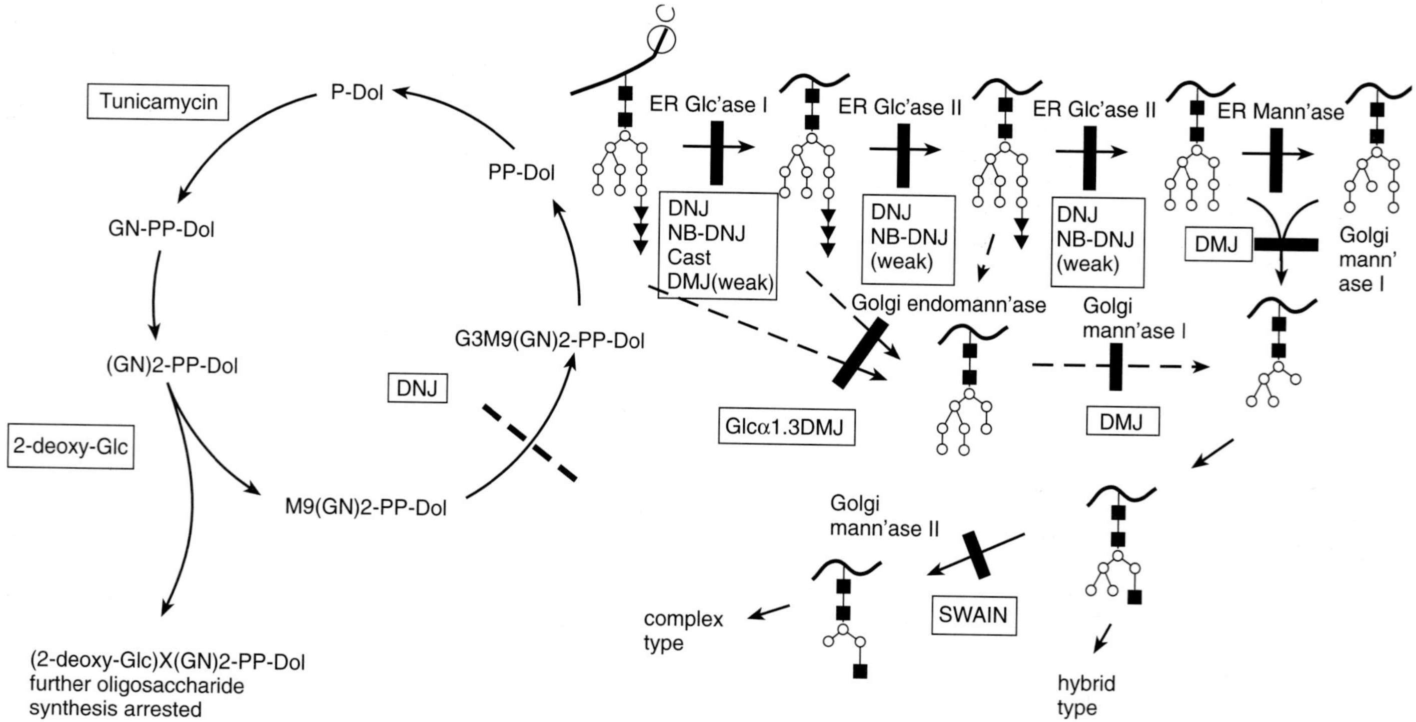

Figure 2. The dolichol lipid cycle and early steps in *N*-linked oligosaccharide biosynthesis. The steps where specific inhibitors inhibit enzymes in the pathway are indicated by black bars and boxed areas. DNJ, deoxynojirimycin; NB-DNJ, *N*-butyldeoxynojirimycin; Cast, castanospermine; DMJ, deoxymannojirimycin; Swain, swainsonine; ER, endoplasmic reticulum; Glc'ase, glucosidase; G/Glc/▲, glucose; M/○, mannose; GN/■, *N*-acetylglucosamine; Mann'ase, mannosidase; P, phosphate; Dol, dolichol; 2-deoxy-Glc, 2-deoxy-glucose.

on a dolichol lipid carrier. The final structure synthesized and co-translationally transferred to the glycosylation sequon is $Glc_3Man_9GlcNAc_2$. The *N*-glycan structure is then subjected to a number of processing reactions in which it is extensively remodelled to form the three major classes of mature *N*-glycans, oligomannose, complex, and hybrid type structures. The processing reactions are initiated in the ER and involve the sequential removal of the α1,2-glucose residue by α-glucosidase I and the subsequent removal of the two α1,3-linked glucose residues by α-glucosidase II. The resulting oligomannose structure is then acted upon by trimming mannosidases in the ER and early Golgi complex. For instance, Golgi mannosidase I can cleave all the α1,2-linked mannose residues to yield a Man_5 structure. It is quite common for cell surface and secreted glycoproteins to retain oligomannose structures such as Man_5 through to Man_9, with the mannosidase trimming arresting at different stages in the processing pathway. However, in the case of complex *N*-glycan biosynthesis, GlcNAc transferase I transfers a GlcNAc residue to the α1,3-linked mannose that forms part of the trimannosyl core. This then permits the action of Golgi mannosidase II in the medial Golgi complex that hydrolyses the terminal α1,3- and α1,6-mannose residues to yield $GlcNAc_1Man_3GlcNAc_2$. Sequentially acting glycosyltransferases then act upon this processed *N*-glycan to form complex type structures through the addition of monosaccharides, which include GlcNAc, Gal, Fuc, and NeuNAc. Hybrid type *N*-glycans are formed when mannosidase II either fails to act or is only partially active.

Several of the enzymes in the *N*-glycan processing pathway can be specifically inhibited. This permits the manipulation of the pathway to generate *N*-glycan structures of uniform type, as it arrests the pathway at one particular stage and the maturation of the oligosaccharide is blocked and can progress no further. The points of action of some inhibitors of *N*-glycan processing enzymes are shown schematically in *Figure 2*. These inhibitors are invaluable tools for probing the roles of *N*-glycans in the structure and function of many glycoproteins. Most of the inhibitors characterized to date are natural products, some of which have been chemically modified to enhance specificity or increase the potency of inhibition (14, 15).

1.3.2 *O*-glycosylation

The biosynthesis of *O*-linked glycans, in contrast to *N*-glycans, occurs by the sequential addition of single monosaccharides to the glycoprotein (16). Initiation of *O*-glycosylation, i.e. the addition of the first monosaccharide to the hydroxy amino acid in the protein, is post-translational and can occur in the ER. This is the case for yeast *O*-glycans, where the first mannose residue added to the protein is transferred via a dolichol-mannose precursor (17). The majority of *O*-glycosylation reactions, however, occur after the protein has been transported to the Golgi apparatus and involve sugar-nucleotide donors.

1.4 Characterization of protein glycosylation

Analysis of the glycosylation of glycoproteins is one of the most complex tasks in modern biochemistry. The complete characterization of the glycosylation of a glycoprotein entails the determination of the occupancy of potential glycosylation sites, the type(s) of glycans present at these sites, and the distribution of oligosaccharide structures at each site. The sequencing of individual glycans requires the determination of monosaccharide composition, linkage positions between monosaccharides, and the anomeric configuration of each structure. There are a number of technical challenges associated with glycosylation analysis when compared with polypeptide or nucleic acid analysis. For example, the occurrence of branching in oligosaccharide structure generates an enormous variety of potential configurations and no single analytical method can deliver the complete characterization of small amounts of oligosaccharide. Also, the fact that oligosaccharides do not contain a natural chromophore means that they must be labelled for any type of sensitive analysis. These difficulties have been tackled by the development of numerous labelling techniques and a battery of separation and analysis methods (many of them dependant on the particular label introduced). Details of these methods are given in the following sections.

2. Is my protein glycosylated?

Examination of the amino acid sequence deduced from the nucleic acid sequence may reveal the presence of potential glycosylation sites in the protein. The sequence for *N*-glycosylation (Asn-X-Ser/Thr, see Section 1.1) is readily detected and, although no analogous sequence exists for *O*-glycosylation, computer sequence analysis programs can predict domains and sequences that are likely to be *O*-glycosylated. Amino acid sequence analysis of the protein or derived peptides can then determine whether asparagine, serine, and threonine residues are occupied (since they will show up as 'blanks' in the cycles of Edman sequencing). Mass spectrometric techniques will also indicate whether glycosylation exists at the protein and the peptide level. Methods such as assaying for carbohydrate by colorimetric methods, proprietary detection methods, and lectin binding do not require expensive instruments and are described here.

2.1 Colorimetric methods

Many colorimetric assays exist for the determination of carbohydrate, e.g. phenol/sulfuric acid, anthrone, and orcinol methods, and they are widely used. The formation of colour in these assays is dependent on the composition of the glycan(s) as each monosaccharide has its own molar response. Therefore, the appropriate monosaccharide standard should be chosen and the measured concentration of carbohydrate in the protein should be expressed relative to that standard. *Protocol 1* describes the phenol/sulfuric acid assay. This assay measures the total neutral carbohydrate content of a sample. It cannot differentiate between monosaccharides but it is reasonably sensitive and no hydrolysis is required. The amount of carbohydrate in the sample is determined by reference to a standard curve constructed simultaneously. Suitable standards are Xyl (pentose) or Man, Gal, and glucose (hexoses).

Protocol 1. Phenol/sulfuric acid assay[a]

Equipment and reagents

- Protein sample to be assayed (containing approx. 0.5–20 μg carbohydrate/200 μl)
- 5% (w/v) phenol
- Concentrated sulfuric acid (analytical grade, s.g. 1.84)
- 13 × 100 mm medium or heavy wall glass tubes
- Standard monosaccharide solution (100 μg/ml)

Method

1. Set up a standard curve (in duplicate) by adding known amounts of the standard monosaccharide to a series of tubes to give a range of concentrations between 0.5–20 μg/tube. Make the volumes up to 200 μl with H_2O.
2. Add 200 μl 5% phenol to 200 μl of the protein sample and standard solutions and mix thoroughly.
3. Add 1 ml concentrated sulfuric acid to each tube. **Caution!** The reaction is exothermic and the solution may boil; wear safety glasses and handle with care.
4. Allow the tubes to cool at room temperature for 10 min. Mix thoroughly. **Caution!** Hot sulfuric acid; wear safety glasses and handle with care.
5. Leave the tubes to cool completely to room temperature.
6. When cool, read the A_{480} for pentoses or A_{490} for hexoses.
7. Use the data from the monosaccharide standards to construct a standard curve. Calculate the carbohydrate content of the protein sample using the standard curve.

[a] From ref. 18.

2.2 Proprietary detection methods

These methods rely on periodate oxidation of the carbohydrate chains to generate reactive groups that are then modified to allow specific high sensitivity detection. An example is the GlycoTrack™ carbohydrate detection kit from Oxford GlycoSciences. Protein mixtures either separated by SDS–PAGE and transferred to membranes (Western blot) or dot blotted directly onto membranes are treated sequentially with periodate and biotin hydrazide. The incorporated biotinyl groups are detected by using a streptavidin–alkaline phosphatase conjugate. Boehringer Mannheim also have detection kits (DIG glycan detection kit) that employs digoxigenin and anti-digoxigenin antibodies in a similar kind of protocol. Both methods claim submicrogram sensitivity.

2.3 Lectin binding

Lectins are carbohydrate-binding proteins and have a number of applications in glycosylation analysis. Each lectin has a specific sugar or oligosaccharide hapten (*Table 1*). Therefore, using a lectin as the detection reagent in a Western blot can not only indicate that a glycan is present on a protein but also provide information on the structures present. *Protocol 2* describes a procedure for detecting glycoproteins that bind the lectin concanavalin A (Con A). The protocol is suitable for both (glyco)proteins separated by polyacrylamide gel electrophoresis and transferred to nitrocellulose (Western blot), and to glycoprotein and glycopeptide samples (e.g. column fractions) applied directly to the nitrocellulose sheet (dot blot). Con A bound to glycoproteins on the blot is detected by the glyco-enzyme horseradish peroxidase (HRP) because Con A is a multimeric lectin with multiple carbohydrate binding sites and it can bind to carbohydrate chains on both the immobilized glycoprotein and HRP.

Table 1. Lectins and their oligosaccharide specificity

Lectin source and abbreviation	Principal specificity	Inhibitory sugar and concentration[a]
Canavalia ensiformis (jack bean, Con A)	Man-α-Man,	200 mM α-Me-Man
Lens culinaris (lentil, LCA)	Man-α-Man[b]	200 mM α-Me-Man
Galanthus nivalis (snowdrop, GNA)	Man-α1,3-Man	100–200 mM α-Me-Man
Triticum vulgaris (wheat germ, WGA)	β1,4-GlcNAc oligomers	500 mM GlcNAc
Erythrina cristagalli (coral tree, ECL)	Terminal β1,4-linked Gal	200 mM lactose
Sambucus nigra (elderberry, SNA)	NeuNAc-α2,6-Gal/GalNAc	500 mM lactose
Maakia amurensis (MAA)	NeuNAc-α2,3-Gal	200 mM lactose
Glycine max (soy bean, SBA)	GalNAc-α1,3-Gal	200 mM GalNAc
Arachis hypogea (peanut, PNA)	Gal-β1,3-GalNAc	200 mM Gal
Aleuria aurantia (orange peel fungus, AAL)	Terminal α-linked Fuc[c]	10–50 mM Fuc
Ulex europaeus (gorse, UEA-I)	Fuc-α1,2-Gal	50–100 mM Fuc

[a] The concentrations given will elute the majority of specifically bound glycoproteins from a lectin affinity column or inhibit the binding of a glycoprotein to the lectin. Weakly bound glycoproteins can be eluted with lower concentrations of inhibitor.
[b] Specificity differs slightly from Con A. Will bind oligosaccharides with core fucose.
[c] Binds core α1,6-fucose in *N*-linked glycans.

Protocol 2. Detection of glycoproteins on nitrocellulose by concanavalin A[a]

Equipment and reagents

- Western or dot blot to be tested
- Fixing solution: glacial acetic acid:propan-2-ol:water (10:25:65, by vol.)
- Tris-buffered sodium chloride solution (TBSalt): 20 mM Tris–HCl pH 7.4 containing 500 mM NaCl
- Blocking solution: 3% (w/v) gelatin (Sigma) in TBSalt
- Detection buffer: TBSalt containing 1% (w/v) gelatin, 1 mM $CaCl_2$, and 1 mM $MnCl_2$
- Con A reagent: 25 μg/ml Con A (Vector Labs) in detection buffer
- TBSalt/Tween solution: TBSalt containing 0.1% (w/v) Tween 20 (Sigma), 1 mM $CaCl_2$, and 1 mM $MnCl_2$
- 50 μg/ml horseradish peroxidase (HRP, Sigma) in detection buffer
- Staining solution: 60 mg 4-chloro-1-naphthol (Sigma) and 60 μl hydrogen peroxide in 100 ml TBSalt—prepared just before use
- α-Methyl mannoside (Sigma)

Method

In the following steps, the volumes of solutions are not critical but use sufficient of each solution to ensure that the blot stays wet at all times.

1. Fix proteins and peptides on the blot by immersing it in fixing solution for 30 min. Then wash it twice with water and three times with TBSalt.
2. Incubate the blot in blocking solution for 1 h at room temperature, or overnight at 4°C, to saturate protein binding sites.
3. Wash the blot once with TBSalt and incubate it with gentle rocking in the Con A reagent for 1 h at room temperature.
4. Rinse the blot twice with TBSalt and then wash four times, for 15 min, with TBSalt/Tween solution.
5. Repeat step 3 with the HRP reagent.
6. Repeat step 4.
7. Develop the blot with the staining solution, ensuring that the sheet stays completely immersed throughout. Purple bands appearing within 5–10 min indicate a positive result, i.e. Con A has bound to glycoprotein(s) on the blot and has been detected by the HRP.
8. Rinse the blot twice with water and dry gently. Store at room temperature in the dark. Photograph the blot for a permanent record since the bands fade with time.

[a] From ref. 19.

To show that the Con A binding is specific and to control for non-specific protein binding, process a replicate blot by *Protocol 2* but include 200 mM α-methyl mannoside in all the reagents in steps 3–6. This will specifically inhibit Con A binding to carbohydrate. Strong Con A binding requires 200 mM α-methyl mannoside to inhibit formation of complexes. Weak binding can be inhibited by including 20 mM α-methyl glucoside in the wash steps after incubation with Con A. To detect all the proteins on the Western blot and thus allow a comparison with Con A binding, stain a replicate blot, after fixing, with Coomassie blue or Amido black.

3. Study of whole protein glycosylation

3.1 Detection of *N*-glycosylation

This can be done in a number of different ways. Two of these methods are:

- inhibition of *N*-glycosylation in a tissue culture expression system using tunicamycin,
- treatment of isolated protein with peptide:*N*-glycosidase F (PNGase F) to specifically release the *N*-glycans.

In the former method, the electrophoretic mobility in SDS–PAGE (and, therefore, the molecular weight) of the unglycosylated protein generated in the presence of tunicamycin is compared with the mobility of the protein derived from untreated cultures (glycosylation status unknown). If glycosylated, the molecular weights of the two proteins should differ, with the tunicamycin treated cultures giving a protein species of increased electrophoretic mobility. In the second method, the use of PNGase F, which is an amidase that cleaves between the innermost GlcNAc and the asparagine residue of all classes of *N*-glycans (except some plant and insect *N*-glycans which have a core α1,3-linked fucose that prevents PNGase F cleavage), leads to de-*N*-glycosylation of the protein. The electrophoretic mobility in SDS–PAGE of the protein can be compared, before and after digestion, in the same way as for tunicamycin.

An average *N*-glycan structure has an approximate molecular weight of 2 kDa and so a rough estimate can be made of the number of *N*-glycans associated with the protein. This can be compared with the number of potential *N*-glycosylation sites predicted by the primary amino acid sequence of the protein. *Protocols 3* and *4* describe the use of tunicamycin for inhibiting glycosylation of proteins and the use of PNGase F, respectively.

Protocol 3. Tunicamycin treatment

Equipment and reagents

- Eukaryotic cells expressing the protein of interest, and the appropriate tissue culture facilities for their growth
- Tran ^{35}S-label (> 1000 Ci/mmol, ICN-Flow)
- Methionine- and cysteine-free medium (ICN-Flow)
- Tunicamycin (Sigma)
- Phosphate-buffered saline pH 7.2 (PBS): 0.14 M NaCl containing 2.7 mM KCl, 1.5 mM KH_2PO_4, and 8.1 mM Na_2HPO_4; sterilize this by autoclaving or by passage through a 0.2 μm filter

- Lysing buffer: PBS containing 2% (w/v) Triton X-100 (Sigma), 50 μg/ml *N*-tosyl-L-phenylalanine chloromethylketone (Sigma), 50 μg/ml *N*-*p*-tosyl-L-lysine chloromethylketone hydrochloride (Sigma), and 200 μg/ml phenylmethylsulfonyl fluoride (PMSF, Sigma)[a]
- Sterilization equipment: autoclave or sterile filter (0.2 μm)
- SDS–PAGE gel, electrophoresis equipment, and reagents
- Autoradiography film and film processing equipment
- Materials for immunoprecipitation (ref. 41)

Method

1. Harvest the cells expressing the protein of interest and wash them three times with sterile PBS by resuspension and centrifugation at 1000 *g* for 10 min.
2. Resuspend the cells (at 1×10^7 cells/ml) in methionine/cysteine-free medium with or without tunicamycin (1 μg/ml)[b] and incubate them for 1 h at 37 °C.
3. Add 100 μCi/ml Tran ^{35}S-label and incubate the cells for a further 4 h.
4. If the protein of interest is secreted from the cells as a soluble protein, remove the cells (step 1) and retain the medium.[c] If the protein is cell-associated (e.g. a transmembrane protein) lyse the cells with 100 μl lysing buffer.
5. Isolate the protein from the tissue culture medium or the cell lysate by immunoprecipitation (see ref. 41).
6. Determine the relative molecular weights of the untreated and tunicamycin-treated protein by SDS–PAGE and autoradiography.

[a] This buffer is only needed if the protein of interest is cell-associated rather than soluble.
[b] Tunicamycin is cytotoxic and this toxicity varies with cell type. Therefore, it is advisable to titrate the tunicamycin concentration (0.1–5 μg/ml) with the cells being used.
[c] Concentrate the medium 10 x, if necessary, using a proprietary ultrafiltration system.

Protocol 4. Treatment with PNGase F

Equipment and reagents

- Purified protein for analysis
- PNGase F (*N*-glycanase, New England Biolabs or Oxford GlycoSciences): 266 U/ml in incubation buffer
- Incubation buffer A: 20 mM sodium phosphate pH 7.5 containing 50 mM EDTA, 0.02% sodium azide
- 10% (w/v) NP-40 (Sigma)
- Denaturation buffer A: incubation buffer A containing 0.5% (w/v) SDS and 5% (v/v) 2-mercaptoethanol
- SDS–PAGE gel, electrophoresis equipment, and reagents
- Western blotting equipment and reagents (if required)

Protocol 4. *Continued*

Method

1. Denature the purified protein (1–50 μg) by heating it at 100°C for 10 min in 20 μl denaturation buffer A.[a]
2. Cool the sample.
3. Take a 10 μl aliquot into a fresh tube and add 2.5 μl of 10% NP-40[b] and 7.5 μl PNGase F (2 U). To the other 10 μl of sample add 2.5 μl of 10% NP-40 and 7.5 μl incubation buffer A.
4. Incubate both samples for 18 h at 37 °C.
5. Analyse the samples with and without PNGase F digestion by SDS–PAGE.
6. Depending on the experiment, stain the gel with silver or Coomassie blue, or perform a Western blot to detect the protein of interest.

[a] Denaturation of the protein prior to enzyme treatment will increase the efficiency of the deglycosylation.
[b] This prevents inactivation of the PNGase by SDS.

3.2 Monosaccharide composition

Determination of the monosaccharide composition of a glycoprotein gives valuable information on the glycosylation by identifying monosaccharides in the sample and determining their relative and absolute molar proportions. This provides useful clues to the structure of the oligosaccharides (as outlined in Section 1). Compositional analysis can also detect the presence of unusual monosaccharides. After complete acid hydrolysis or methanolysis of the glycoprotein, constituent monosaccharides can be separated and analysed by a number of techniques. The most commonly used of these are gas chromatography (GC), as alditol acetate (see *Protocol 5*) or trimethylsilyl derivatives; HPLC, as conjugates labelled with a fluorophore by reductive amination; and high performance anion exchange chromatography (HPAEC, Dionex). The latter can be performed directly on the acid hydrolysate after neutralization and requires no derivatization or labelling. However, this technique requires specialized equipment as it uses a pulsed amperometric detector (PAD).

The glycosidic bonds between monosaccharides differ in their susceptibility to acid hydrolysis and the released monosaccharides differ in their susceptibility to destruction by acid. It is therefore often necessary to compromise the conditions to get maximum release with minimum destruction. The trifluoroacetic acid (TFA) hydrolysis method for neutral sugars described in *Protocol 5* will not give quantitative cleavage of GlcNAc–GlcNAc and GlcNAc–Asn bonds but does give good recovery of pentose and deoxyhexose residues (e.g. Xyl and Fuc). Conversely, the HCl hydrolysis method gives quantitative release and recovery of GlcNAc but Xyl and Fuc are destroyed. It is advisable to use both hydrolysis methods if you suspect your glycoprotein contains both amino and neutral sugars. Also, depending on the monosaccharide composition, a time course of hydrolysis may be required to correct for any losses. After hydrolysis, the monosaccharides are re-*N*-acetylated (if required), reduced, and per-*O*-acetylated to produce the alditol acetate derivatives that will be analysed by GC. *Protocol 5* describes all of these procedures.

Protocol 5. The alditol acetate method for determining the monosaccharide composition of a glycoprotein[a]

Equipment and reagents

- Purified glycoprotein for analysis
- Glass screw-cap vials with Teflon-faced seal (Pierce or Supelco)
- 2 M trifluoroacetic acid (TFA)
- 4 M HCl
- Propan-2-ol
- Heating block or oven
- Compressed N_2 or argon gas (dry)
- *myo*-Inositol (Fluka or Sigma)
- Standard sugar mixture containing 1 mg/ml of each of the monosaccharides that might be present in the sample to be analysed: the minimum suggested mixture is, Glc, Gal, Man, Fuc, Xyl, GlcNAc, and GalNAc
- Centrifugal evaporator (Savant SpeedVac)
- Acetic anhydride
- Saturated solution of sodium hydrogen carbonate
- 1 M ammonia solution containing 10 mg/ml sodium borohydride
- Glacial acetic acid
- 10% (v/v) acetic acid in methanol
- Methanol
- Pyridine
- Toluene
- Dichloromethane
- Heating block
- Gas chromatograph fitted with an SP-2380 fused-silica capillary column (20 m × 0.25 mm i.d., Supelco) and flame ionization detection[b]

A. *Acid hydrolysis*

1. Place duplicate samples of glycoprotein (10–100 μg of carbohydrate) in screw-cap vials.
2. In two other vials place 20 μl of the standard sugar mixture (20 μg of each standard monosaccharide) and 20 μg *myo*-inositol.
3. To all samples, add 2–20 μg (in proportion to the carbohydrate content of the sample) of *myo*-inositol as internal standard.
4. Dry all of the samples in the centrifugal evaporator.
5. Process one glycoprotein sample(s) and one standard sugar mixture for neutral sugars by adding 0.5 ml of 2 M TFA.
6. Flush the vials with dry nitrogen or argon and seal them tightly using screw-caps with Teflon-faced seals.
7. Heat the vials at 100°C for 4 h or 121°C for 2 h.
8. Cool the vials to room temperature and evaporate the TFA under a stream of dry nitrogen or argon, or in the centrifugal evaporator.
9. To ensure complete removal of the TFA, add 250 μl propan-2-ol to the dried samples and re-evaporate. Repeat this step.

Protocol 5. *Continued*

10. Process the other glycoprotein sample(s) and standard sugar mixture for amino sugars by adding 0.5 ml of 4 M HCl and proceeding as in step 6.[c]
11. Heat at 100°C for 18 h.
12. Process these samples as in steps 8 and 9 but use H_2O in place of the propan-2-ol.

B. *Derivatization*

If the sample is thought to contain amino sugars, carry out part B, step 1. Otherwise proceed directly to part B, step 2.

1. To each of the dried hydrolysates, add 5 μl acetic anhydride and 100 μl saturated sodium hydrogen carbonate. Mix gently and incubate at room temperature for 10 min. Add another 5 μl acetic anhydride, mix gently, and continue the incubation at room temperature for a further 20 min. Stop the reaction by heating at 100°C for 2 min. Cool to room temperature and evaporate to dryness as in part A, step 8.
2. Dissolve the dried mixture in 250 μl of 1 M ammonia solution containing 10 mg/ml sodium borohydride. Incubate the vials at room temperature for 2 h or overnight.
3. Add glacial acetic acid to the solution a drop at a time until all bubbling stops.[d]
4. Evaporate to dryness as in part A, step 8. Add 0.5 ml of 10% (v/v) acetic acid in methanol to the dried samples and re-evaporate. Repeat this step twice with 10% (v/v) acetic acid in methanol and once with methanol.[e]
5. Add 100 μl each of acetic anhydride and pyridine, close the vials tightly, and heat them at 100°C for 2 h (or at 121°C for 20 min) to per-*O*-acetylate the alditols.
6. Cool the vials to room temperature and evaporate to dryness as in part A, step 8. Evaporate twice with 100 μl toluene to remove the last traces of the reagents.
7. Dissolve the dried mixture in 0.5 ml dichloromethane. Add 0.5 ml water and mix thoroughly. Centrifuge at 1000 *g* to separate the phases. Transfer the lower organic phase to a clean vial with a glass Pasteur pipette or syringe. Be careful not to transfer any of the aqueous phase.
8. Evaporate the organic phase to dryness as in part A, step 8 and redissolve the residue in 20 μl dichloromethane.

C. *Separation and analysis*

1. Set the initial oven temperature of the GC apparatus at 190°C.
2. For each sample from part B, step 8 inject 1 μl of sample. Develop the

chromatogram by holding the temperature at 190°C for 2 min, and then increasing the temperature at 3°C/min to 260°C. Hold at this temperature for 10 min.

3. Identify the alditol acetates and therefore the component monosaccharides from the retention times of the standards. Quantify them by reference to the internal standard, *myo*-inositol, taking into account the response factor for each standard monosaccharide.

[a] From ref. 20.
[b] The GC can also be part of a GC-MS instrument. This offers further confirmation of the identification of the monosaccharide components by acting as the GC detector and analysing the fragmentation ions produced by each alditol acetate peak.
[c] Part A, steps 10–12 can be omitted if it is known that the glycoprotein sample does not contain amino sugars.
[d] This breaks down any unreacted sodium borohydride and acidifies the solution.
[e] This removes borate ions produced by the breakdown of the sodium borohydride.

3.3 Lectin binding analysis

Lectins, as described in Section 2.3, bind to specific carbohydrate structures. Their usefulness in the present context is that they can be used to separate glycoproteins by affinity chromatography. Many lectins with different specificities are known and some of the more useful ones are listed with their specificities in *Table 1*. *Protocol 6* describes affinity chromatography using the lectin concanavalin A (Con A). It also serves as a guide for lectin affinity chromatography separations using lectins other than Con A. Lectins usually have particular properties, apart from the obvious one of their hapten, that affect the way they are used in chromatography. Factors such as the flow rate, temperature, or column dimensions can dramatically affect separations. Established procedures in the literature normally take these properties into account.

Protocol 6. Concanavalin A affinity chromatography

Equipment and reagents

- Tris-buffered saline (TBS): 20 mM Tris–HCl pH 8 containing 150 mM NaCl
- TBS containing 1 mM $CaCl_2$ and 1 mM $MnCl_2$
- Glycoprotein sample for analysis (in 0.5 ml TBS)
- 20 mM α-methyl glucoside in TBS
- 200 mM α-methyl mannoside in TBS
- Immobilized concanavalin A: Con A–agarose (Vector Labs or Sigma) or Con A–Sepharose (Pharmacia)
- 2 ml chromatography column (e.g. Bio-Rad Econo-column 7 × 50 mm)

Method

1. Pack the column with immobilized Con A and wash it with 10 column volumes of TBS containing 1 mM $CaCl_2$ and 1 mM $MnCl_2$.

Protocol 6. *Continued*

2. Apply the glycoprotein sample (in 0.5 ml of TBS) to the column. Wash the sample into the column with 0.5 ml TBS and then elute with 5–10 column volumes of TBS at a flow rate of 0.5 ml/min. Collect 0.5 ml fractions throughout.
3. Change the eluent to 20 mM α-methyl glucoside and collect a further 5 column volumes.
4. Change the eluent to 200 mM α-methyl mannoside and collect a further 5 column volumes.
5. Wash the column with 10 column volumes of TBS containing 1 mM $CaCl_2$ and 1 mM $MnCl_2$ to regenerate it. Store the packed column at 4°C in TBS containing 0.1% sodium azide or other bacteriostat.
6. Measure the protein concentrations of the fractions by reading the A_{280}, or assay for protein using a standard assay procedure.
7. Pool fractions containing protein and desalt them (e.g. by dialysis or chromatography on Sephadex G25) prior to analysis by SDS–PAGE or a functional assay.

In this particular fractionation, a glycoprotein may elute in one or more of four ways: unbound (flows straight through the column); weakly bound but not requiring hapten to elute (retarded by the column and eluted with buffer); bound but eluted with weak hapten (eluted with 20 mM α-methyl glucoside); strongly bound (eluted with 200 mM α-methyl mannoside).

3.4 Susceptibility to endoglycosidase H

Differential sensitivity to endoglycosidase H (Endo H) digestion provides a simple, rapid method for determining if the *N*-glycans present on a glycoprotein are oligomannose/hybrid or complex structures, or if the protein is modified with both types of structure. Endo H selectively cleaves the β1,4 linkage between the two GlcNAc residues of the chitobiose core of oligomannose and hybrid type structures, provided that the α1,6-mannose linked to the β-mannose is substituted by an additional mannose residue. All the early biosynthetic intermediates prior to the action of Golgi mannosidase II (*Figure 2*) are sensitive to Endo H. Complex type *N*-glycans are, therefore, not sensitive to Endo H cleavage.

Endo H digestion analysis can be performed on the following types of sample:

- purified glycoprotein
- Tran ^{35}S-labelled glycoprotein (*Protocol 3*)
- crude culture medium or cell lysates (*Protocol 3*)

Whenever possible work with a purified protein. Considerably more Endo H may be required with crude culture medium or cell lysates than with a purified or partially purified protein. After Endo H digestion, the protein of interest is normally separated by SDS–PAGE and detected by Coomassie blue or silver stain, or Western blotted and detected by probing the blot with a specific antibody probe (see ref. 41). Alternatively, if it is radiolabelled it may be immunoprecipitated, run on SDS–PAGE, and detected by fluorography (*Protocol 3*).

Protocol 7. Digestion with endoglycosidase H

Equipment and reagents

- SDS–PAGE gel, electrophoresis equipment, and reagents
- Western blotting apparatus (if the protein is to be immunodetected)
- 50 mM sodium citrate pH 5.5
- Endo H (Boehringer Mannheim or Oxford GlycoSciences): make this up at 0.1 U/ml in 50 mM citrate buffer pH 5.5
- Denaturation buffer B: citrate buffer pH 5.5 containing 0.5 mg/ml SDS and 0.1 M 2-mercaptoethanol
- PMSF: 10 mg/ml in methanol (Sigma)
- Purified glycoprotein for analysis

Method

1. Place 5 μg of protein in each of two tubes. Add 24 μl denaturation buffer B (giving a substrate:SDS ratio of 1:1.2) and heat at 100 °C for 10 min to denature the proteins. Cool the samples to room temperature.
2. To one sample add 35 μl Endo H and 1 μl PMSF to give a final digestion volume of 60 μl.[a] To the other sample, add 35 μl citrate buffer pH 5.5 and 1 μl PMSF. **Note:** the final SDS concentration must not exceed 0.2 mg/ml or it will inactivate the Endo H.
3. Incubate the samples at 37 °C for 18 h.
4. Add an equal vol. of SDS–PAGE sample buffer and analyse the samples by SDS–PAGE. Western blot the gel and detect the protein of interest with a suitable antibody probe if required (ref. 41).

[a] The final Endo H concentration is approx. 0.06 U/ml. As an approximate guide the optimal digestion conditions for Endo H are in the range of 0.02–0.4 U/ml.

3.5 Other methods

Direct analysis of protein glycosylation can be obtained by electrospray and matrix-assisted laser desorption mass spectrometry of the intact protein. Characteristics such as the degree of glycosylation site occupancy and the relative proportions of separate populations of glycoproteins within the sample can be determined by the estimation of molecular mass. Different glycoforms of the protein will have distinct masses that can be detected and analysed. The calculated mass of the protein from its amino acid sequence can then be subtracted to give a measure of the mass of the attached carbohydrate. Based on the residue masses of individual monosaccharides the number and partial composition of the glycans can be deduced. Glycoforms can also be separated utilizing the high efficiency of capillary electrophoresis (21).

4. Analysis of glycosylation sites

Proteases such as trypsin, chymotrypsin, and endoproteinase Glu-C (V8 protease) have been used extensively in protein sequencing because their cleavage sites are determined by one or a restricted number of amino acid residues. These enzymes produce a predictable pattern of peptides from a polypeptide. If the polypeptide is glycosylated then some of the peptides will be glycopeptides. Proteolytic cleavage of a glycoprotein and isolation of the glycopeptides allows the characterization of both the glycans and the peptide sequence to which they are attached.

4.1 Protease mapping of glycopeptides

Knowledge of the amino acid sequence of the glycoprotein and the location of the potential glycosylation sites in that sequence are prerequisites for glycosylation site analysis. The cleavage pattern of the protein with highly specific proteases can then be predicted. A strategy for proteolytic cleavage is chosen to digest the polypeptide into peptides and glycopeptides so that each predicted glycopeptide will contain only a single glycosylation site. This may require more than one enzyme to ensure that single sites are isolated. An example of proteolytic digestion with trypsin is described in *Protocol 8*. The glycopeptides can then be separated by HPLC as described in *Protocol 9*, and analysed to determine their amino acid composition and sequence (Chapter 1), monosaccharide composition (Section 3.2), and glycan structures (see Section 5).

Protocol 8. Tryptic cleavage of glycoprotein

Equipment and reagents

- Glycoprotein for analysis (lyophilized)
- Denaturation buffer C: 0.5 M Tris–HCl pH 8.6 containing 2 mM EDTA and 6 M guanidine–HCl
- Dithiothreitol (DTT)
- Iodoacetic acid
- 0.1% and 0.01% (w/v) TFA
- 0.1 M ammonium hydrogen carbonate pH 8.6
- Sequencing grade trypsin (preferably Promega)[a]
- 1 M NaOH
- Glass screw-cap vials with Teflon-faced seal (Pierce or Supelco)
- Compressed N_2 or argon gas (dry)
- Freeze-drier

A. *Reduction and S-carboxymethylation of disulfide bonds*[b]

1. In a glass screw-cap vial, dissolve the lyophilized glycoprotein in denaturation buffer C (final protein concentration 5–20 mg/ml), mix well, but take care not to aerate. Flush the vial with dry nitrogen or argon and seal it tightly using screw-caps with Teflon-faced seals. Incubate for at least 30 min at 50°C.
2. Add solid DTT (50-fold molar excess over total cysteine in the protein). Mix to dissolve, reflush the vial with gas, and continue the incubation at 50°C for 2–4 h (or overnight at room temperature).
3. Cool the mixture to room temperature and add a twofold molar excess of iodoacetic acid (over DTT) in an equal volume of denaturation buffer C. Check that the pH of this solution is 8.6 and if necessary adjust the pH with 1 M NaOH.
4. Monitor the pH of the reaction mixture for the first 10 min maintaining it at 8.6 by addition of 1 M NaOH. Incubate the sample at 50°C for 15 min or at room temperature for 30 min. Dialyse the solution exhaustively against distilled water and then lyophilize.

B. *Digestion with trypsin*

1. Dissolve the dried sample in the minimum volume of 0.1% TFA then dilute it with ammonium bicarbonate buffer to give a 1–2 mg/ml solution of protein.
2. Add trypsin (dissolved in 0.01% TFA) to give a protease to protein weight ratio between 1:50 and 1:100, and incubate the digestion mixture at 37°C for 6 h or overnight. Add a second aliquot of trypsin and continue the incubation to give a total incubation time of 24–30 h. Stop the reaction by freezing at –20°C.
3. Remove the ammonium bicarbonate by repeated lyophilization. Separate the peptide and glycopeptide mixture by HPLC (*Protocol 9*) for subsequent analysis.

[a] Promega sequencing grade trypsin is modified to prevent autolysis and therefore lasts longer in the digest. This means that one can use a lower ratio of protease to protein. Also, for high sensitivity applications, there are no peptides in the digest derived from the trypsin itself. An alternative source of sequencing grade trypsin is Boehringer Mannheim.
[b] From ref. 22.

Protocol 9. Separation of tryptic peptides and glycopeptides by HPLC

Equipment and reagents

- Tryptic digest of glycoprotein of interest (*Protocol 8*, part B, step 3)
- HPLC grade acetonitrile (low UV) containing 0.1% TFA
- 0.1% (w/v) TFA
- Reversed-phase HPLC column (e.g. Vydac 218TP54, 4.6 × 250 mm,[a] Hichrom)
- Gradient HPLC system with UV detection

Protocol 9. *Continued*

Method

1. Dissolve the mixture of tryptic peptides in an appropriate volume of 0.1% TFA to give approx. a 2% solution.
2. Equilibrate the reversed-phase column with 0.1% TFA at a flow rate of 1 ml/min.
3. Inject the peptide mixture (or an aliquot) onto the column and develop the separation with a linear gradient of acetonitrile from 0–60% at 1%/min.
4. Monitor the eluate at 206, 214, or 225 nm.[b]
5. Collect continuous fractions or collect only peaks by watching the absorbance trace and diverting the effluent to one tube per peak.
6. Analyse the fractions to identify glycopeptides.

[a] For separation of multi-milligram amounts of digested protein, a larger column such as the Vydac 218TP510 is recommended.
[b] Simultaneous monitoring at 280 nm can be useful to detect peptides that contain aromatic amino acids.

4.2 Glycopeptide identification and analysis

To identify the glycosylated peptides in the proteolytic digest, methods for the detection of carbohydrate, such as those described in Sections 2 and 3, can be used. For example, if the original protein bound Con A, then a dot blot using the method in *Protocol 2* would detect the glycopeptides that bound to Con A. Compositional assays (Section 3.2) will detect the component monosaccharides in glycopeptides. A quicker method, more suitable to large numbers of samples, is to use amino acid analysis techniques to detect amino sugars which will be present in all *N*-glycans and in animal *O*-glycans. *Protocol 10* describes the analysis of peptides for hexosamines.

Protocol 10. Analysis of peptides for hexosamine[a]

Equipment and reagents

- Peptides for analysis, separated by HPLC as described in *Protocol 9*
- 6 M HCl (Sequenal grade, Pierce)
- 6 × 50 mm glass tubes (Pyrex or equivalent)
- PicoTag amino acid analysis workstation and reagents (Waters)
- Standard hexosamine mixture: 5 μmol/ml glucosamine and galactosamine
- α-Aminobutyric acid (ABA): 1 μmol/ml
- Compressed N_2 or argon gas (dry)
- C18 reversed-phase HPLC column (Spherisorb ODS-2, 3 μm beads, 250 × 4.6 mm, Hichrom)
- Solvent A: 0.14 M sodium acetate pH 5.7 containing 1.2% (v/v) triethylamine and 4% (v/v) acetonitrile
- Solvent B: 60% (v/v) acetonitrile
- Gradient HPLC system with UV detection

Method

1. Prepare duplicate standards by adding 10 μl of the hexosamine mixture and 25 μl ABA to glass tubes.
2. Prepare duplicate samples by adding 100 μl of each peptide sample to glass tubes. Add 5 μl ABA to each sample tube to act as the internal standard. Place the sample and standard tubes in a Waters PicoTag reaction vial and dry them using the Waters PicoTag workstation.
3. Prepare the samples and standards for hydrolysis with HCl using the Waters PicoTag workstation following the manufacturer's instructions.
4. Hydrolyse the samples and standards by heating the reaction vial at 100°C for 6 h using the Waters PicoTag workstation.
5. Derivatize the amino acids and hexosamines in the hydrolysate with phenyl isothiocyanate by the Waters PicoTag method according to the manufacturer's instructions.
6. Equilibrate the reversed-phase column with solvent A at a flow rate of 1.5 ml/min at 50°C.
7. Inject the samples onto the column and separate the phenylthiocarbamyl derivatives with the following elution gradient: 0–2% solvent B for 2.4 min, 2–20% solvent B for 5.6 min, 20–25% solvent B for 2.4 min, 25–100% solvent B for 0.4 min, and 100% solvent B for 1.5 min.
8. Identify and quantify the hexosamines in each sample by reference to the hydrolysed standard hexosamines and the ABA internal standard.

[a] From ref. 23.

Having identified the glycosylated peptides, these can be analysed for their amino acid composition and sequence to ascertain which peptide sequence they correspond to in the protein. The intact glycopeptides can be analysed by mass spectrometry to determine the mass of the glycans, the types of glycan, and their heterogeneity as described in Section 3.5. The glycan portion can be characterized by their monosaccharide composition (see Section 3.2 and *Protocol 5*) and by the methods for glycan analysis described in Section 5.

5. Analysis of glycan structure

5.1 Glycan release

Chemical release of *O*-glycans commonly utilizes the β-elimination reaction under alkaline conditions. This is generally done in the presence of excess reducing agent to prevent degradation of alkali-sensitive linkages (24) but this precludes labelling of the reducing terminus of the glycan. Hydrazine has been used extensively to release *N*-linked oligosaccharides and the technique has now been extended to release of *O*-glycans (25, 26). Conditions of hydrazinolysis can be chosen to release both types of glycans (*N* + *O*) or only *O*-glycans. If it is not known whether the glycoprotein or glycopeptide of interest has only one or both types of glycosylation, by employing different hydrazinolysis conditions on separate samples, and analysing the resulting glycan profiles, the type(s) of glycosylation can be determined. If glycans are found under *N* + *O* conditions and none are found under *O* only conditions then the sample contains only *N*-glycans. If identical glycans are found under both *N* + *O* and *O* only conditions then the sample contains only *O*-glycans. If, however, glycans are seen under the *N* + *O* conditions that are not found under *O* only conditions then both types of glycan are present. The Glyco-Prep 1000 instrument (Oxford GlycoSciences) uses hydrazinolysis for glycan release and includes a series of chromatographic steps to isolate the glycans free of peptide. The major advantage of chemical methods is that they are non-specific in their release. However, enzymatic methods may offer advantages when dealing with labile substituents. The alternative to release is to isolate glycopeptides or glycosyl amino acids from the glycoprotein. This is protein-dependent, but can offer advantages when chemical release is too harsh and enzymatic release is not feasible. The protein can be digested with proteases or, in the case of hydroxyproline-containing plant glycoproteins, hydrolysed with strong base.

Protocol 11. Hydrazinolysis[a]

Equipment and reagents

- Glycoprotein (exhaustively dialysed against H_2O or 0.1% TFA) or glycopeptide (from *Protocol 9*)
- Hydrazine: essentially anhydrous, water < 1% (v/v) (Aldrich)[b]
- Toluene (anhydrous)
- Acetic anhydride
- HPLC grade water
- Saturated solution of sodium hydrogen carbonate, at 0°C
- Dowex AG50-X12 resin (H^+ form): washed with 3 × 2 vol. H_2O
- Whatman 3MM chromatography paper (washed with H_2O and dried)
- Butan-1-ol:ethanol:H_2O (4:1:1, by vol.)
- Copper reagent: 1 mM copper (II) acetate in 1 mM acetic acid
- Octan-2-ol
- Glass screw-cap vials with Teflon-faced seal (Pierce or Supelco)
- Glass syringe with gas-tight plunger (Hamilton)
- Heating block or oven
- Freeze-drier
- High vacuum pump with liquid nitrogen vapour trap
- Chromatography tank for descending paper chromatography
- 5 ml plastic Luer-lock syringe
- 0.5 μm PTFE syringe filter (3 mm diameter)

A. *Hydrazine treatment*

1. Lyophilize the glycoprotein or glycopeptide sample (0.1–5 mg) in the reaction vial for at least 24 h.
2. In a fume-cupboard, add 200 μl hydrazine with a dry syringe to the sample *immediately* on its removal from the freeze-drier and seal the vial tightly.[c]

3. Mix gently to wet all the dried sample and heat the vial at 95°C for 4 h to release *N*- and *O*-linked glycans or at 60°C for 5 h to release only the *O*-linked glycans.
4. Cool the reaction vial to room temperature. Remove the hydrazine by evaporating under high vacuum at room temperature. Remove the final traces of hydrazine by evaporating with anhydrous toluene (3 × 200 μl).

B. *N-acetylation*

1. Place the dried reaction mixture (from part A, step 4) on ice and add 100 μl/mg protein (minimum 100 μl) of ice-cold saturated sodium hydrogen carbonate followed *immediately* by 50 μl/ml solution (minimum 5 μl) of acetic anhydride. Mix gently to avoid foaming and leave the reaction on ice for 10 min. Warm vial to room temperature (control frothing if it occurs by adding a drop of octan-2-ol). Add the same volume of acetic anhydride again and continue the incubation at room temperature for 20–30 min. This step *N*-acetylates all the amino groups in the mixture.
2. After the *N*-acetylation step, mix the solution thoroughly, but carefully to avoid losses due to frothing, and apply the sample to a column of Dowex AG 50 resin (2 ml resin/ml of reaction solution, minimum 200 μl). Rinse the vial twice with 0.5 ml water and apply the rinsings to the column. Wash the column with 3 column volumes of water. Collect all the effluent from the column and evaporate it to dryness at room temperature.

C. *Isolation of released glycans*

1. Redissolve the dried mixture (from part B, step 2) in a minimum volume of water and apply the sample to water-washed strips of 3MM paper.[d]
2. Separate the glycan pool from the peptides by descending paper chromatography with butan-1-ol:ethanol:H_2O (4:1:1) for 16–24 h. Disaccharides and larger glycans will remain at the origin. Dry the strips and elute the glycans from the origin with water and evaporate them to dryness at room temperature.
3. Redissolve the dried glycans in 0.5 ml of copper reagent and incubate at room temperature for 1 h.
4. Remove the plunger from a 5 ml plastic syringe, fit it with the PTFE filter, and suspend it over a collection tube. Suspend a 0.5 ml column of Dowex AG 50 resin over the syringe barrel.
5. Apply the reaction solution to the resin column, rinse the tube twice with 0.25 ml water, and apply the rinsings to the column. Wash the

Protocol 11. *Continued*

column with 3 column volumes of water. Collect all the effluent from the column in the syringe barrel. Using the syringe plunger push the glycan solution through the PTFE filter into the collection tube. Evaporate to dryness at room temperature.

6. Add 0.5 ml water to the residue and re-evaporate to remove traces of acetic acid. Redissolve the glycans in 0.5 ml water and store the solution at –20°C.

[a] From ref. 26.
[b] Preparation of anhydrous hydrazine is also described briefly in ref. 26.
[c] The lyophilized sample and the hydrazine should not be exposed to the atmosphere for more than a few minutes as they will start to absorb water.
[d] If the hydrazinolysis conditions used in part A, step 3 were for release of only *O*-linked glycans, it is advisable to perform a reversed-phase clean-up step, to remove *N*-glycopeptides, before the paper chromatography. See *Protocol 12*, steps 3–5, or ref. 26.

As pointed out in Section 3.1, PNGase F cannot release some *N*-glycans from plant and insect glycoproteins. A similar enzyme from almond (peptide:*N*-glycosidase A, PNGase A) can release these structures but it is inefficient with glycoprotein substrates, only showing full activity with glycopeptides. Therefore, this enzyme is not used for protein analysis. However, when releasing *N*-glycans for further study, an intact protein is no longer a priority so PNGase A is the enzyme of choice for glycoproteins expressed in plant and insect cells. *Protocols 12* and *13* describe the release of *N*-glycans using PNGase F and PNGase A. The release of *N*-glycans using PNGase F (*Protocol 12*) is a similar procedure to that used earlier in the chapter for deglycosylation of *N*-glycoproteins (*Protocol 4*), except that after the incubation, instead of analysing the deglycosylated protein, it is removed from the sample and the released glycans are isolated for further analysis.

Protocol 12. Release of *N*-glycans using PNGase F

Equipment and reagents

- Glycoprotein or glycopeptide for analysis
- PNGase F enzyme, buffers, and reagents (see *Protocol 4*)
- C18 reversed-phase pre-packed columns (Sep-Pak, Millipore)
- Methanol
- HPLC grade water
- Freeze-drier

Method

Use HPLC grade water throughout.

1. Denature the purified glycoprotein (100–500 μg) in 100 μl denaturation buffer as described in *Protocol 4*, step 1.

2. Cool the sample and add 25 μl 10% NP-40 and 20 U PNGase F in 75 μl incubation buffer. Incubate for 18 h at 37 °C.
3. Prepare the reversed-phase columns by washing with 5 column volumes of methanol and then 5 column volumes of water.
4. Apply the sample to the column. Rinse the reaction vial twice with 100 μl water and add the rinsings to the column. Wash the column with 3 column volumes of water. Collect all the effluent and evaporate it to dryness.
5. Redissolve the eluted glycans in a minimum volume of water and desalt them by ion exchange (see *Protocol 17*) or gel filtration chromatography. Store the desalted glycans in 0.5 ml water at −20 °C.

Protocol 13. Release of *N*-glycans using PNGase A

Equipment and reagents

- Glycoprotein (100–500 μg)
- PNGase A (Seikagaku or Boehringer Mannheim)
- 0.1 M sodium citrate/phosphate buffer pH 5
- C18 reversed-phase columns, methanol, and HPLC grade H_2O (see *Protocol 12*)

Method

1. Digest the glycoprotein with protease to generate glycopeptides.[a]
2. If using the Seikagaku enzyme, it is supplied dry (10 mU/vial). Make a 20 mU/ml solution by adding 500 μl citrate/phosphate buffer to the vial. If using the Boehringer enzyme, it is supplied as 100 μl of a 50 mU/ml solution in 50 mM citrate/phosphate buffer pH 5 containing 50% (v/v) glycerol. Micro-dialyse or spin-ultrafilter with 0.1 M citrate/phosphate buffer to remove the glycerol. Dilute the enzyme to 20 mU/ml by adding citrate/phosphate buffer.
3. Add sufficient PNGase A solution to the dried peptide mixture to make the peptides approx. 10 μmol/ml and incubate the mixture at 37 °C for 18 h.
4. Recover the released glycans as described in *Protocol 12*, steps 3–5.

[a] Use a protease such as trypsin, chymotrypsin, or pepsin that will generate a reasonable number of peptides but not make them too small. *Protocol 8* can be used as a guide for the procedure but, as quantitative cleavage is less critical for this protocol, the denaturation steps can be omitted. Use a volatile or dilute buffer for the proteolytic digestion to ensure that it will not interfere with the pH of the PNGase reaction, step 3.

5.2 Glycan labelling

Glycan or glycopeptide pools can be separated directly using techniques that resolve on the basis of the carbohydrate. However, detection methods for unlabelled glycans are insensitive (with the notable exception of PAD). More often, the glycans are labelled prior to separation and the labelling enables sensitive methods of detection to be used such as radioactivity or fluorescence. PAD, radioactivity, and fluorescence all allow detection in the low picomole range. Radioactive labelling, although sensitive, is going out of favour due to environmental restrictions on use and disposal of radioactive materials. The principal method of fluorophore labelling is to use reductive amination to couple an amine-containing compound to the carbonyl group at the reducing terminal end of the oligosaccharide. The fluorescent dye, 2-aminobenzamide (2-AB), introduced by Oxford GlycoSciences is useful as it does not introduce a charged group into the labelled glycan and is compatible with all liquid chromatography methods. The Signal™ labelling kit from Oxford GlycoSciences (Code K-404) utilizes this reagent in a convenient package, complete with reagents and instructions. However, the clean-up cartridges provided in the kit are not as efficient as the chromatography 'clean-up' in the non-kit method described in *Protocol 14*.

Protocol 14. Labelling of glycans with 2-aminobenzamide[a]

Equipment and reagents

- Glycans (salt-free) as prepared in *Protocols 11–13*
- 2-Aminobenzamide (2-AB, Sigma)
- DMSO containing 30% (v/v) glacial acetic acid
- Sodium cyanoborohydride (Aldrich)
- Polypropylene tubes (0.5 ml Eppendorf or similar)
- Whatman 3MM chromatography paper (3 × 10 cm)
- Chromatography tank for ascending paper chromatography
- Butan-1-ol:ethanol:H_2O (4:1:1, by vol.), HPLC grade H_2O, oven or heating block, freeze-drier (see *Protocol 11*)

Method

1. Dry the glycans in 0.5 ml polypropylene tubes.
2. Make up *fresh* labelling reagent by dissolving 2-AB in 30% acetic acid, 70% DMSO to 0.35 M final concentration. Add solid sodium cyanoborohydride to 1 M final concentration.
3. Add 5 μl labelling reagent to the dried glycans. This volume of reagent will label up to 50 nmol glycans; scale it up if required.[b]
4. Cap the tubes and heat them at 65°C for 2 h.
5. Cool the tubes to room temperature and apply the contents in one spot to a strip of Whatman 3MM paper and dry thoroughly. Separate the glycans from the reagents by ascending paper chromatography using butan-1-ol:ethanol:H_2O until the solvent front is 1 cm from the top of the strip (~ 45 min). The unreacted dye and reagents move with the solvent front and the labelled glycans will remain at the origin.

6. Elute the glycans from the origin with water, filter, and evaporate them to dryness as described in *Protocol 11*, part C.
7. Redissolve the glycans in a minimum of water and store at –20°C.

[a] Based on ref. 27.
[b] Labelling is independent of the glycan concentration over the range 5–100 μM (27).

After *Protocol 14*, step 7 the glycan pool is ready for analysis by Bio-Gel P-4 chromatography, HPLC, or other methods (see Section 5.3).

5.3 Glycan separation

The separation of oligosaccharides based on their charge is commonly performed by ion exchange chromatography. The majority of charged substituents on oligosaccharides are acidic and, therefore, anion exchange methods predominate, but they vary in both the column matrix and the eluent used. Strong and weak anion exchangers are used with functional groups such as QAE (strong) and DEAE (weak). The matrix can be soft gel, silica, or polymer based. Eluents can use sodium phosphate, sodium acetate, ammonium acetate, or ammonium formate buffers. The ammonium ion buffers have the advantage that they are volatile and the glycans can be easily freed of buffer by evaporation.

Gel electrophoresis techniques have also been developed recently for charge (and size) separation of fluorophore labelled glycans (28) but they will not be discussed here.

Bio-Gel P-4 chromatography has been used to separate oligosaccharides on the basis of their hydrodynamic volume. The columns can be calibrated with dextran hydrolysate; this consists of glucose and isomalto-oligosaccharides with increasing degrees of polymerization. The elution position of an unknown oligosaccharide can then be expressed in glucose units (g.u.). The component monosaccharides of a glycan have characteristic, and generally additive, sizes. Hexose residues are equivalent to 0.7–1 g.u. and *N*-acetylhexosamine (HexNAc) residues to 2 g.u. (29). The fluorescent conjugates formed with 2-AB (*Protocol 14*) cause a downward shift of ~ 2.5 g.u. for each oligosaccharide, but this is easily corrected for. When pure water is used as eluent for Bio-Gel P-4 columns, negatively charged oligosaccharides are excluded from the matrix and elute at the void volume of the column. For this reason only neutral glycans are generally separated by this method. If it is necessary to separate negatively charged glycans by Bio-Gel P-4 chromatography then a dedicated column running in buffer (e.g. 50 mM sodium acetate or ammonium formate pH 4.5) or 1% acetic acid is required. The size in g.u. of an unknown oligosaccharide is then indicative of its composition and structure. This, combined with a composition of the intact glycoprotein or the glycan pool, allows the first interpretations of the structure of the component oligosaccharides. As gel

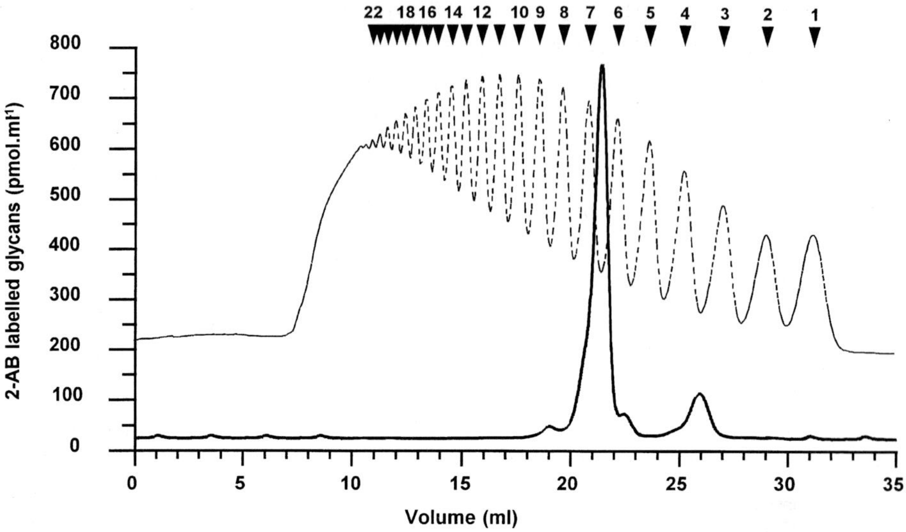

Figure 3. Gel filtration profile of the 2-AB labelled glycans from horseradish peroxidase on an Oxford GlycoSciences RAAM 2000 GlycoSequencer. The glycans were released using hydrazinolysis and labelled with 2-AB as described in *Protocols 11* and *14*, respectively. The dashed line shows the elution profile of the dextran hydrolysate internal standard. The arrows and numbers at the top of the panel indicate the elution position of glucose and isomalto-oligosaccharides in the dextran hydrolysate. This internal calibration (scale at top) allows the elution positions of the labelled oligosaccharides to be expressed in glucose units (g.u.). Thus, the major peak eluting at 21.5 min is 6.5 g.u. in size and corresponds to the plant modified type glycan, Manα3(Manα6)(Xylβ2)Manβ4GlcNAcβ4 (Fucα3)GlcNAc, depicted in *Figure 1*.

filtration separation is based on hydrodynamic volume, there is the possibility that several glycan structures can have the same size in g.u. Sometimes this is coincidental and unrelated to similarities in structure but more often it arises through the existence of structural isomers, i.e. the same number and type of monosaccharides but in a different sequence.

Protocol 15 describes high resolution Bio-Gel P-4 gel filtration chromatography of fluorescently labelled glycans using a system built from individual components. The RAAM 2000 GlycoSequencer (Oxford GlycoSciences) incorporates all the hardware (with the exception of a fraction collector) required for *Protocol 15*. An example of a separation on the GlycoSequencer is shown in *Figure 3*. As described above, since the gel is run in pure water, it is essential that the glycan preparation for analysis (e.g. from *Protocol 14*) is salt-free. *Protocol 15* also includes a desalting procedure using ion exchange chromatography.

Mass spectrometry techniques can give an accurate measure of the size and composition of the components of a glycan mixture (31). The mass of each oligosaccharide indicates the number and type of monosaccharide units.

These methods cannot discriminate individual hexose and hexosamine isomers but, as with Bio-Gel P-4 chromatography, a knowledge of the size of oligosaccharide can give a very good idea of its composition. There is less chance of fortuitous mass similarities with mass spectrometry, but again structural isomers cannot be resolved.

Protocol 15. Bio-Gel P-4 gel filtration chromatography of glycans[a]

Equipment and reagents

- Glycans to be separated (*Protocol 14*)
- HPLC grade water: degas this using a vacuum pump
- Dextran hydrolysate: 15 mg/ml (Oxford GlycoSciences)[b]
- Bio-Gel P-4 (–400 mesh) (Bio-Rad)
- Two water-jacketed glass chromatography columns (1.5 × 100 cm) with flow adapters and a packing reservoir (Pharmacia)
- Circulating water-bath or column heaters
- Refractive index detector (ERMA)
- Fluorescence detector (Jasco)
- 1.5% (v/v) triethylamine
- Liquid chromatography system consisting of a low pulsation pump, sample injector, pressure gauge, pre-column in-line filter, and fraction collector (see ref. 30 for a schematic)
- Vacuum pump (water or diaphragm pump)
- Dowex AG50-X12 (H^+ form) and Dowex AG1-X8 (acetate form): wash these well with water before use
- Plastic mini-column (Bio-Spin from Bio-Rad)
- Plastic syringes (Luer lock, 5 ml)
- PTFE syringe filters (0.5 μm, 3 mm diameter)

NB: use degassed HPLC grade water throughout the following steps.

A. *Preparation of the Bio-Gel P-4 columns*

1. Suspend 200 g Bio-Gel P-4 in 3 litres water and allow the beads to swell overnight.[c]
2. Pour off the supernatant and fine the swollen gel thoroughly (at least five times, but seven to ten times is recommended) by resuspending the slurry in 3 litres water, allowing the gel to settle for 3 h, and removing the supernatant containing the suspension of small particles.[d]
3. Resuspend the fined gel in water to give a 1:1 slurry. Degas the slurry at 55°C for 1 h using the vacuum pump.
4. Assemble the column with packing reservoir and clamp it in a vertical position. Stopper the bottom of the column and fill it with a 10–20 cm column of water. Fill the rest of the column and the reservoir with gel slurry and allow it to settle overnight under gravity.
5. Start pumping water through the column at 0.2 ml/min and continue for 24 h.
6. Increase the flow rate to generate 60–75 p.s.i. back pressure (1–2 ml/min). As the column packs, maintain the pressure at this level by gradually lowering the flow rate. The column should stabilize at this pressure, flowing at ~ 0.6 ml/min. If it does not, it may be because the gel was not fined sufficiently or the batch of gel has slightly

Protocol 15. *Continued*

different flow properties. Unpack the column and try to repack at 50 p.s.i. If this fails then unpack the column and re-fine the gel.

7. Remove the packing reservoir, fit the flow adapter, and invert the column. Connect the column inlet and outlet tubing to open reservoirs of water (to allow liquid expansion without the risk of the column running dry or air siphoning into the column).
8. Heat the column to 55°C and maintain it at this temperature. When the expansion has finished, start pumping at 0.2 ml/min.
9. Connect the column to the liquid chromatography system and continue pumping at 0.2 ml/min until the refractive index detector shows a stable baseline.
10. Test the resolution of the column by running 50 μl dextran hydrolysate standard. Detect the components of the dextran hydrolysate with the refractive index monitor.
11. Repeat steps 4–10 for the second column. It is important for column resolution and flow properties that each column is pressure-packed individually before connecting them.
12. Connect the columns in series and pump at 0.2 ml/min. The total column back pressure should be $<$ 50 p.s.i. Re-run the dextran hydrolysate standard.
13. When not in use for separations the columns must be kept at 55°C and kept running but the flow rate can be reduced to 50 μl/min.

B. *Desalting the glycans prior to separation*

1. Prepare a 50% suspension of Dowex AG 50 (H^+) in H_2O and add approx. 400 μl of the resin suspension to a mini-column. Allow it to drain and then wash it with 5 ml 1.5% triethylamine. Wash three times with 1 ml of water.
2. Prepare a 50% suspension of Dowex AG 1 (acetate) and carefully add approx. 400 μl of the resin suspension to the mini-column above the AG 50. Allow the column to drain and wash it with 1 ml water.
3. Remove the plunger from a 5 ml plastic syringe, fit it with a PTFE filter, and suspend it over a collection tube. Suspend the tandem ion exchange mini-column over the syringe.
4. Dissolve the labelled glycans in 100 μl water and apply the mixture to the column. Wash the column with 0.5 ml water. The effluent will collect in the syringe barrel. Repeat the wash four times, continuing to collect the effluent. Remove the mini-column. Using the syringe plunger, push the deionized glycan solution through the filter into the collection tube.
5. Evaporate the glycan solution to dryness.[e]

C. *Separation of labelled glycans*

1. Dissolve the dried glycans in 100 μl water and add 50 μl dextran hydrolysate.
2. Check that the Bio-Gel columns are running at 0.2 ml/min. Load all the sample mixture onto the columns through the sample injector without introducing any air bubbles.
3. Run the columns at 0.2 ml/min to separate the mixture. Collect 1 ml fractions. A double column takes approx. 20 h to run.
4. Detect the components of the dextran hydrolysate with the refractive index monitor and the labelled glycans with the fluorescence detector. The elution positions of the fluorescent glycan peaks are expressed in glucose units (g.u.) by reference to the glucose oligomers in the dextran hydrolysate.
5. Pool the fractions containing glycan peaks and evaporate them to dryness.[e]

[a] Based on refs 29 and 30.
[b] Dextran hydrolysate may also be prepared as described in ref. 30.
[c] This should provide enough gel for two to three columns (1.5 × 100 cm).
[d] Fining is a compromise. If you remove too much small material, then there is the risk that the column will run well but have poor resolution. Conversely if the gel is not sufficiently fined, then the column will have a high back pressure and be prone to blocking.
[e] If desired, the glycans may be redissolved in 100 μl H_2O and stored at –20°C for later analysis.

Numerous other separation methods such as reverse-phase or amino phase HPLC, HPAEC, and capillary electrophoresis, are all capable of separating an oligosaccharide mixture to give a profile of the components (8) but the separation is not based solely on size. This was seen as a drawback of these techniques for the primary separations of unknown mixtures because they are not as directly interpretable as size separations and external standardization with known compounds was required. However, they are more sensitive to the composition and structure of oligosaccharides than size separations and the characteristics of these methods are extremely useful for the separation and purification of structural isomers. A recent paper by Guile *et al.* (32) describes an HPLC separation that can be calibrated with dextran hydrolysate and in which the relative contributions of component monosaccharides to the elution position of glycans was determined. This method is given in *Protocol 16*. This is a normal phase separation that exploits the small differences in hydrophilicity between different glycans. The glycans bind to the amide column at low concentrations of water in the mobile phase and are eluted by an increasing gradient of water. Potential ion exchange problems with the matrix are overcome by buffering the aqueous component of the mobile phase to pH 4.4, the pK_a of the amide functional groups on the column. An example of an HPLC separation is shown in *Figure 4*.

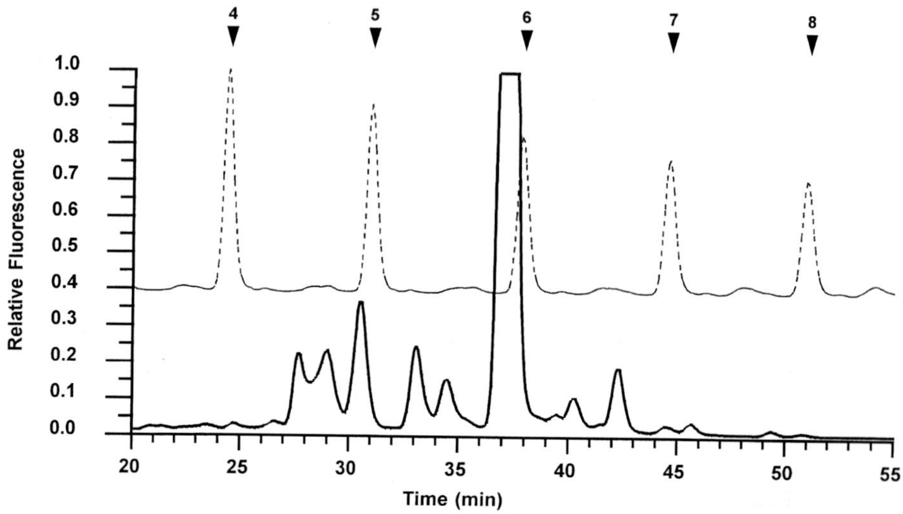

Figure 4. Normal phase HPLC separation of the 2-AB labelled glycans from horseradish peroxidase on an Oxford GlycoSciences GlycoSep N column as described in *Protocol 16*. The glycans were released using hydrazinolysis and labelled with 2-AB as described in *Protocols 11* and *14*, respectively. The dashed line shows an elution profile of 2-AB labelled dextran hydrolysate used as an external standard. The arrows and numbers at the top of the panel indicate the size in g.u. of the isomalto-oligosaccharide at that elution position. The major overloaded peak at ~ 37 min corresponds to the 6.5 g.u. peak in *Figure 3*. In this separation it elutes at 5.9 g.u.

Protocol 16. HPLC separation of 2-AB labelled glycans[a]

Equipment and reagents

- Glycans to be separated (*Protocol 14* or *15*)
- Solvent C: 250 mM ammonium formate pH 4.4—make this using formic acid and ammonia solution (Aristar grade or equivalent)
- HPLC grade acetonitrile
- GlycoSep N HPLC column (Oxford GlycoSciences)
- HPLC system with fluorescence detection (λ^{Ex}, 330 nm; λ^{Em}, 420 nm)
- Centrifugal evaporator (e.g. Savant SpeedVac)

Method

1. Equilibrate the GlycoSep column with 20% solvent C, 80% acetonitrile at 0.4 ml/min at 30°C.
2. Dissolve the glycan sample in a minimum volume of 90% acetonitrile (< 100 μl).
3. Inject the sample onto the column and separate with the following elution gradient: 20–53% solvent C for 132 min, 53–100% solvent C for 3 min, and then increase the flow rate with 100% solvent C to 1 ml/min over 2 min. Continue with 100% solvent C for 5 min at 1 ml/min.

Reduce the flow rate to 0.4 ml/min and re-equilibrate the column with 20% solvent C before injecting the next sample. The total run time is 180 min.

4. Monitor the eluate continuously and collect the glycan peaks. Evaporate them to dryness in the centrifugal evaporator to remove the volatile buffers. Analyse the glycans as appropriate to the study.

[a] From ref. 32.

By using lectin affinity chromatography (Section 3.3), oligosaccharides with similar chromatographic properties can be separated based on specific differences in their sugar sequence. Also serial lectin affinity chromatography can be used to fractionate a mixture of glycans into a number of pools based on their reactivity. With the known specificity of the lectins used (*Table 1*), a lectin binding pattern can be built up which indicates the presence of certain types of oligosaccharide structure or specific sequences in the glycans (33).

5.4 Structural analysis of glycans

Methylation (or linkage) analysis of a glycan (34) will indicate which monosaccharide residues are present, which of these residues are terminal, and give the substitution (linkage) positions of each residue. This technique cannot produce information on the ordering of the linkages or their anomeric configuration. Assumptions based on known sequences can be used to interpret the result of methylation analysis but with large or complicated oligosaccharides additional structural information is necessary for a full sequence assignment.

Methods employing glycosidic enzymes are capable of providing considerably more sequence data than chemical methods. There are now a large number of exoglycosidases that are well characterized and highly specific. Digestion of oligosaccharides with these enzymes gives the identity of the terminal monosaccharide, the linkage to the next residue, and the anomeric configuration of the linkage, for each cycle of the analysis. Branching and other substitutions can also be defined. Sequential enzyme treatments with analysis between each treatment are then used to determine the sequence of the oligosaccharide (35, 36). The method for exoglycosidase digestion is illustrated with jack bean α-mannosidase in *Protocol 17*. Other exoglycosidases may have different pH optima and require different digestion conditions.

Sequential use of several exoglycosidase can provide valuable information concerning the structure of a glycan. *Figure 5* illustrates this type of analysis. However, the number of cycles in sequential enzymatic digestion leads to progressive sample loss, and the need to isolate and characterize the product

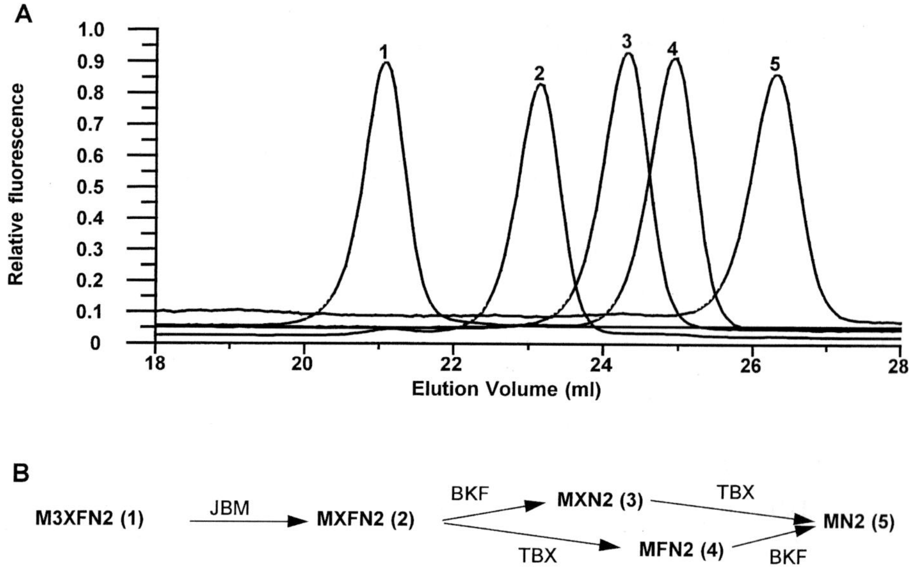

Figure 5. Sequential enzymatic digestion of the major glycan of horseradish peroxidase. (A) The decrease in size on gel filtration of the starting material, 1, after each enzyme treatment. (B) The enzymes and their digestion sequence, and the glycan structures corresponding to the peaks in (A). M3XFN2, Manα3(Manα6)(Xylβ2)Manβ4GlcNAcβ4(Fucα3)GlcNAc; MXFN2, Xylβ2Manβ4GlcNAcβ4(Fucα3)GlcNAc; MXN2, Xylβ2Manβ4GlcNAcβ4GlcNAc; MFN2, Manβ4GlcNAcβ4(Fucα3)GlcNAc; MN2, Manβ4GlcNAcβ4GlcNAc; JBM, jack bean α-mannosidase; BKF, bovine kidney α-fucosidase; TBX, tobacco β-xylosidase.

after each cycle makes this method laborious. A technique has recently been described which uses defined mixtures of enzymes in combination instead of singly and sequentially. The reagent array analysis method (RAAM) (37) involves dividing the oligosaccharide to be analysed into equal aliquots which are then incubated with mixtures of exoglycosidases that differ in their composition. The products of the individual reactions are combined and analysis is performed by a single chromatographic separation. The RAAM technique, therefore, overcomes the limitations of sequential enzyme digestion. The mixtures of exoglycosidases generate fragments of the oligosaccharide and will act until a linkage that is resistant to all the enzymes in the mixture is reached. By omitting different enzymes from each mixture, different fragments are produced in the individual incubations; these are termed stop-points. The separation of the recombined incubation mixtures generates a pattern of fragments that is characteristic of the oligosaccharide. This pattern can be interpreted by computer software that deduces the sequence of the original oligosaccharide.

Protocol 17. Exoglycosidase digestion with jack bean α-mannosidase

Equipment and reagents

- Glycan for analysis (*Protocol 15* or *16*)
- Jack bean α-mannosidase (Oxford GlycoSciences)[a]
- Incubation buffer B: 0.1 M sodium acetate pH 5 containing 2 mM Zn^{2+} (5 × buffer is supplied with the enzyme by Oxford GlycoSciences)
- Dowex AG3-X4A (OH^- form)
- Affi-Gel Blue (Bio-Rad)
- HPLC grade water
- Dowex AG50-X12 (H^+ form), Dowex AG1-X8 (acetate form), plastic mini-columns, syringes (3 ml), and PTFE syringe filters (see *Protocol 15*)
- Incubator or heating block

Method

1. Reconstitute 2 U of the dry enzyme in 80 µl incubation buffer B to give a 25 U/ml solution.
2. Add the enzyme solution to the dried glycan to give approx. 10 µM glycan solution.
3. Incubate the mixture for 18 h at 37 °C. Stop the reaction by heating at 100 °C for 2 min or by freezing at –20 °C.
4. Wash the Dowex resins and Affi-Gel Blue well with water and prepare 50% suspensions. Add 200 µl of each suspension to the mini-column in the following order: Affi-Gel Blue, Dowex AG 3, Dowex AG 50, Dowex AG 1. Allow the column to drain between each addition and layer each carefully to avoid disturbing the bed below. Wash the assembled column with 1 ml water.
5. Remove the plunger from a 3 ml plastic syringe, fit it with a PTFE filter, and suspend it over a collection tube. Suspend the ion exchange column over the syringe.
6. Add the enzyme reaction mixture (step 3) to the column. Rinse out the tube with two aliquots of 100 µl water and apply the rinsings to the column. The effluent will collect in the syringe barrel. Wash the column with 0.5 ml water. Repeat the wash three times, collecting all the effluent. Remove the mini-column. Filter the glycan solution as described in *Protocol 15*.
7. Evaporate the glycan solution to dryness. Analyse the enzyme digest by your method of choice (e.g. *Protocol 15* or *16*) or redissolve the residue in 100 µl water and store at –20 °C for later analysis.

[a] Sigma and Boehringer Mannheim also supply jack bean α-mannosidase but as an ammonium sulfate suspension. Dialyse this suspension against incubation buffer B in a similar way to that described for the Boehringer PNGase A in *Protocol 13*, step 1. Dilute the enzyme to 25 U/ml and follow the protocol from step 2 onwards.

6. Manipulation of protein glycosylation

It may be useful to alter the structures of *N*-glycans present on a recombinant protein for the following reasons:

- To investigate the role of the *N*-glycans in the folding, stability, or functions of a glycoprotein.
- Because you need to remove *N*-glycans, for instance as a starting point for crystallography trials.
- To generate a uniform population of glycoforms.
- To target the protein to particular carbohydrate receptors, for instance, drug targeting *in vivo*.

Whatever the reasons for modifying the carbohydrate, there are several methods which can be applied, some of which are listed below.

(a) Expression of the glycoprotein in a prokaryotic expression system will result in a non-glycosylated protein. However, this can result in misfolding and aggregation and so needs to be evaluated for each protein of interest.

(b) Expression of the glycoprotein in a eukaryotic expression system in the presence of tunicamycin results in a non-glycosylated protein (see Section 3.1 and *Protocol 3*). The tunicamycin concentration requires careful titration to find an inhibitory dose with minimal cytotoxicity. This must be determined for each cell system used (evaluate concentrations in the range of 0.1–2 μg/ml).

(c) Site-directed mutagenesis in a eukaryotic expression system, where the asparagine residue in the glycosylation sequon is mutated to another amino acid. Again no glycosylation will occur. This is a powerful technique for exploring the role of individual glycosylation sites.

(d) Expression of a recombinant protein in a mutant cell line (38). These are frequently processing enzyme mutants and so the protein will be glycosylated but processing of the glycans will be arrested at a certain stage in the pathway (39). The precise structures resulting depend on the nature of the mutant.

(e) Exoglycosidase digestion. This allows the glycans to be modified by subjecting the oligosaccharide to exoglycosidase digestion to trim back the glycan to the desired structure. This is carried out after protein expression and purification. A good knowledge of the structures present is essential if this is to be carried out with minimal wastage of enzymes and complete glycan sequence information is, therefore, desirable. An example of the application of this technique to drug targeting is the trimming of the *N*-glycans on glucocerebrosidase to target this enzyme to macrophages for the treatment of type 1 Gaucher's disease (40).

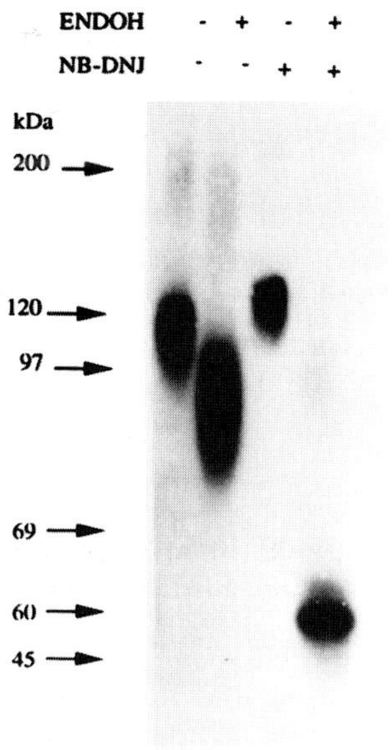

Figure 6. Western blot of affinity purified recombinant HIV gp120 generated either in the presence or absence of the *N*-butyl derivative of DNJ (NB-DNJ), an inhibitor of processing α-glucosidases I and II. The gp120 samples were treated with or without Endo H. The 'wild-type' molecule has the predicted molecular weight of 120 kDa and is partially sensitive to Endo H as approximately 50% of its *N*-glycans are oligomannose structures. The gp120 generated in the presence of 2 mM NB-DNJ migrates as a slightly larger species, relative to the untreated gp120, due to the retention of the glucosylated oligomannose type precursor *N*-glycans. When this material is subjected to Endo H digestion, all of the *N*-glycans are released leaving the polypeptide backbone of 60 kDa. (Photograph kindly provided by Khairina Tajul-Arifin.)

(f) Glycosidase inhibitors. Addition of chemical inhibitors of enzymes in the *N*-glycan processing pathway to the culture medium during protein expression in a eukaryotic expression system results in the inhibition of steps in *N*-glycan biosynthesis. This generally involves the use of glucosidase or mannosidase inhibitors.

The use of the α-glucosidase inhibitors deoxynojirimycin (DNJ) or castanospermine, and trimming mannosidase inhibitors such as deoxymannojirimycin (DMJ), are described in *Protocol 18*. DNJ and castanospermine both result in the *N*-glycans remaining in the glucosylated oligomannose precursor form whereas DMJ results in non-glucosylated oligomannose structures. The glycan in both cases will be fully sensitive to Endo H digestion (*Protocol 7*) and, therefore, the glycoprotein can be readily deglycosylated. An example of this effect is shown in *Figure 6*.

Protocol 18. Inhibition of *N*-glycan processing in eukaryotic cells

Equipment and reagents

- SDS–PAGE gel, electrophoresis apparatus, and reagents
- Castanospermine, DNJ, or DMJ (Oxford GlycoSciences or Toronto Research Biochemicals): prepare a 200 mM stock solution in water, sterilize the solution by passage through a 0.2 μm filter, and store at −20 °C[a]
- Eukaryotic cells expressing the glycoprotein of interest and facilities for their growth in tissue culture
- Sterile filters (0.2 μm)
- Haemocytometer
- PBS and lysing buffer (*Protocol 3*)
- Materials for Endo H digestion (*Protocol 7*)

Method

1. Harvest the cells expressing the glycoprotein of interest from the logarithmic phase of growth. Count them using a haemocytometer.
2. Culture the cells in fresh tissue culture medium at a density of 5×10^4 cells/ml.
3. Add the appropriate volume of the sterile inhibitor solution to the cells in culture. The inhibitory range for these compounds is generally 0.5–2 mM.[b]
4. Incubate the cells in the presence of the inhibitor for three days without medium change.[c]
5. Depending on whether the glycoprotein of interest is secreted into the medium or retained within the cells, process the medium or cell lysate as described in *Protocol 3*.
6. Determine the relative sensitivity of the glycoprotein to Endo H using SDS–PAGE analysis (*Protocol 7*).

[a] The inhibitors are stable for at least 12 months under these conditions.
[b] It is advisable to check viability and growth rates in the presence of the drugs the first time you work with them and each time you use a new cell system. Set up several concentrations of the inhibitors with the cells to evaluate the optimal inhibitor concentration and to assess cell viability. These compounds are generally well tolerated by eukaryotic cells but there are exceptions (for instance some neuronal cell lines).
[c] The inhibitors are very stable and are not metabolized by eukaryotic cells. There is no need to add additional compound during the culture period.

References

1. Rademacher, T. W., Parekh, R. B., and Dwek, R. A. (1988). *Annu. Rev. Biochem.*, **57**, 785.
2. Parekh, R. B. (1991). *Curr. Opin. Struct. Biol.*, **1**, 750.
3. Goochee, C. F., Gramer, M. J., Andersen, D. C., Bahr, J. B., and Rasmussen, J. R. (1992). In *Frontiers in bioprocessing II* (ed. P. Todd, S. K Sikdar, and M. Bier), p. 199. American Chemical Society, Washington.

4. Sharon, N. and Lis, H. (1993). *Eur. J. Biochem.*, **218**, 1.
5. Varki, A. (1993). *Glycobiology*, **3**, 97.
6. Fukuda, M. and Kobata, A. (ed.) (1993). *Glycobiology: a practical approach.* IRL Press, Oxford.
7. Dwek, R. A., Edge, C. J., Harvey, D. J., Wormald, M. A., and Parekh, R. B. (1993). *Annu. Rev. Biochem.*, **62**, 65.
8. Lennarz, W. J. and Hart, G. W. (ed.) (1994). *Methods in enzymology*, Vol. 230. Academic Press, London.
9. Hounsell, E. F. (ed.) (1993). *Glycoprotein analysis in biomedicine. Methods in molecular biology*, Vol. 14. Humana Press, Totowa.
10. Fukuda, M. and Hindsgaul, O. (ed.) (1994). *Molecular glycobiology.* IRL Press, Oxford.
11. Goochee, C. F. and Monica, T. (1990). *Bio/Technology*, **8**, 421.
12. Mackiewicz, A. and Mackiewicz, K. (1995). *Glycoconjugate J.*, **12**, 241.
13. Kornfeld, R. and Kornfeld, S. (1985). *Annu. Rev. Biochem.*, **54**, 631.
14. Elbein, A. D. (1987). *Annu. Rev. Biochem.*, **56**, 497.
15. Winchester, B. and Fleet, G. W. J. (1992). *Glycobiology*, **2**, 199.
16. Schachter, H. and Brockhausen, I. (1992). In *Glycoconjugates: composition, structure and function* (ed. H. J. Allen and E. C. Kisailus), p. 263. Marcel Dekker, New York.
17. Lehle, L. and Tanner, W. (1974). *Biochim. Biophys. Acta*, **350**, 225.
18. Dubois, M., Gilles, K. A., Hamilton, J. K., Rebers, P. A., and Smith, F. (1956) *Anal. Chem.*, **28**, 350.
19. Faye, L. and Chrispeels, M. J. (1985). *Anal. Biochem.*, **149**, 218.
20. Merkle, R. K. and Poppe, I. (1994). In *Methods in enzymology* (ed. W. J. Lennarz and G. W. Hart), Vol. 230, p. 1. Academic Press, London.
21. Rudd, P. M., Scragg, I., Coghill, E., and Dwek, R. A. (1992). *Glycoconjugate J.*, **9**, 86.
22. Konigsberg, W. (1972). In *Methods in enzymology* (ed. C. H. W. Hirs and S. N. Timasheff), Vol. 25, p. 185. Academic Press, London.
23. Costa, J., Ashford, D. A., and Pinto Ricardo, C. P. (1996). *Planta*, **198**, 221.
24. Iyer, R. N. and Carlsson, D. M. (1971). *Arch. Biochem. Biophys.*, **142**, 101.
25. Patel, T., Bruce, J., Merry, A. H., Bigge, J. C., Wormald, M. R., and Parekh, R. B. (1992). *Biochemistry*, **32**, 679.
26. Patel, T. P. and Parekh, R. B. (1994). In *Methods in enzymology* (ed. W. J. Lennarz and G. W. Hart), Vol. 230, p. 57. Academic Press, London.
27. Bigge, J. C., Patel, T. P., Bruce, J. A., Goulding, P. N., Charles, S. M., and Parekh, R. B. (1995). *Anal. Biochem.*, **230**, 229.
28. Jackson, P. (1994). In *Methods in enzymology* (ed. W. J. Lennarz and G. W. Hart), Vol. 230, p. 280. Academic Press, London.
29. Yamashita, K., Mizuochi, T., and Kobata, A. (1982). In *Methods in enzymology* (ed. V. Ginsberg), Vol. 83, p. 105. Academic Press, London.
30. Kobata, A. (1994). In *Methods in enzymology* (ed. W. J. Lennarz and G. W. Hart), Vol. 230, p. 200. Academic Press, London.
31. Dell, A., Reason, A. J., Khoo, K.-H., Panico, M., McDowell, R. A., and Morris, H. R. (1994). In *Methods in enzymology* (ed. W. J. Lennarz and G. W. Hart), Vol. 230, p. 108. Academic Press, London.

32. Guile, G. R., Rudd, P. M., Wing, D. R., Prime, S. B., and Dwek, R. A. (1996). *Anal. Biochem.*, **240**, 210.
33. Mir-Shekari, S. Y., Ashford, D. A., Harvey, D. J., Dwek, R. A., and Schulze, I. T. (1997). *J. Biol. Chem.*, **272**, 4027.
34. Geyer, R. and Geyer, H. (1994). In *Methods in enzymology* (ed. W. J. Lennarz and G. W. Hart), Vol. 230, p. 86. Academic Press, London.
35. Jacob, G. S. and Scudder, P. (1994). In *Methods in enzymology* (ed. W. J. Lennarz and G. W. Hart), Vol. 230, p. 280. Academic Press, London.
36. Prime, S., Dearnley, J., Ventom, A. M., Parekh, R. B., and Edge, C. J. (1996). *J. Chromatogr. A*, **720**, 263.
37. Edge, C. J., Rademacher, T. W., Wormald, M. R., Parekh, R. B., Butters, T. D., Wing, D. R., *et al.* (1992). *Proc. Natl. Acad. Sci. USA*, **89**, 6338.
38. Stanley, P. (1989). *Mol. Cell. Biol.*, **9**, 377.
39. Davis, S. J., Puklavec, M. J., Ashford, D. A., Harlos, K., Jones, E. Y., Stuart, D. I., *et al.* (1993). *Protein Eng.*, **6**, 229.
40. Sato, Y. and Beutler, E. (1993). *J. Clin. Invest.*, **91**, 1909.
41. Colyer, J. (1999). In *Protein expression: a practical approach* (ed. S. J. Higgins and B. D. Hames), p. 226. Oxford University Press, Oxford.

5

Lipid modification of proteins

NIGEL M. HOOPER and R. A. JEFFREY McILHINNEY

1. Introduction

Numerous proteins are now known to be covalently modified with one or more lipid species. Indeed, it has been estimated that between 10–50% of all proteins within mammalian cells may be modified by lipid. Such modifications not only alter the structure of the protein but are also clearly important for the correct functioning of the individual proteins. Cytosolic proteins can be modified with *myristic acid* (myristoylation), on their N-terminus or by the addition of a prenyl group (farnesyl or geranylgeranyl) (*prenylation*), on Cys residues which often lie towards the C-terminus. In addition, both cytosolic and membrane spanning proteins can be modified by the attachment of *palmitic acid* (palmitoylation) to Cys residues. It is now becoming clear that some proteins are acylated with two different lipid modifications, e.g. by myristoylation and palmitoylation or by prenylation and palmitoylation. Although integral proteins of the plasma membrane have traditionally been considered as being anchored by membrane spanning hydrophobic polypeptides, over 150 such proteins are now known to be anchored in the membrane by a *glycosylphosphatidylinositol* (GPI) moiety. This complex glycan structure is attached to the C-teminal amino acid of the mature protein and serves to tether the protein in the outer leaflet of the bilayer.

Although analysis of cDNA-derived amino acid sequences can provide a clue as to whether a protein is likely to be modified with lipid, ultimately direct biochemical analysis is required to confirm this. In this chapter we provide information on how to ascertain whether a protein is modified with lipid.

2. Protein acylation

2.1 General considerations

The modification of cell proteins by the covalent attachment of saturated fatty acids, termed protein acylation, is now recognized to be a widespread phenomenon (1–5). Whilst there is evidence that some proteins may have stearic acid attached to them (6) the predominant fatty acids found covalently attached to cell proteins are myristic and palmitic acid, respectively C14 and C16 saturated fatty acids. Despite these two fatty acids differing by only two carbons in chain length there is clear evidence that they are processed via different metabolic pathways and by different enzymes (1–5). Attachment of myristic acid occurs co-translationally as evidenced by its sensitivity to protein synthesis inhibitors such as cylcoheximide. Once attached, it remains bound to the protein for a long time. This contrasts with the post-translational attachment of palmitic acid, which can show rapid turnover following attachment to the target protein (7–9).

2.2 Enzymology

Myristoyl-CoA: protein *N*-myristoyltransferase (NMT; EC 2.3.1.97) is the enzyme responsible for the attachment of myristic acid to proteins. It catalyses the transfer of myristic acid from myristoyl-CoA to the N-terminal amino acid of the target protein, which must be a glycine residue. This results in the fatty acid being amide bonded to the α-amino group of this amino acid. Whilst an N-terminal glycine is obligatory, the other sequence requirements for protein *N*-myristoylation are less certain resulting in a vague consensus sequence for protein *N*-myristoylation (*Table 1*). NMT can make use of a limited range of other fatty acids, including shorter chain and unsaturated fatty acids, to modify proteins, but in general this only occurs when these fatty acids form the majority of the acyl-CoA pool in a cell or tissue, as in the retina (10). For all practical purposes, therefore, NMT can be regarded as essentially specific for myristic acid. Yeast NMT is the best characterized myristoyltransferase and is a single chain polypeptide with an apparent molecular weight of about 50 kDa on SDS–polyacrylamide gels (11). The human enzyme has also been cloned and an active form of this identified, which has a similar molecular weight (12).

In contrast to NMT, the enzyme responsible for the attachment of palmitate to proteins is much less well characterized. Activities capable of performing the palmitoylation of viral proteins have been demonstrated in cell microsomes (13) and cell plasma membranes (14), but the enzyme itself has been difficult to purify. Equally unknown are the sequence determinants necessary for the attachment of palmitate. Most palmitoylated cell and viral proteins are acylated near their C-termini at cysteine residues within 20–30 residues of the inner face of the plasma membrane. However, it has recently become apparent that for some of the src-related protein kinases, the heterotrimeric G protein α-subunits, and GAP-43, palmitoylation can take place on cysteine residues close to the N-terminus (15, 16). Indeed some of these proteins are subject to dual acylation, having an N-terminal myristate as well as the palmitate. A possible motif for N-terminal palmitoylation could be myristoyl-GCXX(C)X but the precise sequence requirements for palmitoylation of this group of proteins remains uncertain.

Table 1. General consensus sequences for N-terminal myristoylation by mammalian myristoyl-CoA:protein *N*-myristoyltransferase

Consensus sequences recognized by mammalian NMTs conform to the following general sequence:

$$NH_2\text{-}\mathbf{G}_1X_2X_3X_4\mathbf{S}_5X_6X_7X_8\text{-}$$

where the permitted residues, based on the sequences of known myristoyl-proteins as described by Rudnick *et al.* (3), are:

Amino acid position	Permitted residue (single letter nomenclature)
1	G is obligatory
2	C, A, L, N, Q, V, S, G
3	T, Y, F, Q, V, A, L
4	T, Y, F, Q, V, A, L, P
5	S preferred
6	C, N, Q, V, S, G, T, Y, F, A, L, R, K
7	K, S, L, E, V, G, P, N, F, D, Q, T
8	K, S, L, E, V, G, P, N, F, D, Q, T

The glycine at position 1 is obligatory. Serine is often found at position 5, and increases the affinity of yeast NMT for substrate peptides, but other amino acids occur here in mammalian myristoyl-proteins. Tryptophan and tyrosine have not been found at positions 7 or 8 nor has proline been found at position 6. Generally, acidic residues are not found at position 4.

3. Identification of acylated proteins

3.1 General points

Given the rather vague nature of the consensus sequences for both palmitoylation and myristoylation, it is difficult from inspection of a particular protein sequence to predict whether that protein is likely to be acylated. Therefore, it is necessary to show directly that a protein of interest is acylated. If the protein is available in reasonable quantities (e.g. 100 pmol) then several methods exist to show that it is myristoylated. However, these methods nearly all depend on mass spectrometric analyses and the availability of the appropriate instrumentation (17–20).

The alternative strategy for identifying acylated proteins has been to label cells containing the protein of interest with radioactive fatty acids and to immunoprecipitate the protein(s). The immunoprecipitate is then analysed by SDS–PAGE followed by fluorography to determine if the protein has incorporated the labelled fatty acid. The advent of molecular biology means that it is now more common to transfect cell lines with the cDNA coding for the protein of interest and then follow the same basic strategy. This has the additional advantage that the expression of the introduced sequence is high in such systems and the labelling of the protein with fatty acids is correspondingly easier to detect. Because of the possibility of conversion of the fatty acids to other metabolites, it is necessary to show that the incorporated label is indeed a fatty acid, and this can be done by identifying the labelled material released following acid hydrolysis of the protein. Alternatively, the presence of myristoyl-glycine in the protein can be determined following exhaustive proteolytic digestion of the protein. Methods for these analyses are given in the following sections.

More recently it has become possible to exploit the co-expression of NMT and the protein of interest in *Escherichia coli* (which does not contain NMT) to determine if a protein is myristoylated (21). Since few proteins are labelled following incubation of *E. coli* with radioactive myristic acid, even in bacteria expressing NMT, the heterologous labelled myristoyl-protein is clearly visible following fluorography of an SDS–polyacrylamide gel of the *E. coli* lysate. However, this approach depends on the availability of a suitably cloned form of the target protein. Also, because it makes use of a system in which there is a very high level of expression of both the enzyme and the substrate protein, there is a risk that a protein, normally a poor substrate for NMT, may be anomalously myristoylated under these conditions. This methodology is clearly described in another volume in the Practical Approach series (22).

Alternatively, the target protein can be myristoylated *in vitro* using purified NMT derived from *E. coli*, and radiolabelled myristoyl-CoA (23). Whilst the stoichiometry of the resulting myristoylation can be low and varies with different proteins (*Table 2*), the sensitivity of the method is high. The concentrations of the NMT and target protein can be altered to reduce the likelihood of artefactual myristoylation. The method depends on a source of the purified or partially purified target protein which in many cases can be obtained by expression in bacteria. Since the resulting myristoylated protein is radioactively tagged, this method also allows the consequences of myristoylation on the properties of the protein to be examined.

Table 2. Stoichiometry of protein myristoylation *in vitro* using NMT expressed in *E. coli*

Target protein	**Myristoylation (pmol myristic acid incorporated/pmol protein)**[a]
cAMP-dependent protein kinase (catalytic subunit)	0.01–0.03
Creatine kinase II	0.36–0.7
MARCKS[b]	0.5–0.7
MARCKS-related protein	0.5–0.7

[a] The ranges given cover the values obtained in at least three experiments. All of the proteins were purified from *E. coli* transfected with the corresponding cDNA apart from creatine kinase II, which was purified from sea urchin sperm (59).
[b] MARCKS, myristoylated alanine-rich C-kinase substrate (60).

For palmitoylated proteins, there is no similar *in vitro* or reconstituted enzyme system. However, transfection of mammalian or insect cell lines with suitable expression vectors containing sequences encoding the target proteins has improved the ease of demonstration of protein palmitoylation due to the high levels of protein produced in these systems. Nowhere has this been more clearly demonstrated than in the use of baculovirus systems for studying the palmitoylation of G protein coupled receptors and G proteins themselves (9, 24).

3.2 Labelling cultured cells with fatty acids

Protocol 1 is suitable for any cell line which expresses the protein under study. It also requires little adaptation for use with insect cells used for baculovirus expression of proteins. The culture medium should be that which is optimal for the growth of the cells under investigation, and should be supplemented with 5 mM sodium pyruvate to minimize catabolism of the added fatty acids. We have found that the cocktail of protease inhibitors used in *Protocol 1* is generally satisfactory to prevent proteolytic degradation of proteins. However, other combinations of protease inhibitors may be required for some proteins. On no account should the fatty acid be dried down in a polypropylene tube or container; fatty acids have a high affinity for plastic and much of the radioactivity will remain on the walls of the tube. This should also be borne in mind when subsequently analysing the proteins.

The short incubation time used in these experiments is essential to minimize the metabolism of the radioactive fatty acid. The degree to which this happens depends on the cell line, but chain lengthening of myristate (C14) to both palmitate (C16) and stearate (C18) should be expected, whereas palmitate will tend to give rise to mainly stearate. Because of this, it is necessary to show that the residue incorporated is indeed the fatty acid used in the labelling of the cells. The amount of radioactive fatty acid used can be increased from that given in *Protocol 1* by four- or fivefold. This will increase the extent of protein labelling, but this should be balanced against the increased expense of the label and the attendant radioactive waste disposal problems.

Protocol 1. Labelling of cell lines with fatty acids

Equipment and reagents

- [9,10-^{3}H]fatty acids (at least 30 Ci/mmol)
- Glass test-tube and supply of nitrogen gas for drying down the radioactive fatty acid
- Ethanol
- Serum-free medium for cell culture (with 5 mM sodium pyruvate)
- 25 cm^2 flask of appropriate cells grown to near confluence
- Tris-saline: 50 mM Tris–HCl pH 7.4, 0.1 M NaCl, 0.5% NaN_3
- Lysis buffer: Tris-saline containing 1% (v/v) Triton X-100
- 1,10-phenanthroline (1,10-P; Sigma): 2 mg/ml in DMSO[a]
- Dichloroisocoumarin (DCI; Sigma): 2.2 mg/ml in DMSO[a]
- E-64 (Boehringer Mannheim): 1.8 mg/ml[a]

Protocol 1. *Continued*

Method

1. Dry down the radioactive fatty acid in a glass test-tube under nitrogen in a fume-cupboard. Redissolve it in ethanol to give a final concentration of 10 mCi/ml.
2. Dilute the radioactive fatty acid to 100 μCi/ml in warm serum-free medium.
3. Remove the medium from the cells and wash them with 5 ml warm serum-free medium. Remove the medium.
4. Add the medium containing the radioactive fatty acid to the cells. For a 25 cm^2 flask, 1–2 ml medium is adequate. Adjust this volume appropriately for culture vessels of different sizes.
5. Incubate the cells at 37 °C for 4 h.
6. Discard the culture medium safely and wash the cells twice with 5 ml Tris-saline.
7. Lyse the cells by gently shaking the flask at 4 °C for 5 min in 1 ml ice-cold lysis buffer containing 10 μl of each of the stock 1,10-P, DCI, and E-64 solutions.
8. Pipette the medium gently to release any adherent cells.
9. Centrifuge the lysate at 3000 *g* for 10 min at 4 °C. Remove and retain the supernatant for analysis of the acylated protein.

[a] These stock solutions are used at 1/100 dilution to inhibit proteases. Store them at –20 °C.

3.3 Analysis of acylated proteins

Acylated proteins in cell lysates (*Protocol 1*) are first immunoprecipitated with a specific antibody followed by SDS–PAGE of the immunoprecipitate. Fluorography of the gel is used to locate the labelled protein for extraction of the fatty acid residue for subsequent analysis.

3.3.1 SDS–PAGE and fluorography

The lysis medium (*Protocol 1*) is generally applicable to most cells and proteins, but may have to be modified to achieve the optimal immunoprecipitation conditions. This must be established in pilot experiments. For SDS–PAGE most laboratories use the discontinuous gel buffer system described by Laemmli (25). However, sample preparation prior to electrophoresis is critical, especially for palmitoylated proteins. This is due to the susceptibility of the thioester linkage to reducing agents and the tendency of such proteins to aggregate when heated (7). For acylated proteins, dithiothreitol at a final concentration of 10 mM should be used rather than 2-mercaptoethanol in the sample buffer and the sample heated only at 70 °C for 2 min.

Following electrophoresis, the gel will require impregnation with fluor in order to detect the radioactive protein. Two methods exist for this: the salicylate method (*Protocol 2*) and the PPO method (*Protocol 3*). Both methods need the film to be pre-flashed for maximum sensitivity and for quantification of the results (26). The PPO impregnation method (*Protocol 3*) gives a greater sensitivity of detection of the incorporated fatty acid compared to salicylate impregnation (27). However, the latter is cheaper and has the additional advantage that it does allow the fatty acid attached to the protein to be identified. In fact the salicylate method must be followed if the identified protein is to have the attached fatty acid analysed (see *Protocols 4* and *6*).

Protocol 2. Salicylate fluorography of gels

Equipment and reagents

- SDS–PAGE gel for analysis
- 1 M sodium salicylate pH 6.8[a]
- Platform shaker
- Vacuum gel drier
- Whatman 3MM paper
- X-ray cassette, Kodak X-AR5 film, and other materials for autoradiography

Method

1. Following electrophoresis, place the gel directly into 100 ml[b] of 1 M salicylate solution and incubate with gentle agitation on a shaker for 1 h.
2. Dry the gel on to Whatman 3MM paper under vacuum at 80 °C.
3. Place the dried gel in an X-ray cassette with the gel against Kodak X-AR5 film and leave at –70 °C to expose.
4. Develop the X-ray film according to the manufacturer's instructions.

[a] Adjust the pH with HCl.
[b] The volumes are for a 20 cm^2, 1.5 mm thick gel and should be adjusted for smaller or larger gels.

Protocol 3. PPO fluorography of gels

Equipment and reagents

- SDS–PAGE gel for analysis
- DMSO
- 22% (w/v) 2,5-diphenyloxazole (PPO) in DMSO
- Equipment and autoradiography materials listed in *Protocol 2*

Method

1. Place the gel in 100 ml DMSO and shake it for 30 min.

Protocol 3. *Continued*

2. Remove the DMSO, replace with fresh DMSO, and continue the incubation for a further 30 min.
3. Place the gel in 100 ml[a] of 22% (w/v) PPO in DMSO and shake it for 1 h.
4. Remove the PPO solution and wash the gel twice with distilled water.
5. Shake the gel in 200 ml distilled water for 1 h.
6. Dry the gel and autoradiograph it as described in *Protocol 2*, steps 2–4.

[a] The volumes are for a 20 cm^2, 1.5 mm thick gel and should be adjusted for smaller or larger gels.

In addition to salicylate and PPO, other commercial fluors are available, including EN^3HANCE, Enlightening, and Entisfy (Dupont-NEN), which can be used to detect labelled bands on SDS–PAGE. These have the advantages of not needing as many solvent changes as the PPO method.

3.3.2 Identification of the fatty acid linkage

The nature of the fatty acid linkage to a protein can be determined after electrophoresis by first running duplicate samples of the specific immunoprecipitate in adjacent gel tracks. Following electrophoresis the tracks are separated. One is incubated in 1 M Tris–HCl pH 7.5 for 1 h. The other track is incubated in 1 M hydroxylamine pH 7.5 for 1 h. Each is then washed with distilled water before being prepared for fluorography as described in either *Protocol 2* or *Protocol 3*. If the fatty acid is linked via a thioester (normally palmitic acid) then the hydroxylamine treated gel track will show a loss of label compared to the Tris–HCl treated track. Amide-linked myristic acid is not released following hydroxylamine treatment.

3.3.3 Identification of the fatty acid residue

To identify the fatty acid attached to the protein, it is necessary to be able to align the autoradiograph (*Protocol 2*) with the dried gel and then excise the labelled protein band. This is most easily achieved by applying an ink spot containing 0.5 μCi/ml of a ^{14}C-labelled compound to the backing paper of the gel. Alternatively, commercially available autoradiography markers may be used. The fatty acid is then extracted from the excised gel slice as described in *Protocol 4*. The released fatty acids can be identified using thin-layer chromatography (TLC) usually on C18 derivatized TLC plates (*Protocol 5*). If the necessary apparatus is available, fatty acids can also be identified using high pressure liquid chromatography (HPLC) on C18 reverse-phase columns. This is faster than TLC and more sensitive. Isocratic elution of suitable columns is sufficient to separate the main fatty acids and this can be done fairly economically using the Apex Octadecyl 5 μm, 25 cm × 4.6 mm column supplied by Jones Chromatography as described in a previous volume of this series (28).

Protocol 4. Release of fatty acids from gel slices

Equipment and reagents

- SDS–PAGE gel for analysis (from *Protocol 2*)
- Glass vials with tightly-fitting lids which can be sealed (Reacti-Vials; Pierce Chemical Co.)
- 6 M HCl
- 10 M NaOH
- Chloroform:water (1:1, v/v)
- Chloroform
- Standard fatty acid solution: myristic acid, stearic acid, and palmitic acid each at 1 mg/ml in ethanol
- Vacuum dryer and supply of nitrogen gas

Method

1. Use the X-ray film (*Protocol 2*) to locate the labelled band on the dried gel. Cut it out of the gel using a clean razor blade.
2. Remove the paper backing and wash the gel slice several times for 10 min at room temperature with distilled water to remove the fluor.
3. When all the fluor is removed and the gel is clear, cut the slice into 2–3 mm^2 pieces and dry them in a glass vial under vacuum.
4. Add 0.7 ml 6 M HCl to the gel pieces, seal the vial, and incubate at 100 °C for 4 h.
5. Allow the hydrolysate to cool and add 0.3 ml 10 M NaOH.
6. Add 3.7 ml chloroform:methanol. Vortex several times over a period of 5 min, and then centrifuge the vial for 10 min at 800–1000 *g* to separate the phases.
7. Using a Pasteur pipette remove the lower layer into a fresh glass vial.
8. Add 50 μl of the standard fatty acid solution to the upper (aqueous) phase and re-extract it with a further 2 ml chloroform.
9. Pool the two chloroform phases and dry under nitrogen.
10. Redissolve the residue in a small volume (100–200 μl) chloroform. Store this material at –20 °C under nitrogen until analysis.

Protocol 5. TLC of fatty acids

Equipment and reagents

- Extracted fatty acid for analysis (*Protocol 4*)
- C18 derivatized TLC plates, 20 cm^2 (KC18 from Whatman or RP18 from Merck)
- TLC tank with lid
- [9,10-^{3}H]myristate, palmitate, and stearate (each at least 30 Ci/mmol)
- Acetonitrile:acetic acid (90:10, v/v)
- Suitable fluor (e.g. EN^{3}HANCE; Dupont New England Nuclear) or a 10% (w/v) solution of PPO in diethyl ether
- Materials and equipment for autoradiography (see *Protocol 2*)

Protocol 5. *Continued*

Method

1. Spot 20 μl of the extracted fatty acid (*Protocol 4*) onto a C18 TLC plate and allow to dry.
2. Repeat the application until all the sample is loaded.
3. In adjacent positions, spot the radioactive fatty acid standards (approx. 50 000 d.p.m. of each fatty acid per spot).
4. Develop the plate with acetonitrile:acetic acid (90:10, v/v). Allow the solvent to reach the top of the 20 cm plate.
5. Dry the plate in a fume-cupboard for 30 min.
6. Spray the plate with fluor and dry it.
7. Expose the plate to X-ray film as described in *Protocol 2* to identify the fatty acid residues.[a]

[a] An alternative is to divide each track into 0.5 cm sectors, scrape the TLC medium from each sector into a scintillation vial, and determine the radioactivity by scintillation counting.

3.4 Myristoylated proteins

Should the protein of interest be myristoylated then the presence of myristoyl-glycine at the N-terminus can be demonstrated by pronase digestion of the protein in the gel slice (*Protocol 6*), and subsequent identification of myristoyl-glycine by reverse-phase HPLC (*Protocol 7*). The HPLC system described in *Protocol 7* will elute the myristoyl-glycine at ~ 66% acetonitrile and the myristic acid at 80%. The position of the myristoyl-glycine can be determined by running either a non-radioactive standard and monitoring the eluate at 214 nm or a radioactive standard prepared with [^{3}H] or [^{14}C]myristic acid as described in ref. 28.

Protocol 6. Pronase digestion of proteins in gel slices

Reagents

- SDS–PAGE gel for analysis (see *Protocol 2*)
- 50 mM Tris–HCl pH 7.2
- Pronase (Sigma): 2 mg/ml in 50 mM Tris–HCl pH 7.2
- Toluene
- Diethyl ether
- 1 M HCl
- Vacuum drier and nitrogen gas supply

Method

1. Rehydrate the gel slice and remove the fluor as described in *Protocol 4*, steps 1–3, but do not dry them.
2. Add 0.7 ml 50 mM Tris–HCl pH 7.2 to the gel pieces, 200 μl pronase solution, and one drop of toluene (as a bacteriostat). Cap the tube and incubate it at 37 °C for 24 h.

3. Add a further 200 μl pronase and fresh toluene, and incubate for another 24 h.
4. Repeat step 3.
5. Remove the solution from the gel pieces and acidify it with 0.2 ml 1 M HCl.
6. Extract the acidified digest with 3 × 2 ml diethyl ether and pool the extracted material.
7. Dry the pooled extract under N_2 and store the residue at –20°C for analysis as in *Protocol 7*.

Protocol 7. Identification of myristoyl-glycine

Equipment and reagents

- C18 reverse-phase HPLC column (e.g. Apex Octadecyl 5 μm, 25 cm × 4.6 mm, Jones Chromatography)
- 0.1% trifluoroacetic acid in distilled water
- HPLC system capable of delivering a gradient of acetonitrile at 1 ml/min
- Extract for analysis (*Protocol 6*)
- Diethyl ether
- Acetonitrile:0.1% trifluoroacetic acid (90:10, v/v)
- Fraction collector
- Scintillation counter and materials for determination of radioactivity

Method

1. Equilibrate the HPLC column with 0.1% trifluoroacetic acid.
2. Redissolve the residue from the diethyl ether extract (*Protocol 6*) in a small volume (100 μl) of diethyl ether.
3. Keep a small aliquot (approx. 10%) of the sample for scintillation counting in order to calculate the yield from the column. Load the rest of the sample onto the column.
4. Elute the column with a gradient of 0.1% trifluoroacetic acid rising to 90% acetonitrile:0.1% trifluoroacetic acid over 30 min at a flow rate of 1 ml/min. Collect 1 min fractions.
5. Determine the position of the eluted radioactivity by scintillation counting: assay 0.2 ml samples from each fraction.

3.4.1 *In vitro* myristoylation of proteins

To perform *in vitro* myristoylation, radioactive myristoyl-CoA and purified NMT are needed, together with a supply of purified or partially purified target protein. The former is available commercially from Amersham International or can be made by the method described in *Protocol 8* which gives a 65–85% conversion of radioactive myristic acid into a product that is stable for many months. Purified NMT is not commercially available but small amounts of the human enzyme, tagged with a polyhistidine sequence to facilitate purification, can be obtained from R. A. J. McIlhinney's laboratory for research purposes. Two other laboratories may be prepared to supply the enzyme since they also have it expressed in *E. coli*. These are the laboratories of Prof. J. I. Gordon (Dept. Molecular Biology and Pharmacology, Washington University School of Medicine, Washington University Medical Centre, 660 South Euclid Avenue, St. Louis, Missouri 63110, USA) (22, 29), and Dr K. Sharma (Dept. Pathology and Saskatoon Cancer Centre, College of Medicine, University of Saskatchewan, Royal University Hospital, Saskatoon, Saskatchewan S7N 4H4, Canada) (30). The protein to be examined for *in vitro* myristoylation should also be expressed in *E. coli* because the bacterium does not possess NMT. Therefore, the N-terminus will be available to be myristoylated. To aid purification, the protein can be tagged but obviously the tag should not be at the N-terminus. The substrate protein does not need to be pure, but does need to be the major protein in the sample. The procedure for *in vitro* myristoylation is described in *Protocol 9*.

Protocol 8. Production of ^{3}H-labelled myristoyl-CoA

Equipment and reagents

- [^{3}H]myristic acid (30–60 Ci/mmol)
- Compressed nitrogen gas
- Freeze-drier
- 20 mM Tris–HCl pH 7.5 containing 0.2 mM EGTA and 10 mM $MgCl_2$
- 10% Triton X-100 in distilled water
- 20 mM lithium CoA
- 50 mM ATP
- 3 U/ml acyl CoA synthetase (Sigma)
- Sep-Pak C18 cartridges (Waters Chromatography Division, Millipore)
- Methanol
- Plastic disposable 5 ml and 10 ml syringes
- 10 mM ammonium acetate pH 6.3
- Acetonitrile
- 10 mM sodium acetate pH 6.3

Method

1. Carefully reduce the volume of 500 μCi [^{3}H]myristic acid to near dryness in a glass tube under nitrogen in a fume-cupboard.
2. Add 100 μl Tris–HCl buffer and 1 μl of 10% Triton X-100. Vortex gently and transfer to a 1.5 ml microcentrifuge tube.
3. Add 20 μl of 20 mM lithium CoA and 20 μl of 50 mM ATP followed by 60 μl of the acyl CoA synthetase. Mix by pipetting and incubate at 30 °C for 2 h.
4. Prepare the Sep-Pak cartridge by passing 5 ml methanol through it with a syringe and then 10 ml of 10 mM ammonium acetate buffer pH 6.3.
5. When the CoA synthetase incubation is completed (step 3), detach the syringe from the Sep-Pak cartridge, pipette the reaction mixture onto the cartridge, and allow it to be drawn in. Re-attach the syringe and wash the cartridge with 5 ml of 10 mM ammonium acetate buffer pH 6.3 using a syringe. Discard these (and all subsequent) washes to radioactive waste.

6. Wash the cartridge with 5 ml of 10 mM ammonium acetate buffer pH 6.3 containing 10% acetonitrile, and then with 5 ml of the same buffer containing 20% acetonitrile.
7. Elute the [^{3}H]myristoyl-CoA with 5 ml of 10 mM ammonium acetate buffer pH 6.3 containing 60% acetonitrile collecting the eluate in a glass test-tube.
8. Wash the Sep-Pak cartridge with 10 ml acetonitrile and discard this wash to radioactive waste.
9. Freeze-dry the eluted [^{3}H]myristoyl-CoA and redissolve it in 1 ml sodium acetate buffer. Adjust the concentration to 200 μCi/ml and freeze it in aliquots. Store these at −20 °C.

Protocol 9. *In vitro* myristoylation

Equipment and reagents

- Purified NMT (100 μg/ml)
- Purified or partially purified target protein (100 μg/ml to 1 mg/ml)
- [^{3}H]myristoyl-CoA (30–60 Ci/mmol, 200 μCi/ml) (*Protocol 8*)
- Incubation buffer: 50 mM Tris–HCl pH 7.5 containing 0.5% Triton X-100)
- Equipment and reagents for SDS–PAGE (see ref. 61)

Method

1. Set up three 1.5 ml microcentrifuge tubes labelled 1–3 and add the reagents as shown below:

Tube number	1	2	3
NMT	–	5 μl	5 μl
protein (100 μg/ml)	10 μl	–	10 μl
[^{3}H]myristoyl-CoA	10 μl	10 μl	10 μl
incubation buffer	30 μl	35 μl	25 μl

2. Incubate the tubes at 30 °C for 1 h.
3. Add 50 μl gel sample buffer, heat at 100 °C for 2 min, and load 50 μl of each sample on to the SDS–polyacrylamide gel.
4. Run the gel and subject it to fluorography as described in *Protocol 2* or *3*.

The fluorograph should take only 24 hours to develop if there is significant myristoylation of the target protein. Tubes 1 and 2 should not give a signal. Tube 3 should have a radioactively labelled band with the M_r of the target protein. Note that prolonged exposure of the fluorograph can reveal a band which is due to the formation of a myristoyl-CoA NMT complex (23), but this will also be apparent in the track for tube 2 (NMT alone). The amounts of reagents given in *Protocol 9* are those used in our laboratories to test potential substrates of NMT. However, it should be noted that the ratio of NMT to protein used is 1:2 by weight and this is a relatively high enzyme to target protein ratio. In addition, the concentration of radioactive myristoyl-CoA in the reaction will be 0.5–1 μM under these conditions and this is well below the K_m of NMT for myristoyl-CoA (7.5 μM). The reason for first trying these conditions is that the specific activity of the radiolabel is high and allows a highly sensitive detection of labelled product. If the protein has been myristoylated then the conditions can be altered and optimized to examine the levels of myristoylation of the protein. It is advisable to confirm that the protein has been N-terminally myristoylated using the methods described in *Protocols 6* and *7*. If stoichiometric myristoylation is desired then the specific activity of the [^{3}H]myristoyl-CoA should be adjusted to 3000 d.p.m./pmol and the final concentration of myristoyl-CoA in the reaction should be 10–20 μM. The incubation time can be lengthened if necessary to obtain stoichiometric myristoylation, and the Triton X-100 concentration reduced or eliminated if it interferes with subsequent functional analysis of the myristoylated protein.

4. Glycosyl-phosphatidylinositol membrane anchorage of proteins

4.1 Structure of GPI anchors

GPI membrane anchors are present in organisms at most stages of eukaryotic evolution, including protozoa, yeast, slime moulds, invertebrates, and vertebrates (31). To date a diverse range of some 150 or more GPI-anchored proteins have been described, including protozoal surface coat proteins, receptors, adhesion molecules, ectoenzymes, and differentiation antigens. Structural determination of the GPI anchors on proteins from protozoa, yeast, and mammals has revealed the common core structure of:

Ethanolamine-PO_4-6Manα1-2Manα1-6Manα1-4GlcNH_2α1-6*myo*-inositol-1-PO_4-lipid.

The lipid moieties range from ceramide in most yeast and slime mould GPI-anchored proteins, to diacylglycerol in protozoa, and (predominantly) 1-alkyl-2-acylglycerol in mammalian proteins. The conserved core structure may possess a variety of side chain modifications that appear to be either protein and/or tissue-specific (32).

4.2 Signals for attachment of a GPI anchor to a protein

To date all GPI-anchored proteins are present either on the outer leaflet of the plasma membrane or on an intracellular membrane with the same disposition (i.e. in the lumen of the endoplasmic reticulum or secretory vesicles). Thus, all GPI-anchored proteins possess two signal sequences in their nascent polypeptide chain. The first is a hydrophobic cleavable N-terminal signal sequence that directs the protein into the lumen of the rough endoplasmic reticulum. The second is another hydrophobic peptide of some 17–30 amino

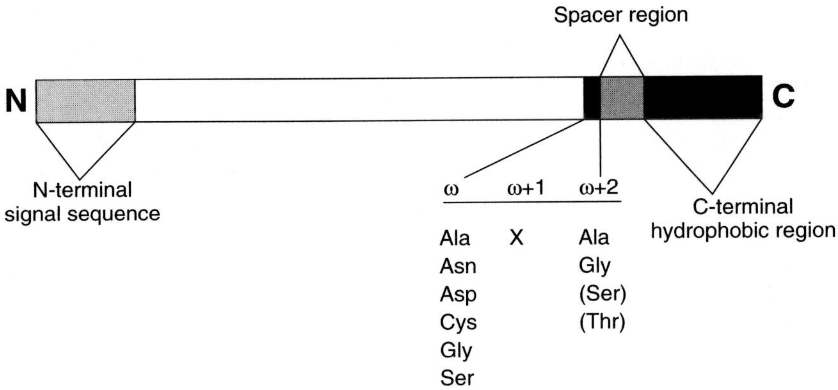

Figure 1. Schematic representation of a typical nascent GPI-anchored protein. The ω residue is the site of cleavage of the polypeptide chain and addition of the preformed GPI anchor by the putative transamidase. X represents any amino acid except Pro or Trp.

acid residues that lies at the very C-terminus of the nascent chain (*Figure 1*). A hydropathy plot of the nascent amino acid sequence of such proteins clearly reveals these two hydrophobic regions at either end of the polypeptide. The C-terminal hydrophobic sequence is preceded by a consensus sequence (ω, ω + 1, and ω + 2) for GPI anchor addition (*Figure 1*). Cleavage of the polypeptide chain occurs on the C-terminal side of the ω residue with concomitant addition of the preformed GPI anchor to the newly exposed COOH group. The preformed GPI anchor is built up by sequential addition of the individual sugars and ethanolamine phosphate to a phosphatidylinositol molecule on the cytoplasmic side of the rough endoplasmic reticulum. The completed precursor is then 'flipped' across the membrane of the rough endoplasmic reticulum prior to addition to the protein (33, 34).

Analysis of native GPI-anchored protein sequences and extensive site-directed mutagenesis of these residues in alkaline phosphatase and decay accelerating factor (35) has shown that the ω residue is restricted to amino acids with small side chains (Ala, Asn, Asp, Cys, Gly, or Ser), whereas ω + 1 can be any residue except for Pro and Trp, and ω + 2 is usually Gly or Ala, or occasionally Ser or Thr. The site of GPI anchor attachment in a protein can now be predicted in a manner analogous to the prediction of the site of signal peptidase cleavage at the N-terminus of a protein, based on the experimental observations in ref. 36 (summarized in *Table 3*). The hierarchical values for the ω, ω + 1, and ω + 2 sites when multiplied together give the probability of a specific amino acid being the ω site in a nascent protein (*Table 3*). A calculated probability of 1.0 would indicate that GPI anchoring is most probably at the ω residue, while a figure < 1.0 would indicate a lower probability of a GPI anchor occurring at that residue.

A transamidase enzyme(s) appears both to cleave the polypeptide chain

Table 3. Experimentally determined hierarchical values of amino acid substituents at ω, ω + 1, and ω + 2 sites of GPI-anchored proteins[a]

Residue	ω	ω+1	ω+2
Ala	0.4	1	1
Arg	n.d.	0.5	n.d.
Asn	0.8	n.d.	n.d.
Asp	0.4	0.4	0.1
Cys	0.2	0.3	0
Gln	0	n.d.	n.d.
Glu	0	0.4	0
Gly	0.4	n.d.	0.7
His	n.d.	n.d.	0
Leu	0.1	n.d.	n.d.
Lys	0	n.d.	n.d.
Met	0	0.3	n.d.
Pro	0	0	0
Ser	1	0.6	0.3
Thr	0	0.3	0.1
Trp	0	0.1	n.d.
Tyr	0	n.d.	n.d.
Val	0.1	n.d.	0.1

[a] Data are taken from ref. 36; n.d., not determined. For the putative GPI anchor attachment site, Ser-Ala-Ala, the probability of GPI anchorage occurring on the ω residue (Ser) would be 1 × 1 × 1 = 1, i.e. most probable, whilst for the sequence Asp-Ser-Gly the probability of GPI anchorage occurring on the Asp would be 0.4 × 0.6 × 0.7 = 0.168, i.e. much less probable.

and add on the preformed GPI anchor. A cell-free assay for this activity is available that involves *in vitro* translation of the mRNA encoding the GPI-anchored protein of interest (37), but as yet it is not useful for determining whether or not a target protein is GPI anchored due to the difficulties in identifying the GPI-anchored product.

5. Identification of GPI anchorage

5.1 General points

The identification of a GPI anchor on a protein does not have to rely upon direct structural determination of the anchor using biophysical techniques such as GC-MS and mass spectroscopy which require relatively large amounts of highly purified protein and access to such equipment (38). Techniques exist whereby the presence of such a structure can be determined even if the protein is present in low abundance or is not readily purified, and these are detailed below. With any of these techniques it is desirable to include an appropriate GPI-anchored protein as a positive control in order to assess the reliability of the method. To this end, at least in mammalian systems, alkaline phosphatase is an easily assayed control protein which appears to be GPI-anchored in all its isozymes under normal circumstances. A simple and rapid spectrophotometric assay for this GPI-anchored enzyme is available that uses *p*-nitrophenyl phosphate as substrate (39) and which, for example, can readily be used to assess the release of GPI-anchored proteins from membranes or cells by exogenous phospholipases (40).

5.2 Release of GPI-anchored proteins by bacterial phosphatidylinositol-specific phospholipase C

The susceptibility of a protein to release from the membrane by bacterial phosphatidylinositol-specific phospholipase C (PI-PLC) is probably the simplest and most commonly used criterion to demonstrate the presence of a GPI anchor on a protein. PI-PLCs from *Bacillus thuringiensis* and *B. cereus* are available, often in recombinant form, from a number of suppliers (Sigma, Fluka, Oxford Glycosystems, Boehringer Mannheim, and Calbiochem). Alternative, relatively inexpensive, sources of PI-PLC are commercial preparations of *B. cereus* phospholipase C (Sigma, Cat. No. P6135; Fluka, Cat. No. 79484) which contain a contaminating PI-PLC activity. If necessary, the PI-PLC activity can be separated from the rest of the phospholipase activity by chromatography on Affi-Gel 501 (Bio-Rad) (39). For PI-PLC release experiments, membranes or cells containing the protein of interest are incubated with bacterial PI-PLC and the loss of the protein from the membrane and/or its appearance in the supernatant is monitored following centrifugation as described in *Protocol 10* (40, 41). Alternatively, following incubation with PI-PLC, the membrane sample can be subjected to temperature-induced phase separation in Triton X-114 (42) to separate the intact, amphipathic protein from the cleaved, hydrophilic form that lacks the fatty acids (*Protocols 11* and *12*). This simple and powerful technique readily resolves the amphipathic and hydrophilic forms of both GPI-anchored and transmembrane polypeptide-anchored integral membrane proteins, with the hydrophilic form being recovered in the upper detergent-poor phase and the amphipathic form remaining in the lower detergent-rich phase (43).

Protocol 10. Digestion of membrane proteins with bacterial PI-PLC

Reagents

- Suspension of cells expressing the protein of interest or membrane preparation from them
- Bacterial PI-PLC (from *B. thuringiensis* or *B. cereus;* Boehringer Mannheim, Calbiochem, Fluka, Oxford Glycosystems, Sigma)
- Digestion buffer: PBS (0.13 M NaCl, 20 mM sodium phosphate pH 7.4), *or* 50 mM Hepes–NaOH pH 7.4[a]

Protocol 10. *Continued*

Method

1. Incubate samples of the membrane preparation or cell suspension containing the protein of interest with a range of concentrations of PI-PLC diluted in digestion buffer (e.g. 0.1–100 mU/ml for *B. thuringiensis* PI-PLC) for 1–2 h at 37 °C.
2. Centrifuge cell suspensions at 1000 *g* to pellet the cells, and membrane preparations at up to 100 000 *g* to pellet the membranes.
3. Remove the supernatant and analyse it for the protein of interest, e.g. by immunoprecipitation and SDS–PAGE (ref. 61), or SDS–PAGE followed by Western blotting, or measurement of enzyme activity, as appropriate, or carry out temperature-induced phase separation in Triton X-114 (*Protocol 12*).

[a] PI-PLCs from *B. thuringiensis* and *B. cereus* are active in most conventional buffers and tissue culture media (for a more comprehensive list of inhibitors/activators see ref. 39).

Protocol 11. Preparation of pre-condensed Triton X-114

Reagents

- Triton X-114 (Sigma)
- Butylated hydroxytoluene
- Tris buffer: 10 mM Tris–HCl pH 7.4, 150 mM NaCl

Method

1. Dissolve 16 mg butylated hydroxytoluene in 20 g Triton X-114.[a]
2. Add the detergent solution to 980 ml Tris buffer in a glass beaker, mix, and incubate at 0 °C for 2–4 h.
3. Incubate the clear solution overnight at 30 °C to allow condensation of the detergent to occur.
4. Remove the larger aqueous phase, depleted in detergent, from above the smaller phase, enriched in detergent.
5. Add an equal volume of fresh Tris buffer to the detergent enriched phase, mix the solution, and repeat the condensation (steps 3 and 4). Repeat.
6. After the third condensation, remove the upper aqueous phase, and determine the concentration of detergent in the final lower phase by comparison of its A_{275} with that of the original commercial detergent.[b]
7. Store the pre-condensed Triton X-114 in a dark glass bottle at room temperature.

[a] The butylated hydroxytoluene prevents oxidation of the detergent.
[b] The concentration of detergent in the final lower phase is usually 10–20% that of the original commercial detergent. To measure their A_{275} the original and pre-condensed detergents have to be diluted 1 in 10 000.

Protocol 12. Temperature-induced phase separation in Triton X-114

Equipment and reagents

- Membrane sample digested with PI-PLC (*Protocol 10*)
- Pre-condensed Triton X-114 (prepared as in *Protocol 11*)
- Tris buffer: 10 mM Tris–HCl pH 7.4, 150 mM NaCl
- Tris buffer containing 2% (v/v) pre-condensed Triton X-114
- Centrifuge with a swing-out rotor

Method

1. Dilute the membrane protein sample with an equal volume of Tris buffer containing 2% (v/v) pre-condensed Triton X-114 and incubate it on ice for 5–10 min.
2. Incubate the sample at 30 °C for 3–10 min.[a]
3. Centrifuge the sample at 3000 *g* for 3 min in a swing-out rotor.
4. Remove the larger clear, detergent-poor (aqueous) phase from above the smaller cloudy, detergent-rich phase.
5. Make the detergent-rich phase up to the same volume as the detergent-poor phase with Tris buffer.
6. Assay each phase for the presence of the protein of interest.

[a] The time of incubation will depend on the sample size, e.g. a 0.2 ml sample will require only 3 min, while a 5 ml sample will require 10 min.

Results obtained using even the most highly purified bacterial PI-PLCs must be interpreted with some caution. As no specific inhibitors are available for PI-PLCs, it is difficult to rule out completely the possibilities that contaminating or endogenous phospholipases or proteases have released the protein of interest. Thus, to demonstrate unequivocally that a protein is GPI anchored, additional criteria should be employed, such as the presence of the cross-reacting determinant (CRD) in the GPI anchor or metabolic labelling of the protein with components of the GPI anchor (Sections 5.4 and 5.5). On the other hand the lack of release of a protein by PI-PLC does not necessarily rule out the presence of a GPI anchor. Several GPI-anchored proteins are resistant to the action of PI-PLC due to additional acylation of the inositol ring (44). However, removal of this acyl group by hydroxylamine treatment (0.8–1 M hydroxylamine hydrochloride, 0.1 M triethylamine pH 10–12 for 2–48 h at 4 °C) then renders the protein susceptible to cleavage by PI-PLC (ref. 45 and see *Protocol 17*).

5.3 Differential detergent solubilization

Most mammalian GPI-anchored proteins are relatively resistant to solubilization by detergents with a low critical micellar concentration, such as Triton X-100 and Nonidet P-40 (46, 47). In contrast, they are readily solubilized by detergents with a high critical micellar concentration, such as *n*-octyl-β-D-glucopyranoside (octyl glucoside) and 3-[(3-cholamidopropyl)dimethylammonio]-1-propane-sulfonate (CHAPS). This detergent insolubility appears to be an intrinsic property of the GPI anchor and is due to the clustering of GPI-anchored proteins in glycosphingolipid/cholesterol-rich membrane microdomains (48, 49). In contrast to GPI-anchored proteins, integral proteins with a hydrophobic membrane spanning polypeptide anchor are effectively solubilized by both classes of detergent (46). Thus the pattern of detergent solubilization, in particular the ratio of protein solubilized by octyl glucoside to that solubilized by Triton X-100, can be used to predict the presence of a GPI anchor.

Arising from this observation, we have modified the technique of differential solubilization and temperature-induced phase separation in Triton X-114 to distinguish between those proteins anchored by a GPI moiety and those anchored by a single membrane spanning polypeptide (*Protocol 13*) (43). In this procedure GPI-anchored proteins, including those in which the inositol is directly acylated and therefore are resistant to PI-PLC digestion, e.g. human erythrocyte acetylcholinesterase, are recovered predominantly in the detergent insoluble pellet, while polypeptide-anchored proteins are recovered in the detergent-rich phase. Following PI-PLC cleavage of the GPI anchor, however, the GPI-anchored proteins are then found in the detergent-poor phase. The first two steps in *Protocol 13* (i.e. the differential solubilization in Triton X-114) provide a simple and rapid technique for effectively separating GPI-anchored from polypeptide-anchored proteins.

Protocol 13. Differential solubilization and phase separation in Triton X-114

Equipment and reagents

- Membrane protein sample for analysis
- Pre-condensed Triton X-114 (as prepared in *Protocol 11*)
- Tris buffer: 10 mM Tris–HCl pH 7.4, 150 mM NaCl
- 10 mM Hepes, pH 7.4 with NaOH
- Microcentrifuge with a fixed angle rotor
- Centrifuge with a swing-out rotor (*Protocol 12*)

Method

1. Dilute the membrane protein sample (to a final concentration of 4 mg protein/ml) with 10 mM Hepes pH 7.4, and add pre-condensed Triton X-114 to a final detergent concentration of 2%. Immediately vortex the samples for 1–2 sec and then leave them on ice for 5 min.

2. Centrifuge the samples in a microcentrifuge with a fixed angle rotor at 8800 *g* for 10 min at 4°C.
3. Remove the supernatant and subject it to temperature-induced phase separation as described in *Protocol 12*, steps 2–5.
4. Rinse (but do not resuspend) the detergent insoluble pellet with 0.5–1 ml of 10 mM Hepes pH 7.4. Centrifuge as in step 2.
5. Resuspend the detergent insoluble pellet in sufficient 10 mM Hepes pH 7.4 to make it up to the same volume as the final detergent-rich and detergent-poor phases.
6. Assay each of the phases and the resuspended detergent insoluble pellet for the protein of interest.

5.4 Detection of the cross-reacting determinant

Polyclonal antisera raised in rabbits against the PI-PLC cleaved, soluble form of a GPI-anchored protein often cross-react with other unrelated GPI-anchored proteins. The site of cross-reactivity is the CRD which is cryptic in the membrane-bound amphipathic form of GPI-anchored proteins and is only exposed after cleavage of the GPI anchor with PI-PLC. The major epitope involved in this recognition is the inositol-1,2-cyclic monophosphate that is generated on PI-PLC cleavage of the GPI anchor (*Figure 2*) (50, 51). Thus, the recognition of a protein by an anti-CRD antiserum following PI-PLC cleavage is virtually unequivocal evidence for the presence of a GPI anchor. Both Western blot analysis (described in *Protocol 14*) and ELISA have been successfully used to identify the CRD on proteins (51, 52). However, appropriate controls should be included to verify the specificity of the recognition by the anti-CRD antiserum. Either treat the protein with mild acid (e.g. 1 M HCl) selectively to decyclize the inositol-1,2-cyclic monophosphate (*Protocol 15*) or treat it with nitrous acid to deaminate the glucosamine residue (*Protocol 16*). Both of these procedures destroy the CRD (*Figure 2*). If the protein of interest is suspected to have an acylated GPI anchor that is resistant to PI-PLC cleavage, then a modified procedure for Western blot analysis of the CRD can be used that deacylates the protein with alkali and subsequently cleaves its GPI anchor with PI-PLC following transfer of the protein to nitrocellulose (*Protocol 17*) (53).

The Western blotting procedure described in *Protocol 14* uses any suitable wet blot apparatus, but semi-dry models can also be used. Nitrocellulose membranes (Schleicher and Schuell) can be used in place of the polyvinylidene difluoride (PVDF) membranes specified in the protocol. BSA or gelatin can replace the Tween 20 in the blocking buffer, but do not use milk powder as there are components in this that react non-specifically with the anti-CRD antisera. Prior to Western blotting, separate the proteins of interest by SDS–PAGE using the Laemmli system (25). It is advisable to include a

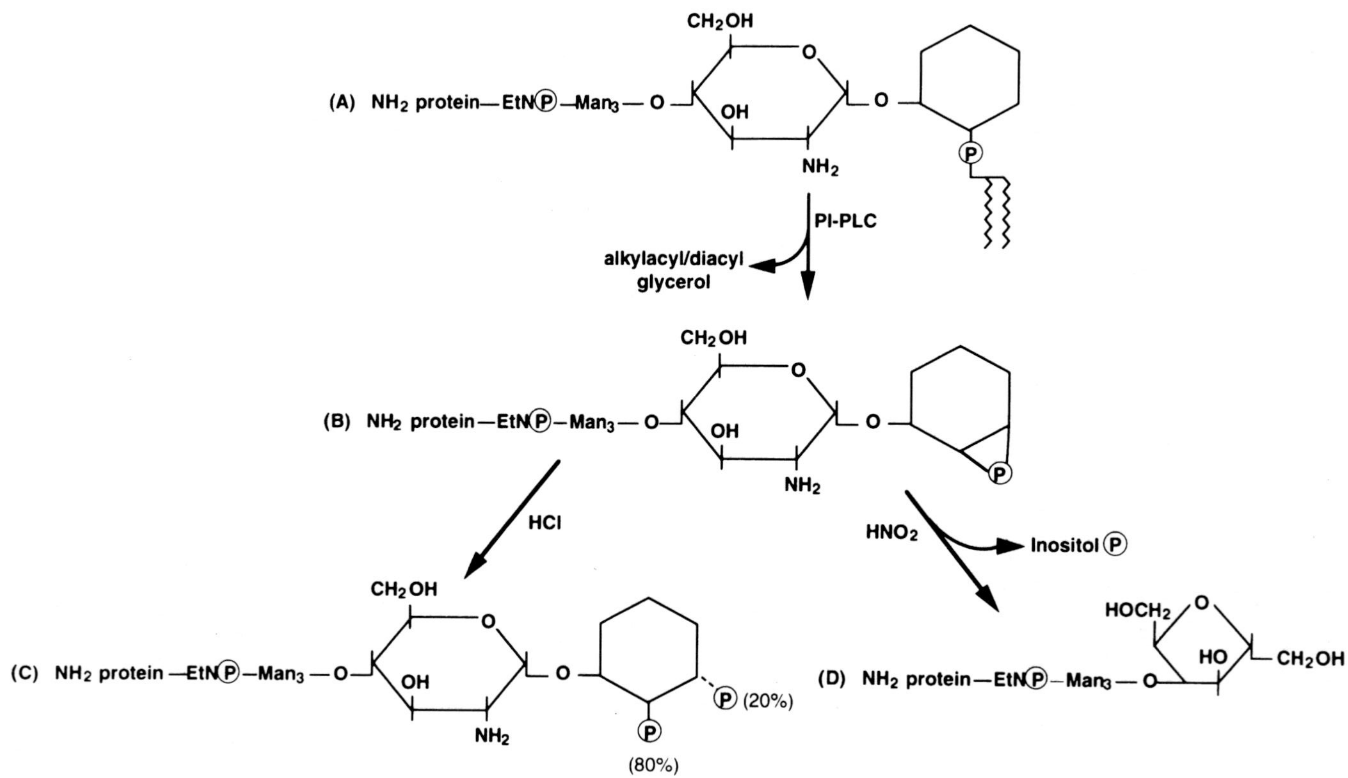

Figure 2. Generation and destruction of the cross-reacting determinant. Cleavage of a GPI anchor (A) with PI-PLC leads to the formation of inositol-1,2-cyclic monophosphate, the CRD (B), which is the major epitope involved in the cross-reactivity of GPI-anchored proteins with anti-CRD antisera. Treatment of this structure with 1 M HCl results in the formation of 80% inositol 1-phosphate and 20% inositol 2-phosphate (C). Treatment with nitrous acid deaminates the glucosamine residue converting it into 2,5-anhydromannose (D) and releasing inositol-1,2-cyclic monophosphate. Both (C) and (D) are unreactive with anti-CRD antisera.

known GPI-anchored protein as a positive control. The soluble form of the trypanosome variant surface glycoprotein (VSG) is available from Oxford Glycosystems, or the soluble form of porcine kidney membrane dipeptidase (leukotriene D_4 hydrolase) is available from Sigma (Cat. No. L2783). Anti-CRD antisera are available commercially (e.g. against trypanosome VSG) from Oxford Glycosystems, or in small amounts for research purposes from one of the authors (N. M. Hooper), in this case raised against porcine or human membrane dipeptidase. To visualize the anti-CRD antibody it is necessary to use a suitable secondary antibody and appropriate detection method, for example, a peroxidase-conjugated secondary antibody in conjunction with ECL (enhanced chemiluminescent) detection kit (from Amersham International), or an anti-(rabbit IgG) alkaline phosphatase-conjugated antibody followed by colour development with nitroblue tetrazolium/5-bromo-4-chloro-3-indoyl phosphate, or a ^{125}I-labelled goat anti-(rabbit IgG) prior to autoradiography.

Protocol 14. Analysis of the cross-reacting determinant of GPI anchors by Western blotting

Equipment and reagents

- Polyvinylidene difluoride (PVDF) membrane (Immobilon P, Millipore)
- Methanol
- SDS–PAGE gel on which the proteins of interest have been separated
- Western blotting apparatus (wet blot type, e.g. Bio-Rad TransBlot)
- Blotting buffer: 20 mM Tris–HCl pH 8–9, 150 mM glycine, 20% (v/v) methanol
- Buffer A: 20 mM Tris–HCl pH 7.4, 150 mM NaCl
- Blocking buffer: 50 mM Tris–HCl pH 8, 0.5 M NaCl, 0.5% Tween 20
- Anti-CRD antiserum (against trypanosome VSG, from Oxford Glycosystems)
- Buffer B: buffer A containing 0.1% Tween 20
- Buffer C: buffer B containing 0.5 M NaCl
- Secondary antibody and detection system suitable for the anti-CRD antiserum being used (see text)

Method

NB: perform all steps at room temperature unless specified otherwise.

1. Pre-wet the PVDF membrane with methanol according to the manufacturer's instructions. Wash the membrane extensively with distilled water prior to use.
2. Transfer the proteins from the gel to the PVDF membrane at 250 mA for 3–6 h in the blotting apparatus containing blotting buffer.
3. Remove the membrane from the blotting apparatus and soak it in 150 ml buffer A for 5 min with shaking.
4. Replace the solution with 150 ml fresh buffer A and soak the membrane for a further 25 min.
5. Block non-specific binding sites on the membrane by soaking it in blocking buffer for 30 min.

Protocol 14. *Continued*

6. Rinse the membrane in 150 ml buffer A for 5 min.
7. Dilute the anti-CRD antiserum into 5–50 ml blocking buffer and incubate with the membrane at 4°C overnight.
8. Remove the antibody solution and wash the membrane with 150 ml of each of the following in the sequence given:
   (a) Buffer A for 5 min.
   (b) 2 × buffer B for 10 min.
   (c) 2 × buffer C for 10 min.
   (d) Buffer A for 5 min.
9. Incubate the membrane with the secondary antibody and detect the bound antibody using the appropriate detection method according to the antibody supplier's instructions.

Protocol 15. Decyclization of GPI-anchored proteins using mild acid conditions

Reagents

- Protein sample to be analysed
- 2 M HCl
- 2 M NaOH
- 3 M Tris–HCl pH 7.5

Method

1. Add an equal volume of 2 M HCl (typically 10–50 μl) to the protein sample in a 1.5 ml microcentrifuge tube.
2. Mix and incubate at room temperature for 30 min.
3. Add an equal volume of 2 M NaOH and 10 μl of 3 M Tris–HCl pH 7.5 to neutralize the sample.
4. Analyse the sample by Western blotting as described in *Protocol 14.*

Protocol 16. Deamination of GPI-anchored proteins with nitrous acid

Reagents

- Protein sample for analysis
- 0.25 M sodium acetate pH 4.0, containing 0.25 M $NaNO_2$; make this fresh daily
- 0.25 M sodium acetate pH 4.0, containing 0.25 M NaCl
- Acetone, cooled to –20°C
- 3 M Tris–HCl pH 7.5

Method

1. Add duplicate aliquots of the protein sample to 1.5 ml microcentrifuge tubes. To one aliquot add 10 vol. of 0.25 M sodium acetate pH 4.0, containing 0.25 M $NaNO_2$ to deaminate the GPI anchor. To the other aliquot add 10 vol. of 0.25 M sodium acetate pH 4.0, containing 0.25 M NaCl (control reaction). Mix each sample and incubate at room temperature for 3 h.
2. Precipitate the proteins in each sample by adding 1 ml acetone (pre-cooled to –20 °C). Mix each sample and leave at –20 °C for 10 min.
3. Centrifuge the samples at 8800 *g* for 15 min at 4 °C.
4. Remove the supernatants and resuspend each pellet in 40 μl water and 10 μl of 3 M Tris–HCl pH 7.5.
5. Analyse the proteins in each sample by Western blotting as described in *Protocol 14*.

Protocol 17 describes a procedure for the detection of PI-PLC-resistant GPI anchors by a Western blotting procedure that is a variation of *Protocol 14* and is based on ref. 53. The same considerations apply to this form of Western blotting as specified for *Protocol 14*.

Protocol 17. Detection of PI-PLC-resistant GPI anchors by Western blotting

- SDS–PAGE gel on which the proteins of interest have been separated
- PVDF membrane, prepared as described in *Protocol 14*, step 1
- Western blotting apparatus (semi-dry type)
- Buffers A, B, and C, anti-CRD antiserum, secondary antibody, and detection system as listed in *Protocol 14*
- 50 mM NaOH in 20% (v/v) propan-1-ol
- Blocking buffer: 50 mM Tris–HCl pH 7.4, 0.15 M NaCl, 1 mM EDTA, 0.05% (v/v) Nonidet P-40, 3% (v/v) BSA
- Bacterial PI-PLC (*Protocol 10*)
- 10 mM Hepes–NaOH pH 7.4

Equipment and reagents

Method

NB: perform all steps at room temperature unless otherwise specified.

1. Transfer the proteins from the SDS–PAGE gel to the PVDF membrane at 40 mA per gel for 1 h in the semi-dry blotting apparatus.
2. Remove the membrane from the blotting apparatus and soak it in 150 ml buffer A for 15 min with shaking.
3. Incubate the membrane with 50 mM NaOH in 20% (v/v) propan-1-ol for

Protocol 17. *Continued*

1 h.

4. Block non-specific binding sites on the membrane by soaking it in blocking buffer overnight at 4°C.
5. Rinse the membrane in 150 ml buffer A for 5 min.
6. Incubate the membrane overnight with 0.1 mU/ml bacterial PI-PLC in 50 mM Hepes–NaOH pH 7.4.
7. Detect the PI-PLC-resistant GPI anchors on the membrane using anti-CRD antiserum and appropriate secondary antibody system as described in *Protocol 14,* steps 6–9.

5.5 Metabolic labelling

If the protein of interest can be expressed in a suitable cell line, then metabolic labelling with components of the GPI anchor can be used to detect the presence of a GPI anchor. For details see another volume in this series (54). The most widely used substrate for labelling GPI anchors is [^{3}H] or [^{14}C]ethanolamine, although other components such as inositol or fatty acids can be used. Following an appropriate labelling period the cells are harvested and the protein of interest immunoprecipitated following either PI-PLC release or detergent solubilization (with octyl glucoside or CHAPS) of the membranes. The presence of the radiolabel in the protein can then be assessed by SDS–PAGE followed by fluorography (see *Protocols 2* and *3*). When using a radiolabelled fatty acid, treatment with PI-PLC (*Protocol 10*) or nitrous acid (*Protocol 16*) should be performed to confirm that the fatty acid is in a GPI anchor and not some other lipid modification. Although there is no selective degradation method available to confirm the location of the ethanolamine label, the only way that ethanolamine can be incorporated into a protein is through a GPI anchor, with the single known exception of the incorporation of ethanolamine into the hydrophilic elongation factor 1α (55).

6. Prenylation

The attachment of prenyl groups, either C15 farnesyl or C20 geranylgeranyl, to numerous cytosolic proteins is now well established (56). The prenyl groups are attached to the protein via thioether linkages to one or more cysteine residues that are usually located towards the C-terminus of the protein. Some proteins destined to be prenylated have a common C-terminal sequence motif of Cys-aliphatic-aliphatic-any amino acid, the so-called CAAX motif. Following attachment of the prenyl group to the cysteine residue, the terminal three amino acids are removed proteolytically and the newly exposed α-carboxyl group is methylated. Other prenylated proteins have alternative motifs, such as Cys-Cys or Cys-X-Cys, in which both cysteine residues are prenylated and the C-terminus carboxyl is methylated (57). Two distinct enzymes have been identified and characterized which add on either farnesyl or geranylgeranyl groups.

As both farnesyl and geranylgeranyl are derived from the cholesterol precursor mevalonic acid, such modified proteins can be identified by labelling cells expressing the protein of interest with [^{3}H]mevalonic acid. If the protein is labelled in this way, the length of the prenyl group attached can be estimated following its cleavage from the protein with methyl iodide and separation by HPLC. The details of these procedures are provided in other volumes in this series (54, 58).

References

1. Towler, D. A., Gordon, J. I., Adams, S. P., and Glaser, L. (1988). *Annu. Rev. Biochem.*, **57**, 69.
2. McIlhinney, R. A. J. (1990). *Trends Biochem. Sci.*, **15**, 387.
3. Rudnick, D. A., McWherter, C. A., Gokel, G. W., and Gordon, J. I. (1993). *Adv. Enzymol.*, **67**, 375.
4. Schmidt, M. F. G. and Burns, G. R. (1989). *Biochem. Soc. Trans.*, **17**, 625.
5. Gordon, J. I., Duronio, R. J., Rudnick, D. A., Adams, S. P., and Gokel, G. W. (1991). *J. Biol. Chem.*, **266**, 8647.
6. Veit, M., Herrler, G., Schmidt, M. F. G., Rott, R., and Klenk, H. D. (1990). *Virology*, **177**, 807.
7. McIlhinney, R. A. J., Pelly, S. J., Chadwick, J. K., and Cowley, G. P. (1985). *EMBO J.*, **4**, 1145.
8. Magee, A. I. and Courtneidge, S. A. (1985). *EMBO J.*, **4**, 1137.
9. Mouillac, B., Caron, M., Bonin, H., Dennis, M., and Bouvier, M. (1992). *J. Biol. Chem.*, **267**, 21733.
10. Dizhoor, A. M., Ericksson, L. H., Johnson, R. S., Kumar, S., Olshevsakya, E., Zozulya, S., *et al.* (1992). *J. Biol. Chem.*, **267**, 16033.
11. Towler, D. A., Adams, S. P., Eubanks, S. R., Towery, D. S., Jackson-Machelski, E., Glaser, L., *et al.* (1987). *Proc. Natl. Acad. Sci. USA*, **84**, 2708.
12. Duronio, R. J., Reed, S. I., and Gordon, J. I. (1992). *Proc. Natl. Acad. Sci. USA*, **89**, 4129.
13. Schmidt, M. F. G. and Burns, G. R. (1989). *Biochem. Soc. Trans.*, **17**, 859.
14. Schmidt, M. F. G., McIlhinney, R. A. J., and Burns, G. R. (1995). *Biochim. Biophys. Acta*, **1257**, 205.
15. Resh, M. D. (1994). *Cell*, **76**, 411.
16. Milligan, G., Parenti, M., and Magee, A. I. (1995). *Trends Biochem. Sci.*, **20**, 181.
17. Goddard, C. and Felsted, R. L. (1988). *Biochem. J.*, **253**, 839.
18. Aitken, A. (1991). In *Lipid modification of proteins: a practical approach* (ed. N. M. Hooper and A. J. Turner), p. 63. IRL Press, Oxford.
19. Neubert, T. A. and Johnson, R. S. (1995). In *Methods in enzymology* (ed. P. J. Casey and J. E. Buss), Vol. 250, p. 487. Academic Press, London.
20. McIlhinney, R. A. J. and Harvey, D. J. (1995). *J. Mass Spectrom.*, **30**, 900.

21. Durino, R. J., Jackson-Machelski, E., Heuckeroth, R. O., Olins, P. O., Devine, C. S., Yonemoto, W., *et al.* (1990). *Proc. Natl. Acad. Sci. USA*, **87**, 1506.
22. Rudnick, D. A., Duronio, R. J., and Gordon, J. I. (1992). In *Lipid modification of proteins: a practical approach* (ed. N. M. Hooper and A. J. Turner), p. 37. IRL Press, Oxford.
23. McIlhinney, R. A. J., Patel, P. B., and McGlone, K. (1994). *Eur. J. Biochem.*, **222**, 137.
24. Veit, M., Nurnberg, B., Spicher, K., Harteneck, C., Ponimaskin, E., Schultz, G., *et al.* (1994). *FEBS Lett.*, **339**, 160.
25. Laemmli, U. K. (1970). *Nature (London)*, **227**, 680.
26. Laskey, R. A. and Mills, A. D. (1975). *Eur. J. Biochem.*, **56**, 335.
27. Chamberlain, J. P. (1979). *Anal. Biochem.*, **98**, 132.
28. McIlhinney, R. A. J. (1992). In *Lipid modification of proteins: a practical approach* (ed. N. M. Hooper and A. J. Turner), p. 15. IRL Press, Oxford.
29. Knoll, L. J., Johnson, D. R., Bryant, M. L., and Gordon, J. I. (1995). In *Methods in enzymology* (ed. P. J. Casey and J. E. Buss), Vol. 250, p. 405. Academic Press, London.
30. Raju, R. V. S., Datla, R. S. S., and Sharma, R. K. (1996). *Protein Expression Purification*, **7**, 431.
31. McConville, M. J. and Ferguson, M. A. J. (1993). *Biochem. J.*, **294**, 305.
32. Brewis, I. A., Ferguson, M. A. J., Mehlert, A., Turner, A. J., and Hooper, N. M. (1995). *J. Biol. Chem.*, **270**, 22946.
33. Englund, P. T. (1993). *Annu. Rev. Biochem.*, **62**, 121.
34. Stevens, V. L. (1995). *Biochem. J.*, **310**, 361.
35. Udenfriend, S. and Kodukula, K. (1995). *Annu. Rev. Biochem.*, **64**, 563.
36. Udenfriend, S. and Kodukula, K. (1995). In *Methods in enzymology* (ed. P. J. Casey and J. E. Buss), Vol. 250, p. 571. Academic Press, London.
37. Kodukula, K., Maxwell, S. E., and Udenfriend, S. (1995). In *Methods in enzymology* (ed. P. J. Casey and J. E. Buss), Vol. 250, p. 536. Academic Press, London.
38. Ferguson, M. A. J. (1992). In *Lipid modification of proteins: a practical approach* (ed. N. M. Hooper and A. J. Turner), p. 191. IRL Press, Oxford.
39. Hooper, N. M. (1992). In *Lipid modification of proteins: a practical approach* (ed. N. M. Hooper and A. J. Turner), p. 89. IRL Press, Oxford.
40. Hooper, N. M., Low, M. G., and Turner, A. J. (1987). *Biochem. J.*, **244**, 465.
41. Brewis, I. A., Turner, A. J., and Hooper, N. M. (1994). *Biochem. J.*, **303**, 633.
42. Bordier, C. (1981). *J. Biol. Chem.*, **256**, 1604.
43. Hooper, N. M. and Bashir, A. (1991). *Biochem. J.*, **280**, 745.
44. Ferguson, M. A. J. (1992). *Biochem. J.*, **284**, 297.
45. Toutant, J.-P., Roberts, W. L., Murray, N. R., and Rosenberry, T. L. (1989). *Eur. J. Biochem.*, **180**, 503.
46. Hooper, N. M. and Turner, A. J. (1988). *Biochem. J.*, **250**, 865.
47. Hooper, N. M. and Turner, A. J. (1988). *FEBS Lett.*, **229**, 340.
48. Brown, D. A. and Rose, J. K. (1992). *Cell*, **68**, 533.
49. Schroeder, R., London, E., and Brown, D. (1994). *Proc. Natl. Acad. Sci. USA*, **91**, 12130.
50. Zamze, S. E., Ferguson, M. A. J., Collins, R., Dwek, R. A., and Rademacher, T. W. (1988). *Eur. J. Biochem.*, **176**, 527.
51. Hooper, N. M., Broomfield, S. J., and Turner, A. J. (1991). *Biochem. J.*, **273**, 301.

52. Broomfield, S. J. and Hooper, N. M. (1993). *Biochim. Biophys. Acta*, **1145**, 212.
53. Guther, M. L. S., Cardoso de Almeida, M. L., Rosenberry, T. L., and Ferguson, M. A. J. (1994). *Anal. Biochem.*, **219**, 249.
54. Masterson, W. J. and Magee, A. I. (1992). In *Protein targeting: a practical approach* (ed. A. I. Magee and T. Wileman), p. 233. IRL Press, Oxford.
55. Whiteheart, S. W., Shenbagamurthi, P., Chen, L., Cotter, R. J., and Hart, G. W. (1989). *J. Biol. Chem.*, **264**, 14334.
56. Casey, P. J. (1995). *Science*, **268**, 221.
57. Khosravi-Far, R., Clark, G. J., Abe, K., Cox, A. D., McLain, T., Lutz, R. J., *et al.* (1992). *J. Biol. Chem.*, **267**, 24363.
58. Gelb, M. H., Farnsworth, C. C., and Glomset, J. A. (1992). In *Lipid modification of proteins: a practical approach* (ed. N. M. Hooper and A. J. Turner), p. 231. IRL Press, Oxford.
59. Quest, A. F., Harvey, D. J., and McIlhinney, R. A. J. (1997). *Biochemistry*, **36**, 6993.
60. Schhiff, E., Schmitz, A., McIlhinney, R. A. J., Manenti, S., and Vagares, G. (1996). *J. Biol. Chem.*, **271**, 26794.
61. Hames, B. D. (ed.) (1998). *Gel electrophoresis of proteins: a practical approach*, 3rd edn. Oxford University Press, Oxford.

6

Proteolytic processing

JOHN M. W. CREEMERS, ELAINE M. BAILYES, IRIS LINDBERG, and JOHN C. HUTTON

1. Introduction

The eukaryotic cell incorporates a series of specialized compartments in which proteolysis proceeds in a regulated and limited manner. This processing is responsible for such diverse processes as sorting and storage of polypeptide hormones, activation of cell surface receptors, presentation to the immune system of self- and foreign antigens, internalization and dissociation of ligands from cognate receptors, and initiation of secretion via non-classical pathways. The same machinery may be exploited by pathogens as in the activation of polio viral polyproteins in the cytosol, the generation of fusigenic HIV and influenza coat glycoproteins, and for the activation of *Pseudomonas* and *Shiga* toxin complexes. The present chapter focuses principally on the secretory pathway taken by proteins which are co-translationally inserted into the lumen of the endoplasmic reticulum and which find their way to the cell surface by vesicle-mediated transport. Specifically excluded are the protein substrates and enzymes which are destined to the lysosome and only mention is made of proteolytic activity which occurs within the endosomal and antigen processing compartments. The latter omission stems in part from lack of understanding of the molecular processes involved but also because the topic is discussed in Chapter 7.

The vast majority of post-translational processing events in the secretory pathway occur via a similar mechanism involving a limited number of enzyme activities, typically one of a series of serine endopeptidases which cleaves substrates at sites marked by clusters of basic amino acids. This is followed by the action of a carboxypeptidase activity that removes the carboxy terminal basic residues exposed by the initial endoproteolytic cleavage. Many polypeptide hormones and neuropeptides undergo a further carboxy terminal amidation catalysed by peptidyl amidating mono-oxygenase in the regulated pathway, an enzyme which uses an exposed carboxy terminal glycine residue as an amide donor.

Seven distinct mammalian endoproteases are now recognized, all of which belong to the subtilisin-like class of serine proteases. All show substantial homology in their catalytic domain and middle domains with yeast kex2p, a Golgi enzyme involved in the processing of the pheromone, alpha-mating factor. Further homology exists upstream of the catalytic domain and all enzymes are produced as pro-proteins which undergo endoproteolytic cleavage necessary for their activation. Removal of the prodomain appears to involve autoproteolytic intramolecular cleavage at one or more sites which are again marked by clusters of basic amino acids. Such cleavages are essential for activation of the enzymatic activity towards exogenous substrates and may be obligatory for the progression of the enzymes from one compartment to another in the secretory pathway. The carboxy termini of the endoproteases are variable and contain one of a number of distinct motifs including transmembrane domains, Cys-rich domains and amphipathic helical segments with putative membrane binding properties. The endoproteases vary in their intracellular localization, furin being localized to the *trans* Golgi network (TGN) through an active retention and retrieval mechanism, other members like prohormone convertases 1 and 2 (PC1 and PC2) being actively sorted to dense-core secretory granules (1, 2). Likewise their tissue distribution is variable; furin, LPC, and PACE4 exhibit an almost ubiquitous localization (3–5); PC1 and PC2 are confined to tissues of neuroendocrine lineage; PC4 is restricted to the testis and ovarian tissue (6). The requirements of PC1 and PC2 for an acidic pH and Ca^{2+} are important factors in the regulation of these enzymes in a cellular context. The V_{max} of the endopeptidases appears to be rate limiting in most cells where they are active, and thus their levels of expression frequently determine the rates of cleavage of potential pro-protein substrates and the nature and bioactivity of the polypeptide products secreted by the cell. The controls on activity and gene expression are clearly very flexible as evidenced by the differential processing of the same substrate by the same enzymes in different tissues or under different physiological conditions. In this context it also has to be considered that the endopeptidases very often work in concert or in a sequential manner. A further layer of complexity is introduced in the case of PC2 by the fact that expression of its activity is closely associated with another protein in neuroendocrine cells, 7B2, a bifunctional molecule which acts as a PC2 binding protein as well as an endogenous inhibitor of the enzyme. Until recently, CPE was the only known member of the secretory pathway carboxypeptidase family. However, new members, carboxypeptidase D and Z, have recently been identified which probably perform a similar catalytic function but in different subcellular localizations.

Post-translational proteolytic events are of key concern to the bioengineer interested in the expression of eukaryotic proteins since these events are critical in determining the efficiency of transport of heterologously expressed proteins through the various intracellular vesicular compartments. They often also determine the molecular form and bioactivity of the secreted product. Not all cells are equivalent in this regard and choices need to be made as to whether a cell type has a regulated pathway and what repertoire of endogenous processing activities it expresses. Few, if any, cultured cells with a regulated pathway of secretion sort and store pro-protein substrates or processing enzymes with comparable efficiencies to their parental cell types. Nevertheless it is often possible to overexpress the processing enzymes even in constitutively secreting cells and achieve virtually stoichiometric conversion of substrates. Likewise it has proven possible for a number of substrates (e.g. prorenin and proinsulin) to change the endopeptidase recognition sequence and achieve processing by a different member of the subtilisin-related endopeptidase in a different cellular compartment.

In some cases there is also the option of expressing the protein in its proform and subjecting it to downstream processing. Furin, which is active in the constitutive pathway, and PC1 and PC2, which can cleave a large number of pro-protein substrates in the regulated pathway, have been extensively studied. However, there remains much to be learned about the kinetic attributes and cell biology of members of the subtilisin-related endopeptidases family. Acting on synthetic substrates, LPC (also known as PC7 and PC8), shows subtle differences in its requirement for basic amino acids at the P4 and P6 position in comparison to furin. PC5A, like PC1 and PC2, has access to the regulated secretory pathway but shows a broader tissue distribution and probably different kinetics towards potential substrates.

The object of this chapter is to present well established procedures that are applicable to many cells, many pro-protein substrates, and most of the members of the pro-hormone convertase family. It is hoped that it will provide the biotechnologist with a set of tools by which processing can be manipulated and at the same time describe techniques which can be applied to more basic research to discover the physiological function of some of the less well characterized enzymes. The processing enzymes, unlike the proteases involved in digestion, lysosomal degradation, blood clotting, or complement cascade, were first discovered as gene products rather than enzymes. Very little is known of their kinetic characteristics or their biophysical properties, a deficiency it is hoped will be redressed in the future by their study as purified proteins rather than through gene transfer experiments in living cells.

The following protocols cover procedures for the assay of the pro-protein convertase activity, methods for generating milligram quantities of the enzymes for more detailed kinetic and structural studies, and an approach for determining activity *in vivo* using a co-transfection procedure that is generally applicable to all enzyme family members and a wide range of biological peptide substrates. Other assay procedures based upon electrophoretic separation of the products generated from radiolabelled substrates have been presented previously in this book series (7). HPLC procedures, although particularly relevant, are not considered here since they generally need to be tailored to particular substrates. Conversely other procedures such as immunoprecipitation, pulse chase radiolabelling, and microsequencing are broadly applicable and are covered elsewhere. Few of the protein convertases mentioned in this chapter are commercially available at present and, apart from carboxypeptidase E and furin, are not readily purified from native tissues because of their low abundance.

2. Immunoadsorbent assay of PC1 and PC2

Immunoadsorbent assays for PC1 and PC2 activity use, as substrates, ^{125}I-labelled proinsulin and des-31,32-proinsulin, respectively. The reaction products and unreacted substrates are then separated using monoclonal antibodies which bind the substrate but not the products. PC1 cleaves proinsulin and des-64,65-proinsulin at the B-chain/C-peptide junction, and in doing so destroys the binding site of monoclonal antibody A6. PC2 cleaves proinsulin and des-31,32-proinsulin at the C-peptide/A-chain junction, and similarly destroys the binding site of monoclonal antibody ANT-1. PC1 is measured by incubating the enzyme with proinsulin followed by the addition of A6 immunoadsorbent to remove all molecules containing an intact B-chain/C-peptide junction. The unbound radioactivity measures the extent of cleavage at the B-chain/C-peptide junction. PC2 is measured by incubating the enzyme with des-31,32-proinsulin, followed by the addition of ANT-1 immunoadsorbent to remove unreacted substrate. The reaction rate with des-31,32-proinsulin is ten times greater than that obtained with proinsulin as substrate and, therefore, the use of des-31,32-proinsulin provides a far more sensitive assay. *Protocol 1* describes the preparation of immunoadsorbents. The enzyme reaction steps of the assay are described in *Protocol 2* and the separation of the reaction products in *Protocol 3*.

Protocol 1. The preparation of cellulose immunoadsorbents

Equipment and reagents

- Glass Petri dish
- Sintered glass funnel
- Vented oven at 125°C (Heraeus)
- Desiccator
- Buchner funnel
- Whatman 41 filter paper
- Whatman CC41 cellulose powder
- Starch iodide paper (Whatman)
- *m*-Nitrobenzyloxymethyl pyridinium chloride (Merck)
- Sodium acetate
- Ethanol
- Benzene
- 20% (w/v) sodium dithionite
- 30% acetic acid
- Ammonium sulfate
- Appropriate antibody preparation (e.g. anti-Ig immunoglobulin from serum or ascites fluid)
- Cupric chloride
- 1 M NaOH (freshly prepared)
- 35% ammonium hydroxide solution
- 10% (v/v) H_2SO_4
- 1 M and 2 M HCl
- 1% (w/v) sodium nitrite
- Urea
- 2-Naphthol
- 0.2 M borate buffer: 12.4 g boric acid, 14.9 g KCl per litre, adjusted to pH 8.2 with NaOH

A. *Preparation of 'Aminocellulose'*

1. Dissolve 0.5 g sodium acetate in 2 ml water, and 1.4 g *m*-nitrobenzyloxymethyl pyridinium chloride in 18 ml ethanol.
2. Mix the two solutions, then add 5 g cellulose powder, and stir to a slurry.
3. Heat to 70 °C in a Petri dish, in a fume-hood, until dry ($\sim$ 30 min), and then for a further 40 min at 125 °C in a vented oven.
4. Wash three times with 200 ml benzene in a sintered glass Buchner funnel, in a fume-hood, and continue to apply vacuum until the powder is dry. Wash with 1 litre water.
5. Resuspend in 150 ml of 20% (w/v) sodium dithionite and mix at 55–60 °C for 30 min.
6. Wash three times with 200 ml water, then twice with 200 ml of 30% acetic acid, and again with water until the 'sulfide' smell disappears.
7. Dry in a desiccator at 20 °C. If lumpy, powder the 'aminocellulose' and store desiccated at 4 °C, where it is stable for several months.

B. *Preparation of antibody*

1. For coupling of anti-Ig immunoglobulin from serum or ascites fluid to cellulose, use an appropriate ammonium sulfate 'cut' (between 40–50%) extensively dialysed against 0.2 M borate buffer.

C. *Preparation of diazocellulose*

1. Dissolve 1.5 g cupric chloride in 5 ml water and add 75 ml of fresh 1 M NaOH with constant stirring.
2. Recover the blue precipitate in a Buchner funnel through two layers of Whatman 41 filter paper, and wash twice with 100 ml water or until the pH of the wash is less than 9. Partially dry on the funnel.
3. Recover the precipitate into a beaker and, inside a fume-hood, dissolve it in 40 ml of fresh 35% ammonia solution to form a saturated solution (deep blue). Stir well for at least 15 min.
4. Dissolve 0.5 g of the powdered 'aminocellulose' prepared in part A in 40 ml of this ammoniacal cupric hydroxide solution. Stir well for at least 15 min. Keep in a fume-hood throughout.
5. Centrifuge ($\sim$ 5000 *g* for 5 min) to remove any excess cupric hydroxide and/or any remaining lumps of 'aminocellulose'.
6. Decant the supernatant into 1500 ml water to produce a light blue solution. Add 10% H_2SO_4 until the solution is colourless or very pale blue and the pH is less than pH 4. The 'aminocellulose' will flocculate to form a white precipitate.

Protocol 1. *Continued*

7. Leave to settle for at least 30 min, then aspirate the bulk of the supernatant. The preparation can be left overnight at 4 °C at this stage if convenient.
8. Wash the precipitate four times with cold water (100 ml aliquots), centrifuging between washes. If the cellulose becomes difficult to spin down, add a drop or two of 1 M HCl.
9. Cool at 4 °C. The cellulose may turn pink.
10. Suspend the 'aminocellulose' in 50 ml of cold fresh 2 M HCl. Add 2 ml of 1% sodium nitrite and check for free oxidant using starch iodide paper which turns black. Then mix for 20 min at 4 °C.
11. Add excess solid urea until the starch iodide test is almost negative, then wash at 4 °C three times with water, and twice with 0.2 M borate buffer. Centrifuge (5000 *g* for 5 min) between washes.
12. Suspend in 25 ml of 0.2 M borate buffer. To test for successful diazotization, add two or three drops of the suspension to three crystals of 2-naphthol in about 3 ml water; it should turn bright orange.

D. *Coupling*

1. Immediately after preparation of the diazocellulose, add an equal quantity of diazocellulose (usually 100 mg/100 mg antibody) to the antibody (from part B) in 0.2 M borate buffer (~ 15 ml total).
2. Mix for 24–48 h at 4 °C in the dark.
3. Centrifuge (5000 *g* for 5 min) and retain the supernatant which can be used again for coupling.
4. Wash the immunoadsorbent three times in 10 ml 0.2 M borate buffer.
5. Resuspend the immunoadsorbent at 5 mg/ml in 0.2 M borate buffer.
6. Measure the amount of protein bound to the immunoadsorbent. Usually 200–300 μg antibody is bound per mg cellulose.

Protocol 2. Enzyme reaction

Equipment and reagents

- 1.5 ml capped microcentrifuge tubes (flip-lid Eppendorf tubes)
- Microcentrifuge with horizontal carriers (Heraeus Biofuge B)
- 30 °C water-bath
- Gamma radiation counter
- Reaction buffer (2 × stock solution): 100 mM sodium acetate pH 5.5, 10 mM $CaCl_2$, 0.2% Triton X-100
- 100 mM ATP-γ-S (optional) (1000 × stock solution)[a]
- Inhibitor cocktail (10 × stock): mix 100 μM E64 (from 20 mM stock solution), 100 μM pepstatin A (from 10 mM stock solution in dimethylformamide), 1 mM TPCK (from 50 mM stock solution)
- 100 mM PMSF in propan-2-ol or dimethylformamide (100 × stock solution)
- ^{125}I-labelled proinsulin or ^{125}I-labelled des-31,32-proinsulin (1 mol ^{125}I/mol peptide) diluted to give 10 000–30 000 d.p.m. in 5 μl reaction buffer (Eli Lilly)[b]
- 1 M Tris pH 8

Method

1. Immediately prior to the assay, mix sufficient reaction buffer, inhibitor cocktail, PMSF, and ATP-γ-S (if required) for the number of reactions to be carried out, such that 40 μl of this assay cocktail will give the required final concentrations in 50 μl total volume.
2. Label duplicate microcentrifuge tubes for each assay to be carried out. To each assay tube add:
   - 5 μl enzyme source
   - 40 μl assay cocktail (from step 1)
   - 5 μl radioactive substrate (^{125}I-labelled proinsulin or ^{125}I-labelled des-31,32-proinsulin) to initiate the reaction
3. Vortex the tubes and spin briefly in horizontal centrifuge racks to bring the tube contents to the bottom.
4. Place in 30 °C water-bath for the incubation time required (typically 30–180 min).
5. Remove the racks to iced water and stop the enzyme reaction by the addition of 10 μl of 1 M Tris pH 8.
6. At this stage, to maximize the accuracy of the assay, the total radioactivity of each tube should be determined, if the geometry of the gamma counter permits. Otherwise, total radioactivity can be determined from 5 μl samples of each tube.

[a] PC1 appears to be susceptible to oxidative damage, catalysed by trace contamination with heavy metals. This inactivation can be reversed by the inclusion of 0.1 mM ATP-γ-S or 50 μM DTT in the assay cocktail (8).

[b] The performance of the assay is dependent on good quality labelled substrate. It can be prepared by one of a number of standard iodination techniques such as Iodogen or chloramine T (9) but should be purified by HPLC (0.1% TFA in acetonitrate gradient) or urea gel electrophoresis (1). The substrate should be at least 95% bindable by the immunoadsorbent since the assay background is the unbindable radioactivity present at the start of the assay.

Protocol 3. Immunoadsorbent separation of reaction products

Equipment and reagents

- Samples for assay from enzyme reactions (*Protocol 2*)
- A6 IgG (for PC1) or ANT-1 IgG (for PC2) cellulose immunoadsorbents prepared as described in *Protocol 1*
- Barbitone buffer: 50 mM sodium barbitone pH 8, 85 mM NaCl, 0.5% (w/v) BSA, 0.1 mg/ml sodium azide
- Gamma radiation counter

Method

1. Wash the immunoadsorbent with barbitone buffer by centrifugation at 5000 *g* for 5 min in order to remove any uncoupled IgG. Resuspend the immunosorbent to 5 mg cellulose/ml in barbitone buffer.

Protocol 3. *Continued*

2. Add 50 μl immunoadsorbent suspension to each assay tube,[a] vortex, and incubate the tubes for ≥ 1 h at room temperature. Due to the flocculent nature of the cellulose and the high molar ratio of IgG to its antigen, little or no further resuspension of the immunoadsorbent is required to achieve maximum binding.
3. Centrifuge the tubes at 10 000 *g* for 5 min to pellet the cellulose.
4. Remove 70 μl of the supernatant and determine its radioactivity in a gamma counter.
5. Calculate the percentage of the total radioactivity (from *Protocol 2*, step 6) recovered in the supernatant. This equals the percentage of substrate hydrolysed by the enzyme.

[a] The amount of immunoadsorbent used in the assay is far in excess of the required binding capacity. To economize on reagents, it is advisable to titrate the immunoadsorbent to determine the minimum quantity to use.

2.1 Calcium-dependence of prohormone convertases

All the members of the mammalian subtilisin-related endopeptidases appear to depend on Ca^{2+} for activity, and the cation plays an important role both in post-translational processing of the enzymes and in regulating the mature protein (10). The presence of endogenous Ca^{2+} necessitates the use of Ca^{2+} buffers if accurate determination of Ca^{2+} requirements for activity are to be made. At the acidic pH values which are often required for enzyme activity, EGTA is not a suitable chelator. A range of calcium concentrations covering the micromolar to millimolar range can be prepared using Ca^{2+} /EDTA buffers as follows:

(a) 2 mM EDTA, 50 mM sodium acetate pH 5.5, 0.1% Triton X-100.
(b) 100 mM $CaCl_2$, 2 mM EDTA, 50 mM sodium acetate pH 5.5, 0.1% Triton X-100 (final pH readjusted to pH 5.5).

Combine the two solutions such that the following concentrations of Ca^{2+} are obtained in the presence of 2 mM EDTA:

Total Ca^{2+} (mM)	Free Ca^{2+} (μM)
0.5	2.6
1	7.6
1.5	21.9
1.7	38.2
2	120
2.5	527
3	1000
4.5	2500
7	5000

3. Expression of prohormone convertases in DG44 CHO cells

This method (*Protocol 4*) is based upon the dihydrofolate reductase (DHFR)-coupled amplification technique in which expression of the desired gene is linked to the amplification of DHFR expression in DHFR-deleted DG44 CHO cells. This gene amplification occurs following stepwise exposure of cell lines to increasing concentrations of methotrexate. The initial selection is based on the ability of the cells to grow in medium lacking nucleosides; only those cells which have successfully taken up the DHFR gene will be able to survive. It should be noted that not all constructs will express well (for example, constructs yielding an unstable protein will not amplify but instead will simply lose expression in the first rounds of amplification). For this reason, expression should be examined following the selection of initial clones and at each amplification step. Use a CMV promoter if possible for your construct; we have had good experiences with rc/CMV, pCDNA3, and pCEP4 from Invitrogen. However, other promoters such as metallothionein will also work. While we have nearly always performed co-transfections of our PC-encoding vectors with a DHFR-encoding vector at a high ratio (to ensure amplification leverage), it is also possible to obtain a dicistronic DHFR-expressing vector containing a weakened DHFR promoter (such as EMC3) and to clone one's cDNA directly into this plasmid. This has the advantage of preserving other antibiotic selection options for later supertransfections (such as for PC2; see below).

The overexpression of PC1 by this method yields an activated protein, i.e. the propeptide is cleaved by the DG44 CHO cells (11). Further carboxy terminal cleavage to a 66 kDa form yields a much more active enzyme, and can be performed *in vitro* (12). PC2 on the other hand is obtained as proPC2 and activation occurs only when the pH drops below 6 (13).

Since expression of active PC2 requires the presence of the helper protein 21 kDa 7B2, enzyme activity cannot be used to screen PC2-expressing DG44 CHO clones and lines. Instead, use Western blotting of conditioned medium to identify a high expressing PC2 cell line which can then be stably super-transfected with a vector encoding 7B2 (for example, using G418 as a selection drug) (13). We found that sequential supertransfection of our highest expressing PC2 DG44 CHO cell line with two vectors (each encoding 21 kDa 7B2 but with either G418 or hygromycin as the selectable markers) resulted in the maximal expression of PC2 activity.

Protocol 4. DHFR-mediated overexpression of PCs in DG44 CHO cells

Equipment and reagents

- Humidified tissue culture incubator at 37 °C, 5% CO_2
- DG44 cells (available from Dr Lawrence Chasin, Department of Biological Science, Columbia University, New York, NY 10027) (these cells have no endogenous DHFR gene expression)
- 0.1 mg/ml of a eukaryotic expression vector encoding DHFR such as pMT2 or EMC3 (Genetics Institute, 87 Cambridge Park Drive, Cambridge, MA 02140)
- 1 mg/ml of the appropriate PC expression vector
- Ham's F12 medium (Life Technologies, Cat. No. 21700-091; powdered form)
- Alpha minus MEM (Life Technologies, Cat. No. 12000-063)
- Optimem (Life Technologies, Cat. No. 22600-134)
- 5 mM methotrexate (also known as amethopterin; Sigma, Cat. No. A6770) in warm Alpha minus MEM—filter through a 0.22 μm filter, aliquot this solution, and store at –20 °C
- PBS (Life Technologies)
- Fetal bovine serum (Irvine Scientific): dialyse 500 ml against at least five changes of 10 litres ice-cold PBS with stirring; include 50 U/ml penicillin and 50 μg/ml streptomycin in the last two changes to guard against bacterial growth. Sterilize by filtration through a 0.22 μm filter, aliquot into 50 ml portions, and store at –20 °C.
- 10 mg/ml gentamycin
- Lipofectin (Life Technologies, Cat. No. 18292-011)
- Roller bottle apparatus at 37 °C (no CO_2 necessary) (optional)
- Cloning rings (Bellco); optional
- Hollow fibre bioreactor (optional) (CellMax, Cat. No. 400-0800/480-008)
- Tissue culture dishes (10 cm diameter) (Corning)
- 24-well microtitre plates (Corning)
- Canted phenolic cap T-25 flasks (Corning)
- Canted phenolic cap T-150 flask (Corning)

Method

1. Plate 0.5×10^6 DG44 CHO cells/10 cm dish in Ham's F12 medium containing 10% fetal bovine serum. Prepare three dishes; one control and two experimental.
2. Next day, wash the dishes twice with 5 ml PBS and then transfect the cells by adding 3 ml Optimem containing 30 μl Lipofectin, 30 μl of 1 mg/ml PC expression vector, and 3–5 μl of 0.1 mg/ml eukaryotic expression vector encoding DHFR. Add 15 μl stock gentamycin solution to the plate since the DNA is not sterile. Treat the control plate identically but omit the PC DNA. Incubate at 37 °C at 5% CO_2 for 4–6 h.
3. To each plate, add 7 ml Ham's F12 medium and 35 μl gentamycin. Incubate overnight at 37 °C, 5% CO_2.
4. Next day replace the medium with Alpha minus MEM, which represents the selection medium (i.e. it lacks nucleosides), containing 10% well dialysed bovine serum and 50 μl gentamycin.
5. Feed the cultures twice weekly (include gentamycin for the first three feeds; it can subsequently be omitted) until robust, well separated clones are visible to the naked eye. This will occur between ten days and three weeks. If prior to colony formation the plate should approach

confluence, split the cell culture 1:100 immediately and keep two plates. Cells in the control plate should be completely dead when colonies appear in the test plates. If not, the reason is probably that the serum has not been dialysed sufficiently.

6. Pick 24 healthy clones using cloning rings (or use the agarose overlay method (14) or other cloning method) and grow them to 50% confluence in a 24-well plate (no well should be more than 80% confluent). Number and score each well as + (least confluent) to +++ (most confluent). Omit from the numbering any well which does not contain cells.
7. Screen clones by adding 500 μl Optimem to each well and incubating overnight. Harvest the conditioned medium in a sterile fashion, replace with growth medium, and return the cells to the incubator. Briefly centrifuge the conditioned medium samples to remove floating cells, and assay either by estimating PC1 enzyme activity (see Section 4) or by Western blotting for PCs (15) following the addition of 0.1 vol. of 10 × concentrated SDS–PAGE sample buffer. Store the remainder of the conditioned medium frozen as a back-up (freezing does not seem to impair PC1 enzyme activity). Note that although PC1 activity can be assayed directly in the conditioned medium (see Section 4), PC2 activity will not be apparent unless 7B2 has previously been transfected into the cells used for the PC2 amplification. If cDNA but no antibodies are available, it is possible (although quite time-consuming) to screen clones using Northern blotting following preparation of mRNA from a 10 cm dish of each clone.
8. Compare the screening results (PC1 activity or Western blot data) to the number of cells in each well, focusing on picking the best expressing well with the least number of cells for further manipulation.
9. Transfer the cells from this best expressing well to a T-25 flask. When this is confluent, harvest the cells, resuspend them in growth medium containing 10% DMSO, and aliquot 90% of the sample into three ampoules. Freeze and store. Begin the methotrexate amplification procedure by transferring the residual 10% of the cells into a T-25 flask in 5 ml of Alpha minus MEM containing 10% well dialysed serum and 5 nM methotrexate. This 5 nM methotrexate-containing medium should be prepared from a stock solution containing 0.5–5 μM methotrexate in Alpha minus MEM and not from the original 5 mM stock, as small variations in the methotrexate concentration can dramatically affect cell growth rates (i.e. error caused by too great dilutions).
10. The cells will take a week to ten days to grow to confluence from this 1:10 split, although they will occasionally express sufficient DHFR to recover quickly. Split the line several times until the cells are growing rapidly in 5 nM methotrexate. When they are able to grow to

Protocol 4. *Continued*

confluence within three days from a 1:10 split, harvest the cells and freeze 90% of them in three ampoules. Culture the residual 10% of the cells in 20–25 nM methotrexate until they are growing well.

11. Make sure that amplification of the desired PC is occurring through an expression check (see step 7) in a test T-25 flask at 70–80% confluence. Incubate with 3 ml Optimem overnight and compare expression in this medium to the expression observed at the previous methotrexate level. Not all increases in methotrexate result in increases in expression, but an increase should be observed over the course of several amplifications.
12. Once a high expressing 50 μM methotrexate-resistant cell line has been established, collection of PC conditioned medium can begin.[a] Seed each of two roller bottles[b] with one T-150 flask of near confluent cells, and grow until the bottles are 70–80% confluent. Then change the medium and replace it with 100 ml Optimem for collection of conditioned medium. Collect this medium after 12–24 h, spin at 2000 *g* for 5 min to pellet floating cells, and store the medium at –20 to –70°C. Depending on the success of the amplification, 100 μg to 3 mg of protein will be present in each 100 ml harvest; contaminating proteins are BSA and transferrin.
13. Collect the conditioned medium on a daily basis to avoid the degradation of secreted proteins. Medium can be collected until expression is lost (this has been repeatedly observed with the PC1 cells but does not appear to occur with the PC2 cells) or the roller bottle becomes overconfluent and the cells detach from the bottle. Between five and ten harvests will generally be possible.

[a] Note that for PC2, expression of *active* enzyme will require stable supertransfection of the 50 μM methotrexate line with a 7B2-encoding vector (16).
[b] Instead of roller bottles, a hollow fibre bioreactor can be used to grow DG44 CHO cells and to collect conditioned medium. While more expensive than roller bottles, the advantage of the bioreactor is that the enzyme is collected in a small volume, thus greatly speeding subsequent purification efforts. We use a CellMax artificial capillary cell culture system and harvest the mature cartridge with about 3–5 ml Optimem each day.

4. Fluorogenic assay for PC1 and PC2

Prohormone convertases cleave prohormone substrates at pairs of basic amino acids, most often a Lys-Arg sequence. Fluorogenic substrates which upon hydrolysis liberate the intensely fluorescent leaving group amino methylcoumarin (amc) represent convenient substrates to assay these enzymes (see *Protocols 5* and *6*). Interestingly, while fluorogenic peptides containing only a pair of basic amino acids are not hydrolysed by members of this family of enzymes, peptides containing an arginine at the –4 position are readily hydrolysed (17). The best commercially available substrate for both PC1 and PC2 is the peptide pyr-Glu-Arg-Thr-Lys-Arg-amc, which is obtained as a powder and made up to a 20 mM stock solution in DMSO. Aliquots are kept frozen until use. A working stock solution is obtained by producing a tenfold dilution in water; this may be refrozen. No assay specific for each PC is currently available; i.e. if a preparation contains both PC1 and PC2, both of these enzymes will act upon the same substrate. However, the action of PC1 can be distinguished from that of PC2 by the use of a PC2-specific inhibitor, the 7B2 CT peptide, which is ineffective against PC1 (18). It should also be noted that a large number of cellular proteinases (such as cathepsin B and cytosolic proteinases) can hydrolyse this fluorogenic substrate but the activity of many of these enzymes can be held in check with an inhibitor cocktail (see *Protocol 5*) although it is impossible to completely control the activity of all of these contaminating enzymes, especially at long incubation times. The purified PC1 and PC2 enzymes are easily lost on surfaces and hence only detergent-containing buffers (such as Brij or Tween) should be used for dilution.

4.1 PC1 assay

The activity of recombinant PC1 exhibits a lag phase, thought to represent a conformational rearrangement, and linear rates of hydrolysis are often not achieved until 30–60 min after the start of the reaction (12). This must be taken into account when measuring the activity of this enzyme using *Protocol 5*. Purified 87 kDa PC1 exhibits a specific activity of about 0.5 μmol/h/mg protein with this substrate.

Protocol 5. Assay for PC1

Equipment and reagents

- 96-well microtitre plate fluorimeter, reading at 380 nm excitation, 460 nm emission (or similar wavelengths)[a]
- Amino methylcoumarin (amc) standard (Peptides International)
- 2 mM pERTKR-amc substrate stock (Peptide International) in water
- 1 M sodium acetate pH 5.5 (keep at 4°C)
- 50 mM $CaCl_2$
- 1% Brij-30
- 10 × inhibitor cocktail (required when the PC1 enzyme is not being assayed in purified form): 10 μM E-64, 10 μM leupeptin, 10 μM pepstatin, and 10 mM TPCK (all from Sigma). Store the cocktail aliquoted at –20°C.
- Polypropylene 96-well microtitre plates (round-bottomed; Costar)
- Reaction buffer stock solution: prepare this by combining 5 ml each of 1 M sodium acetate pH 5.5, 1% Brij-30, 50 mM $CaCl_2$, and 10 × inhibitor cocktail (the inhibitor cocktail is not necessary if purified PC1 enzyme is to be assayed). When diluted during the assay below, this will deliver final concentrations of 0.1 M sodium acetate pH 5.5, 0.1% Brij-30, 1 × inhibitor mix, and 5 mM $CaCl_2$. If prepared without the inhibitor cocktail, it can be stored in the refrigerator (0.02% azide should be added for storage times over two months).

Protocol 5. *Continued*

Method

1. Pipette 20 μl of reaction buffer stock solution into duplicate wells of the 96-well microtitre plate.
2. For each well, add the sample in a volume of up to 25 μl. Duplicates of an appropriate control with the same buffer composition should be included so that the intrinsic fluorescence/water hydrolysis of the substrate alone may be estimated. Typically, 25 μl of conditioned Optimem medium is assayed and plain Optimem is used in the control wells.
3. Bring the volume of each well up to 45 μl with distilled water.
4. Add 5 μl of 2 mM pERTKR-amc substrate to each well. Mix by gently tapping the side of the plate.
5. Cover the plate tightly with Parafilm to prevent evaporation and place at 37 °C (placing another 96-well plate on top of the one containing the samples helps to seal the plate).
6. At 10–60 min intervals (depending on the enzyme concentration), read the fluorescence of the samples using the microtitre plate fluorimeter. Return the plate to the incubator after each reading. The amount of substrate hydrolysed should be kept to below 10% of the total substrate present for accurate quantitation of the enzyme activity, but an activity of at least 20 times the blank should be achieved. Take readings for up to 6 h.[b,c]
7. Plot fluorescence against time, first subtracting the fluorescence in the blank, and estimate PC1 enzyme activity by reference to a standard curve of free amc prepared in the same buffer and at the same temperature. Use the linear portion of the activity curve for activity estimations (i.e. after the lag phase).

[a] If a microtitre plate fluorimeter is not available, assays can be performed in microcentrifuge tubes, quenched with 0.95 ml 1 M acetic acid, and read in a two-sided quartz fluorimeter cuvette in a conventional fluorimeter. The disadvantage of this method is that further incubation following addition of the acid is not possible, so many time points must be set up initially.

[b] Linear rates of hydrolysis are often not achieved until 30–60 min after the start of the reaction (see Section 4.1).

[c] 87 kDa PC1 exhibits much longer rates of linear hydrolysis than PC2 (11).

4.2 PC2 assay

Recombinant PC2 is isolated as a zymogen and also exhibits a short lag phase, in this case due to the necessity for autoactivation of the enzyme. The enzyme can be pre-activated before use by reducing the pH to 5.0 with 2 M acetic acid (first add a buffer if one is not already present) and incubating at 37°C for

30 min. Aliquots of pre-activated enzyme snap-frozen in liquid nitrogen can be stored at –70°C without loss of activity. The procedure for assay of PC2 is described in *Protocol 6*. If a pre-activated enzyme is used, linear hydrolysis rates should be observed soon after the start of the reaction. However, PC2 does not appear to be as stable as PC1 and a drop in the hydrolysis rate may be observed after a few hours. The activity of the enzyme is calculated by reference to a standard curve of free amc prepared in the same buffer. Purified PC2 obtained from the hollow fibre bioreactor of DG44 CHO-PC2-7B2 cells exhibits a specific activity between 2–4 μmol/h/mg protein with the usual substrate.

The only differences between the PC1 (*Protocol 5*) and the PC2 (*Protocol 6*) assays are:

- The use of a lower pH buffer for PC2, sodium acetate at pH 5, since PC2 exhibits a lower pH optimum.
- The inclusion of a parallel sample series containing the 7B2 CT peptide at a final concentration of 1 μM (or higher). Note that if other enzymes are present in the assay and the samples are incubated for long periods of time (i.e. overnight) the inhibition of PC2 by the CT peptide can be overcome by non-specific hydrolysis. The 27 kDa form of 7B2, which contains the CT peptide at its C-terminus, can also be used as a specific PC2 inhibitor; this can be prepared by bacterial expression (19).

Protocol 6. PC2 assay

Equipment and reagents

- Equipment and reagents as specified in *Protocol 5*, but use 1 M sodium acetate pH 5 instead of 1 M sodium acetate pH 5.5 to prepare the reaction buffer stock solution
- 7B2 CT peptide (custom synthesized): 1 mM in water, stored at –20°C in aliquots

Method

1. Pipette samples into the wells of a 96-well microtitre plate as described in *Protocol 5* for PC1. If inhibited samples are to be included, add 1 μl of 1 mM CT peptide to each sample. Take care to avoid contamination of uninhibited samples with the CT peptide since this is a very potent inhibitor of PC2 (IC_{50} about 50 nM).
2. Read the fluorescence of the samples in the fluorimeter at intervals of 10 min for up to 3 h, depending on hydrolysis rates.
3. Plot fluorescence against time, subtracting the fluorescence in the substrate alone at each time point. Use the linear portion of the curve to calculate the activity of the enzyme by reference to the standard curve of free amc.

5. Transient expression with recombinant vaccinia virus V.V.:T7

5.1 Introduction

The principle of the system (20) is based on the infection of eukaryotic cells with a recombinant vaccinia virus (V.V.:T7) expressing the bacteriophage T7 RNA polymerase under the control of the vaccinia virus 7.5K promoter. Subsequent transfection of these cells with one or more plasmids containing the T7 promoter fused to a cDNA of interest will result in the synthesis of T7-directed transcripts, which are efficiently translated. This expression system has been used successfully to study biosynthesis, activity, and specificity of furin (21), PACE4 (22), and LPC (23).

There are several advantages of this system. In general, it allows the rapid analysis of a large number of recombinant proteins without the construction and scaling up of recombinant vaccinia viruses. Furthermore, the infection results in the shut-off of the translation of host cell-specific proteins, making immunoprecipitation of samples from culture medium redundant in most cases. This is of particular interest when specific antibodies are not available. The advantages specifically for the expression of convertases and their substrates are twofold. Recombinant substrates expressed using non-viral expression vectors are most often at least partially processed by endogenous enzymes. In vaccinia virus-infected cells this endogenous activity is reduced to negligible levels (as well illustrated in ref. 21, where this and a non-viral expression system were used). The substrate processing observed in the V.V.:T7 assay can, therefore, be completely attributed to the co-transfected convertase. The second advantage is that for unknown reasons the maturation and activation of convertases seem to be more efficient, e.g. many of the furin mutants described by Creemers *et al.* (21) were found to be active only in the V.V.:T7 expression system.

Protocol 7. Transient expression with recombinant vaccinia virus V.V.:T7

Equipment and reagents

- Vaccinia virus V.V.:T7 (B. Moss, Laboratory of Viral Diseases, NIAID, National Institutes of Health, Bethesda, MD, 20892)
- Recombinant plasmid DNA (containing relevant cDNA cloned downstream of the T7 promoter). Suitable vectors are: pSelect-1 (Promega), pGEM vectors (Promega), and pcDNA3 (Invitrogen). Do not use pBluescript (Stratagene). cDNAs should not contain ATG sequences preceding the start codon.
- Plasmid DNA, but without cDNA insert, to serve as control
- Ultrasonic water-bath (Benson 2200)
- ϕ 35 mm dishes ($\approx$ 10 cm^2) (Sterilin)
- Polystyrene tubes (e.g. Bijou containers, Sterilin)
- DOTAP (liposomal transfection reagent; Boehringer Mannheim)
- PBS (Gibco BRL)
- HBS: 150 mM NaCl, 20 mM Hepes pH 7.4

- Infection medium: DMEM (Gibco BRL) supplemented with 20 μg/ml BSA (Sigma), 100 U/ml penicillin, and 100 μg/ml streptomycin
- Trizma/SDS: 0.5 M Trizma base (Sigma), 6.25% (w/v) SDS
- LSB: 0.25 M Tris–HCl pH 6.8, 50% (v/v) glycerol, 10% (w/v) SDS, 0.005% (w/v) bromophenol blue
- A suitable cell line for transfection: many cell lines can be used, including PK(15), AtT-20, DG44 CHO, COS, and NRK cells
- Radiolabelled amino acids for *in vivo* labelling
- 50 mg/ml BSA
- 100% (w/v) TCA
- SDS–polyacrylamide gel and electrophoresis apparatus (see ref. 24)

Method

1. On day 1, plate the cells in ϕ35 mm plates.
2. On day 2, the cells should be ≈ 80% confluent (i.e. 8×10^5 cells/dish).
3. Wash the cells twice with PBS.
4. Sonicate the vaccinia virus V.V.:T7 stock twice for 2 min each time in the ultrasonic water-bath.
5. Dilute the virus in infection medium to give a total volume equivalent to 0.5 ml/dish and a multiplicity of infection of 5 (i.e. ≈ 4×10^6 plaque forming units/dish or 8×10^6 plaque forming units/ml). Sonicate the diluted virus for 2 min.
6. Add 0.5 ml diluted virus to each dish of cells. Add infection medium without virus to the mock control dish.
7. Incubate the dishes for 1 h at room temperature with occasional agitation.
8. In the meantime prepare the transfection mixture:
   (a) For each dish, add 10 μl DOTAP to 23 μl HBS in a polystyrene tube. In another polystyrene tube, dilute 4 μg recombinant plasmid DNA in HBS in a final volume of 33 μl. This can be either one plasmid or a mixture of equal amounts of two or more different plasmids. The total amount of DNA should, however, always be 4 μg.
   (b) Add the diluted DNA to the diluted DOTAP and mix by pipetting up and down a few times (do not vortex).
   (c) Incubate for 10 min at room temperature.
   (d) Add 2 ml of infection medium.
9. Remove the virus-containing medium from the cells and replace it with the DNA/DOTAP transfection mixture (from step 8). Besides the mock control, also include a T7 control dish which should (after infection) be transfected with the plasmid without insert.
10. Incubate the cells for 16–18 h at 37°C.
11. Label proteins being synthesized *in vivo* by incubating the cells with radioactive amino acids (e.g. 1 h pulse, 4 h chase). Subsequently, separate the cells from the medium and lyse the cells. Carry out an immunoprecipitation for the protein of interest.

Protocol 7. *Continued*

12. Conditioned medium samples can be analysed without further immunopurification. In that case TCA precipitation should be performed as follows:

(a) Collect the medium (1 ml) in a microcentrifuge tube and centrifuge for 10 min at 10 000 *g* at 4°C.

(b) Transfer 0.95 ml of the supernatant to a new microcentrifuge tube, add 1 μl of 50 mg/ml BSA,[a] and vortex. Add 110 μl of 100% (w/v) TCA.

(c) Incubate for at least 1 h on ice and then centrifuge for 10 min at 10 000 *g* at 4°C.

(d) Discard the supernatant and dissolve the pellet in 20 μl Trizma/SDS. Then add 70 μl H_2O and 22.5 μl LSB. 2-Mercaptoethanol (4% (v/v) final concentration) can also be added to reduce disulfide bonds.

(e) Heat the samples for 5 min at 95°C. Load 25 μl on to an SDS–PAGE gel (see ref. 24) and carry out electrophoresis. The mock control should give numerous bands, while the T7 control lane should have only a few discrete viral bands which are also present in all the other lanes except the mock control. Extra bands in the other lanes represent the recombinant proteins encoded by the transfected plasmids. If no extra band can be detected, it might be co-migrating with one of the viral proteins and immunoprecipitation should be performed instead of TCA precipitation.

[a] If the molecular weight of the protein of interest is near that of BSA, ovalbumin can be used instead as carrier.

References

1. Guest, P. C., Arden, S. D., Bennett, D. L., Clark, A., Rutherford, N. G., and Hutton, J. C. (1992). *J. Biol. Chem.*, **267**, 22401.
2. Itoh, Y., Tanaka, S., Takekoshi, S., Itoh, J., and Osamura, R. Y. (1996). *Pathol. Int.*, **46**, 726.
3. Van de Ven, W. J., Voorberg, J., Fontijn, R., Pannekoek, H., Van den Ouweland, A. M. W., and Siezen, R. J. (1990). *Mol. Biol. Rep.*, **14**, 265.
4. Mains, R. E., Berard, C. A., Denault, J. B., Zhou, A., Johnson, R. C., and Leduc, R. (1997). *Biochem. J.*, **321**, 587.
5. Seidah, N. G., Hamelin, J., Mamarbachi, M., Dong, W., Tardos, H., Mbikay, M., *et al.* (1996). *Proc. Natl. Acad. Sci. USA*, **93**, 3388.
6. Seidah, N. G., Day, R., Hamelin, J., Gaspar, A., Collard, M. W., and Chretein, M. (1992). *Mol. Endocrinol.*, **6**, 1559.

7. Hutton, J. C. and Siddle, K. (ed.) (1990). *Peptide hormone secretion: a practical approach*. IRL Press, Oxford.
8. Bailyes, E. M. and Hutton, J. C. (1992). *Biochem. J.*, **286**, 223.
9. Edwards, R. (1990). In *Peptide secretion: a practical approach* (ed. J. C. Hutton and K. Siddle), p. 71. IRL Press, Oxford.
10. Guest, P. C., Bailyes, E. M., and Hutton, J. C. (1997). *Biochem. J.*, **323**, 445.
11. Zhou, Y. and Lindberg, I. (1993). *J. Biol. Chem.*, **268**, 5616.
12. Zhou, Y. and Lindberg, I. (1994). *J. Biol. Chem.*, **269**, 18408.
13. Lamango, N. S., Zhu, X., and Lindberg, I. (1996). *Arch. Biochem. Biophys.*, **330**, 238.
14. Lindberg, I. and Zhou, Y. (1995). In *Methods in neurosciences* (ed. I. Smith, I. Lindberg, and Y. Zhou), pp. 94–108. Academic Press.
15. Vindrola, O. and Lindberg, I. (1992). *Mol. Endocrinol.*, **6**, 1088.
16. Zhu, X. and Lindberg, I. (1995). *J. Cell Biol.*, **129**, 1641.
17. Lindberg, I., Lincoln, B., and Rhodes, C. J. (1992). *Biochem. Biophys. Res. Commun.*, **183**, 1.
18. Lindberg, I., van den Hurk, W. H., Bui, C., and Batie, C. J. (1995). *Biochemistry*, **34**, 5486.
19. Martens, G. J., Braks, J. A., Eib, D. W., Zhou, Y., and Lindberg, I. (1994). *Proc. Natl. Acad. Sci. USA*, **91**, 5784.
20. Fuerst, T. R., Niles, E. G., Studier, F. W., and Moss, B. (1986). *Proc. Natl. Acad. Sci. USA*, **83**, 8122.
21. Creemers, J. W. M., Siezen, R. J., Roebroek, A. J. M., Ayoubi, T. A. Y., Huylebroeck, D., and Van de Ven, W. J. (1993). *J. Biol. Chem.*, **268**, 21826.
22. Creemers, J. W., Kormelink, P. J., Roebroek, A. J., Nakayama, K., and Van de Ven, W. J. (1993). *FEBS Lett.*, **336**, 65.
23. Van de Loo, J.-W. H. P., Creemers, J. W. M., Bright, N. A., Young, B. D., Roebroek, A. J. M., and Van de Ven, W. J. M. (1997). *J. Biol. Chem.*, **272**, 27116.
24. Shi, Q. and Jackowski, G. (1998). In *Gel electrophoresis of proteins: a practical approach* (ed. B. D. Hames), 3rd edn, p. 1. Oxford University Press, Oxford.

7

Protein degradation in mammalian cells

AARON CIECHANOVER and BERND WIEDERANDERS

1. The ubiquitin-proteasome pathway in mammalian cells

1.1 Introduction

Recent experimental evidence implicates the ubiquitin system in the specific and programmed degradation of many key regulatory and short-lived cellular proteins. Among these are mitotic cyclins, transcriptional activators, tumour suppressors, and growth regulators, as well as membrane receptors involved in signal transduction and the immune and inflammatory responses. It is also involved in processing of antigens for presentation via class I MHC molecules. The system can specifically target abnormal, misfolded, and unassembled proteins that may result from mutations, environmental damage, or non-stoichiometric synthesis of subunits of complex proteins. It is clear therefore that ubiquitin-mediated proteolysis plays important roles in the regulation of basic pathophysiological cellular processes such as regulation of cell cycle and division, development and differentiation, malignant transformation, modulation of the immune system, and neurodegeneration. With the multitude of substrates targeted, it is not surprising that recent studies have implicated the system in the pathogenesis of several disease states as well. For example, the inability to degrade the kidney Na^+ channel leads to stabilization of the protein, accumulation of the channel subunits, excessive reabsorption of the cation and water, and a severe form of hypertension (1). Mutations in the gene that encodes E6-AP, a ubiquitin-protein substrate ligase, lead to the Angelman syndrome that is characterized by mental retardation, seizures, and abnormal gait (2). It is predicted that the list of ubiquitin-related pathological derangements will grow significantly in the future.

Degradation of a protein via the ubiquitin pathway proceeds in two discrete and successive steps:

(a) Covalent attachment of multiple ubiquitin molecules to the protein substrate.
(b) Degradation of the targeted protein by the 26S proteasome complex.

Detailed studies have revealed that the ubiquitin pathway (reviewed in refs 3–9, and depicted in *Figure 1*) consists of several components that act in concert. One of these, ubiquitin, an evolutionarily conserved protein, is activated at its C-terminal Gly residue by the ubiquitin-activating enzyme, E1. Following activation, one of several E2 enzymes (ubiquitin-carrier proteins or ubiquitin-conjugating enzymes, UBCs) transfers the activated ubiquitin moiety

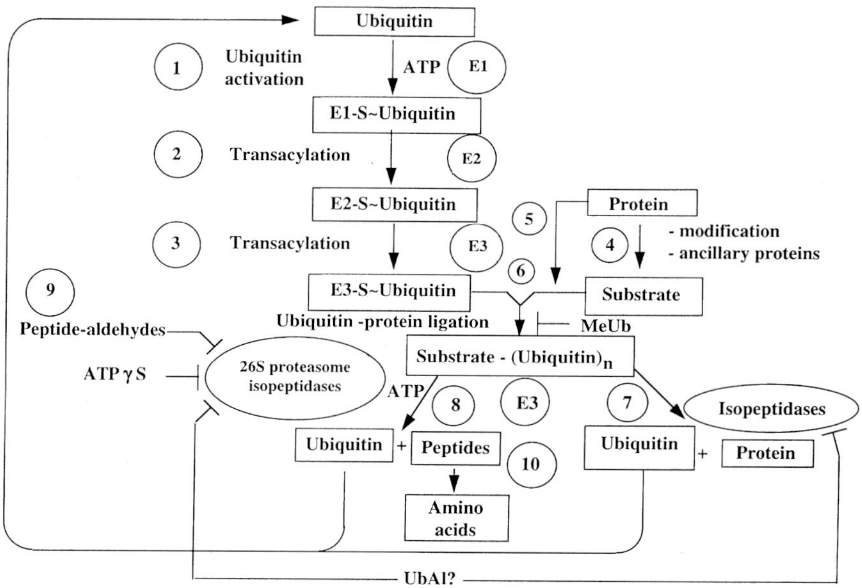

Figure 1. Proposed sequence of events in the degradation of a protein via the ubiquitin-proteasome pathway. Steps of the pathway are denoted by arrows. 1. Activation of ubiquitin by the ubiquitin-activating enzyme, E1. 2. Transfer of activated ubiquitin from E1 to a member of the ubiquitin-carrier protein family, E2 (ubiquitin-conjugating enzymes, UBCs). 3. Transfer of the activated ubiquitin from E2 to a member of the ubiquitin-protein ligases family, E3. 4. Conversion of a protein to a substrate (in certain cases) by a post-translational modification or association with an ancillary protein. 5. Direct recognition (other cases) of a protein substrate by the ubiquitin ligase, E3, without prior modification or association with an ancillary protein. 6. Conjugation of multiple ubiquitin moieties to the E3-bound substrate. 7. Isopeptidases release intact substrate and free ubiquitin from mistakenly conjugated proteins. 8. ATP-dependent degradation of conjugates into peptides mediated by the 26S proteasome; free ubiquitin is released by the activity of 26S proteasome-associated and free isopeptidases. 9. Recycling of the released ubiquitin. 10. Terminal degradation of peptides into free amino acids. Blocked arrows denote specific inhibitors: MeUb, methylated ubiquitin (serves as a chain terminator that cannot generate a polyubiquitin tree); UbAl, ubiquitin aldehyde (inhibitor of certain isopeptidases). Peptide aldehydes such as Acetyl-Leu-Leu-Norleucinal (ALLN; Calpain inhibitor I) or MG132, or the more specific compound lactacystin, inhibit the 20S proteasome; ATPγS promotes conjugation by E1, but not assembly of, and degradation by, the 26S proteasome complex.

from E1 to a member of the ubiquitin-protein ligase family, E3, to which the substrate protein is specifically bound. E3 catalyses the last step in the conjugation process, covalent attachment of ubiquitin to the substrate. The ubiquitin moiety is covalently linked to the ϵ-NH_2 group of a Lys residue of the protein substrate to generate an isopeptide bond. In successive reactions, a polyubiquitin chain is synthesized by processive transfer of additional activated ubiquitin moieties to Lys48 of the previously conjugated ubiquitin molecule. The chain serves, most probably, as a recognition marker for the protease. The binding of the substrate to E3 is specific and implies that E3 enzymes play a major role in recognition and selection of proteins for conjugation.

The ubiquitin system appears to be hierarchical. A single E1, most probably, carries out activation of ubiquitin required for all modifications. Several major species of E2 enzymes have been characterized in mammalian cells (see, for example, refs 3, 10, 11) that are involved in different aspects of cellular metabolism and in the targeting of different substrate proteins. Mammalian E2-14 kDa is involved in conjugating 'N-end rule' substrates (8), whereas UbcH5 and UbcH7 are involved in the degradation of many short-lived proteins such as p53 (10) and c-Fos (12). All these E2 enzymes act in concert with E3 enzymes, and it appears that each E2 can interact with one or more E3 proteins. Few E3 enzymes have been described so far. Mammalian E3α and E3β recognize protein substrates via their free N-terminal amino acid residues ('N-end rule') (3, 8, 9). E6-AP is involved in recognition of p53 (13). A series of unique E6-AP homologous proteins have been identified (14), some of which have an E3 activity and have been implicated in the targeting of specific substrates. Nedd4, for example, is involved in the degradation of the epithelial Na^+ channel (1), whereas Pub1 targets Cdc25 (15). The roles of most of the members of the group and the identity of their cellular substrates is still obscure. It is clear however that the E3 ligases, even more than the E2 ubiquitin-carrier proteins, belong to a large, still growing, family of enzymes. As for their mode of recognition, it is unlikely that each E3 enzyme targets a single substrate. Rather, it is conceivable that a single ligase recognizes several different cellular proteins via a similar, but clearly not identical, structural motif. The mode of recognition of the many substrates of the ubiquitin system appears to be complex. A few proteins may be recognized via their free and 'destabilizing' N-terminal residue ('N-end rule') (3, 8, 9). However, the vast majority of cellular proteins are either acetylated at the N-terminal position or contain free 'stabilizing' N-termini. These proteins are targeted by primary signals that reside downstream and are distinct from the N-terminal residue, by secondary post-translational modification(s) such as phosphorylation, or following association with ancillary proteins. Adding to the complexity of the recognition is the fact that the degradation of many proteins is programmed, is spatially temporal, and occurs only during specific stages of the cell cycle for example.

The stability of other cellular proteins is governed by extracellular and environmental stimuli such as growth promoting factors and stress:

(a) The transcription factor inhibitor IκBα is degraded following phosphorylation of specific Ser residues that results from activation of a specific kinase in response to a whole array of extracellular stimuli such as exposure to viral and bacterial products (16, 17).

(b) Degradation of the tumour suppressor protein p53 is a complex process. In transformed cells, degradation is greatly enhanced following interaction of p53 with the human papillomavirus (HPV) oncoprotein E6 (18). Targeting of the p53 suppressor for degradation by this oncoprotein probably plays an important role in HPV-mediated malignant transformation. In normally dividing cells, p53 is activated by stabilization following exposure of the cell to stress signals such as DNA damage. The activation leads to growth inhibitory responses, including cell cycle arrest. The arrest is believed to provide sufficient time for DNA repair to occur, which maintains the integrity of the genome. Following repair of the damage, p53 is destabilized again.

(c) The programmed degradation of cell cycle regulators such as G1 and mitotic cyclins is essential for transition of the cell from G1 to S or for their exit from mitosis, and occurs only at the G1-S interphase or prior to mitosis.

The degradation of some other substrates of the ubiquitin system requires association with molecular chaperones (19, 20).

Following conjugation, the protein moiety of the ubiquitin-substrate protein adduct is degraded and free ubiquitin is released for reuse. A multisubunit 26S proteasome complex that degrades specifically conjugated proteins in an ATP-dependent mode has been purified and characterized (see *Figure 1*). The 'core catalytic' unit of the protease is a ~ 700 kDa 20S proteasome complex. Two 19S 'cap' complexes associate with the 20S enzyme to generate the 26S dumbbell-shaped complex. These are probably the 'regulatory' subunits that impart specificity (for ubiquitinated proteins) and control. Energy is required to promote assembly of the 26S complex, but also for other functions such as peptide bond cleavage. The 26S proteasome also contains one or several ubiquitin C-terminal hydrolases (isopeptidases) involved in recycling of ubiquitin. Little is known about the regulation of the 26S protease complex. Several activators, inhibitors, and regulators of the 20S complex have been purified and characterized (for recent review articles, see refs 21–25). Using specific inhibitors, it has been shown recently that the 20S enzyme is involved in the degradation of the bulk of cellular proteins, short- and long-lived alike (26). Experimental evidence indicates that the 20S complex acts in the cell as a part of the larger and regulated 26S complex, and not as a distinct ATP-independent protease. Also, it appears that the 26S proteasome is the major proteolytic arm of the ubiquitin system. So far, except for ornithine decarboxylase (ODC) that is degraded by the 26S proteasome in a ubiquitin-independent process (27), all other known substrates of the enzyme must be ubiquitinated prior to their recognition and degradation.

It is evident that the ubiquitin pathway contains many enzymatic factors that act in concert along a well-defined pathway. Some of the factors are 'universal' and are required for the conjugation and subsequent degradation of all protein substrates. These are, for example, ubiquitin, and most probably the ubiquitin-activating enzyme, E1. Others, for example the E2 and E3 enzymes, are specific to certain subsets of substrates. In addition, certain ancillary proteins and modifying enzymes are essential for the degradation of specific protein substrates. Many of these enzymes and components have been purified and characterized, and some have been cloned and can be expressed in bacterial or eukaryotic cells.

In this chapter, space constraints prevent us from covering all of the methodologies that have been used to explore the ubiquitin-proteasome pathway. Rather, we describe important general procedures that will enable researchers to study the conjugation and subsequent degradation of specific protein substrates via this pathway. The preparation of cell extracts for monitoring the ubiquitin-proteasome pathway is described in Section 1.2. Then Section 1.3 describes the fractionation of theses extracts to allow the individual stages in the process to be studied. Preparation of labelled protein substrates is given in Section 1.4, and ubiquitin-mediated conjugation and proteolysis are covered in Sections 1.5 and 1.6, respectively. Finally, Section 1.7 describes the use of specific inhibitors to study the function of the proteasome *in vivo*.

1.2 Preparation of cell extracts for monitoring conjugation and degradation

To conjugate or degrade a protein substrate, one has to utilize the appropriate cell extract. Rabbit reticulocyte lysate contains all the enzymes required for degradation of most proteins and, therefore, can be used in most cases. Unlike lysates from cultured cells, one can obtain reticulocyte lysate in relatively large amounts. Also, the lack of requirement for tissue culture media and sera for its preparation make this lysate significantly cheaper than its counterpart from nucleated cells. *Protocols 1* and *2* describe the preparation of rabbit reticulocyte lysate and extracts from cultured cells respectively. For monitoring degradation of labelled proteins in these crude extracts, it is not necessary to deplete the cells of ATP prior to the preparation of the extract. However, this will be necessary to reconstitute a cell-free proteolytic system, to monitor dependence of the proteolytic process upon the addition of exogenous ubiquitin, or to monitor conjugation of labelled ubiquitin to different substrates. The procedure for depleting cells of ATP is included in *Protocol 1* and *2* but can be omitted if ATP depletion is unnecessary.

Protocol 1. Preparation of rabbit reticulocyte lysate

Equipment and reagents

- New Zealand white rabbits (preferably females) of ~ 2 kg body weight (two to three months old)
- Phenylhydrazine (Sigma)
- PBS pH 7.5
- Krebs Ringer phosphate (KRP) buffer pH 7.5 (120 mM NaCl, 5 mM KCl, 1.2 mM $MgSO_4$, 16 mM sodium phosphate) containing 20 mM 2-deoxyglucose (Sigma) and 0.2 mM 2,4-dinitrophenol (Sigma)
- 0.1 M DTT (stock solution; store at –20 °C)
- Low speed (up to ~ 1800 *g*) refrigerated centrifuge and tubes (50 ml, capped and conical)
- Ultracentrifuge, rotor, and tubes (fixed angle rotor for 35 ml capped tubes)
- Light microscope and methylene blue or brilliant cresyl blue for staining reticulocytes (28)

Method

Note: ensure that you comply with local animal experimentation legislation and animal welfare regulations.

1. Inject rabbits subcutaneously with 10 mg/kg phenylhydrazine (in PBS) on days 1, 2, 4, and 6.
2. Check the induction of reticulocytosis by taking a drop of blood from the ear vein on day 8. Induction of reticulocytosis should be dramatic with reticulocytes representing more than 90% of the circulating red blood cells. To check this, prepare blood films and stain them with methylene blue or brilliant cresyl blue as described in ref. 28.
3. Wash the cells three times with PBS by centrifuging them at 1000 *g* for 10 min, and resuspend the cell pellet in two to three cell pellet volumes of PBS.
4. Remove the white blood cells (buffy coat) by gentle aspiration with a Pasteur pipette.
5. To deplete the cells of ATP, resuspend the pellet in 1 vol. of KRP buffer containing 20 mM 2-deoxyglucose and 0.2 mM 2,4-dinitrophenol.[a]
6. Incubate the cells with gentle shaking for 90 min at 37 °C.
7. Wash the cells twice in PBS (see step 3).
8. Lyse the cells in 1.5 vol. distilled water containing 1 mM DTT.
9. Centrifuge the lysate at 80 000 *g* for 1 h to remove particulate material.
10. Collect the supernatant and freeze it in aliquots at –70 °C.

[a] Omit steps 5–7 if depletion of ATP is not required.

1.3 Fractionation of cell extracts for monitoring conjugation and degradation

Fractionation of the lysate enables one to study ubiquitin conjugation and subsequent proteolytic degradation independently, and to monitor their dependence upon the addition of exogenous ubiquitin and certain E2 enzymes.

Protocol 2. Preparation of extracts from cultured cells

Equipment and reagents

- Cells of choice grown in suspension or monolayer culture, as appropriate
- Hepes/saline buffer: 20 mM Hepes–NaOH pH 7.5, 150 mM NaCl
- Hepes/DTT buffer: 20 mM Hepes–NaOH pH 7.5, 1 mM DTT
- High nitrogen pressure chamber (50 ml volume; Parr Instrument Co.) and nitrogen bottle
- NaF (**caution: toxic**)
- NaN_3 (**caution: toxic**)
- KRP buffer (see *Protocol 1*)
- 0.1 M DTT (stock solution; store at –20°C)
- Centrifuges, rotors, and tubes (see *Protocol 1*)
- KRP buffer containing 20 mM 2-deoxyglucose, 0.2 mM 2,4-dinitrophenol, 20 mM NaF, and 10 mM NaN_3
- Light microscope (an inverted tissue culture microscope is sufficient)

Method

1. Wash the cells three times in Hepes/saline by centrifugation at 1000 *g* for 10 min.
2. Resuspend the cells (10^7 cells/ml) in KRP buffer containing 20 mM 2-deoxyglucose, 0.2 mM 2,4-dinitrophenol, 20 mM NaF, and 10 mM NaN_3.[a]
3. Incubate the cells for 60 min at 37°C with mild shaking to deplete them of ATP.
4. Wash the cells twice in Hepes/saline (as in step 1).
5. Resuspend the washed cells at 10^7–10^8 cells/ml in Hepes/DTT buffer.
6. Cavitate the cells in a high pressure nitrogen chamber at 4°C. For HeLa cells, the best conditions are two cycles of 30 min each at 1000 p.s.i. min. However, these conditions may vary among different cell species. Make sure that most cells are disrupted by visualizing the suspension in a light microscope before and after cavitation. Following disruption, one should observe intact nuclei and cell debris.
7. Centrifuge the homogenate at 3000 *g* for 15 min.
8. Remove the supernatant and centrifuge it at 10 000 *g* for 15 min.
9. Remove the supernatant and centrifuge it at 80 000 *g* for 1 h.
10. Freeze the supernatant in aliquots at –70°C.

[a] Omit steps 2–4 if depletion of ATP is not required.

To prepare lysate fractions, an extract from ATP-depleted cells is resolved on a DEAE cellulose column. When cells are depleted of ATP, all the ubiquitin is released from conjugates by isopeptidases and, in the absence of ATP, reconjugation is prevented. Fractionation of such a lysate over DEAE cellulose (*Protocol 3*) yields *Fraction I*, the unadsorbed flow-through material that contains the free ubiquitin and certain E2 enzymes, and *Fraction II*, the high salt eluate that contains the remaining ubiquitin-conjugating enzymes and the 26S proteasomes.

In cell extracts from which ATP was not depleted, the ubiquitin that is still conjugated to endogenous protein substrates will adsorb to DEAE via the protein substrate and will elute in *Fraction II*. During subsequent incubations, this bound ubiquitin fraction will be released by the activity of isopeptidases and will be available for conjugation to other proteins, including the test substrate under investigation. Therefore, it will be difficult to demonstrate ubiquitin-dependent conjugation and degradation in *Fraction II* that is prepared from an extract from cells that were not depleted of ATP. In addition, the bound ubiquitin fraction, when released, will dilute any added labelled ubiquitin, and thus decrease the detectable signal in ubiquitin adducts formed from exogeneously added labelled (radioactively or tagged) ubiquitin.

Protocol 3. Fractionation of cell lysates by DEAE[a] chromatography

Equipment and reagents

- Diethylaminoethyl cellulose (DEAE cellulose, DE-52; Whatman)
- Ammonium sulfate (enzyme grade; Life Sciences, Inc.)
- Dialysis tubing (12–14 kDa M_r cut-off; Spectrum)
- Buffer A: 5 mM potassium phosphate pH 7.0
- Buffer B: buffer A containing 1 mM DTT
- Buffer C: buffer A containing 1 mM DTT and 20 mM KCl
- Buffer D: 20 mM Tris–HCl pH 7.2, 1 mM DTT
- Buffer E: buffer D containing 1 M KCl
- 0.3 M potassium phosphate pH 7.0
- Chromatography column: the length should be approx. threefold greater than its width; the column volume required depends on the amount and protein concentration of the extract resolved lysate (see step 1, below)
- Reticulocyte lysate (*Protocol 1*) or cell extract (*Protocol 2*) prepared from ATP-depleted cells

Method

Carry out all procedures at 4°C.

1. Swell the DEAE cellulose in 0.3 M potassium phosphate pH 7.0 for several hours. Use enough matrix to adsorb all the proteins in the extract that can be bound. As a rule, use 0.6 volume of swollen matrix per volume of reticulocyte lysate, or 1 ml matrix for every 5 mg (approx.) of protein in a cell extract.
2. Load the matrix into a chromatography column and wash it with 8 column volumes of buffer A, and then with 2 volumes of buffer B.
3. Load the reticulocyte lysate or cultured cell extract.
4. Wash the column with buffer B and collect the flow-through (*Fraction I*). When fractionating reticulocyte lysate, collect only the dark red fraction. When fractionating a cell extract, pool only the fractions with the highest A_{280}. Freeze *Fraction I* in aliquots at –70°C.
5. Wash the column extensively with buffer C. When fractionating a reticulocyte lysate, make sure that all the haemoglobin is eluted.

When fractionating a cell extract, wash until the A_{280} returns to baseline.

6. Elute *Fraction II* with 2.5 column volumes of buffer E.
7. Gently stir *Fraction II* on ice and slowly add ammonium sulfate to saturation (~ 70 g/litre solution). Continue stirring on ice for 30 min.
8. Centrifuge the slurry at 15 000 *g* for 15 min.
9. Resuspend the pellet in 0.2–0.3 volumes of buffer D (relative to the volume of original extract). It may not be possible to dissolve all the proteins; this is not essential since they will dissolve during dialysis (step 10).
10. Dialyse the sample against two changes of buffer D.
11. Remove any particulate material by centrifugation at 15 000 *g* for 15 min.
12. Freeze the supernatant in aliquots at –70 °C.

[a] The method is a modification of the procedure described in ref. 29.

1.4 Labelling of proteolytic substrates

In most cases, monitoring the conjugation and/or degradation of a protein substrate requires it to be labelled, although these processes can be also followed by Western blotting and detection using specific antibodies directed against the test protein. Two methods of labelling protein substrates have proved to be useful: iodination with Na[^{125}I] and biosynthetic incorporation of a labelled amino acid such as [^{35}S]methionine. Radio-iodination (*Protocol 4*) is used mainly when a pure protein is available. The main advantage of the method is the high specific radioactivity that can be attained. The disadvantages, beyond the requirement to use a radioactive isotope, are:

- One needs a pure protein.
- Unless the iodination is carried out using the Bolton-Hunter reagent, the protein can be damaged by the oxidizing agent that is used to oxidize the iodide ion.
- The labelled substrate may be subject to radiochemical damage from isotopic decay.

The other method of labelling utilizes incorporation of [^{35}S]methionine into the protein during its synthesis in a cell-free system from the corresponding mRNA. Although the protein generated is native, the specific activity obtained is relatively low. Also, the labelled protein is not pure, being contained in the crude extract in which it has been synthesized. Furthermore, this extract contains enzymes of the ubiquitin system that may interfere with the reconstitution of a cell-free ubiquitinating system from purified components.

Protocol 4. Radio-iodination of protein substrates

Equipment and reagents

- Purified protein dissolved in buffer: do not dissolve the protein in a buffer that contains free amino or hydroxyl groups, as this may result in iodination of these groups
- 10 mM unlabelled NaI in H_2O
- Na[^{125}I] sodium iodide (specific activity 100–350 mCi/ml; Amersham Pharmacia Biotech, ICN, or NEN)
- Chloramine T (Sigma) freshly dissolved in 50 mM sodium phosphate buffer pH 7.5
- Sodium metabisulfite (Sigma) freshly dissolved in 50 mM sodium phosphate buffer pH 7.5
- Desalting column (low molecular mass exclusion gel filtration matrix; Amersham Pharmacia Biotech or Bio-Rad)
- 1 M sodium phosphate buffer pH 7.5
- 10 mM Tris–HCl pH 7.6, 150 mM NaCl
- 1.5 ml microcentrifuge tubes
- Fraction collector (optional)

Method

1. Equilibrate the desalting column in 10 mM Tris–HCl pH 7.6, 150 mM NaCl.
2. Add the following reagents in the order given to a 1.5 ml microcentrifuge tube. The total volume of the reaction mixture can be between 20–100 μl.
   - sodium phosphate buffer pH 7.5, 100 mM final concentration
   - 10–500 μg protein substrate
   - 50 nmol unlabelled NaI
   - 0.1–2 mCi radiolabelled Na[^{125}I]
   - 10–50 μg chloramine T, in 50 mM sodium phosphate buffer pH 7.5
3. Mix by vortexing and incubate for 1–2 min at room temperature.
4. Add 20–100 μg sodium metabisulfite (twice the amount of added chloramine T) in 50 mM sodium phosphate buffer pH 7.5, and mix.
5. To remove unreacted radioactive iodide, load the mixture onto the desalting column equilibrated in step 1.
6. Collect fractions (either manually or using a fraction collector) each of ~ 10% of column volume. The radioactive protein is typically eluted in fraction 4 (the void volume of the column which is ~ 35% of the total column volume).
7. Store the labelled protein in aliquots at –20 °C.

To label proteins biosynthetically, one can first synthesize the mRNA from a suitable cDNA template cloned in an expression vector, using the appropriate RNA polymerase. Following digestion of the cDNA plasmid, the mRNA can be extracted and translated *in vitro* in a reticulocyte or wheat germ extract. Alternatively, one can use a coupled transcription–translation cell-free extract. Such systems are available commercially (TNT®; Promega). Biosynthesis is carried out basically according to the manufacturer's instructions (30).

When monitoring ubiquitin conjugation and degradation using such *in vitro* translated and labelled proteins, one should bear in mind several principles:

(a) In many cases it is preferable to use a wheat germ extract for investigating the ubiquitin-proteasome pathway. This extract lacks many, although not all, of the mammalian E3 enzymes such as E6-AP (13) and the E3 that is involved in processing of NF-κB (31). Therefore, the protein synthesized in this extract can be used in experiments in which an E3 of interest is used to reconstitute a cell-free system. In contrast, proteins synthesized by the reticulocyte lysate may contain endogenous E3 enzyme(s) that are carried through to the reconstituted system and can be involved in the targeting of the test protein. This may interfere with the examination of the role of an exogenously added E3 enzyme in the conjugation of the translated protein. However, one should remember that the wheat germ extract does contain certain E3s and many enzymatically active homologues of mammalian E2 enzymes. Therefore, caution is still required when monitoring the reconstitution of the pathway with novel E2 and E3 enzymes.

(b) If one suspects interference from the components of the wheat germ extract or where translation in the wheat germ is not efficient, the more efficient reticulocyte translation extract can still be used and the interfering enzymes can be inactivated. All known conjugating enzymes (E1, E2s, and certain E3s) contain an essential functional -SH group and thus can be inactivated by the alkylating agent *N*-ethylmaleimide (NEM). After translation, incubate the extract with 10 mM NEM at room temperature for 10 min. Then neutralize the NEM with 7.5 mM DTT. This procedure can denature and inactivate the substrate protein as well and render it either resistant or more susceptible to the activity of the ubiquitin system. In most cases however, the substrate can still be utilized and it reproduces faithfully the behaviour of the native substrate.

(c) Crude extracts containing *in vitro* labelled proteins can be used to follow the degradation of these proteins without any further processing. This can be done by monitoring their disappearance by SDS–PAGE followed by fluorography or PhosphorImaging. However, this may not be always possible, as the degradation rates of certain proteins are low and it may be difficult to follow the disappearance of a labelled protein band in a gel. In this case, it will be necessary to monitor the release of TCA soluble radioactive material, but first the excess unincorporated labelled methionine present in the cell-free extract must be removed by dialysis or by fractionating it on DEAE exactly as described for fractionation of lysates (*Protocol 3*). The vast majority of the labelled proteins will elute in *Fraction II*, while the unincorporated labelled amino acid will be eluted in *Fraction I*. On the rare occasions that the labelled protein also elutes in *Fraction I*, changing the pH of the loading buffer may lead to its adsorption.

1.5 Conjugation of proteolytic substrates

To demonstrate that the degradation of a certain protein proceeds in a ubiquitin-dependent manner, it is essential to demonstrate the existence of intermediates in the process, i.e. ubiquitin-protein adducts. Typically, incubation of the labelled protein in a complete cell extract in the presence of ATP will lead to the formation of high molecular mass adducts that can be detected following resolution of the mixture in SDS–PAGE. To increase the amount of the adducts, one can use two approaches, independently or simultaneously:

(a) The non-hydrolysable ATP analogue, adenosine-5′-*O*-(3-thiotriphosphate), ATP-γ-S, can be used instead of ATP. The ubiquitin-activating enzyme, E1, can catalyse activation of ubiquitin in the presence of the analogue, as it utilizes the α-β high energy phosphoester bond of the nucleotide that is also present in the analogue. In contrast, assembly and activity of the 26S proteasome complex requires the β-γ phosphoanhydride bond that is absent from ATP-γ-S (32). Caution should be exercised, however, when utilizing ATP-γ-S; often, phosphorylation of the target protein is required for the ubiquitin ligase to recognize and conjugate it with ubiquitin. In these cases, the ATP-γ-S cannot substitute for ATP.

(b) An additional approach to increase the amount of generated conjugates in a cell-free system is to use ubiquitin aldehyde (UbAl), a specific inhibitor of certain ubiquitin C-terminal hydrolases, the isopeptidases (33).

UbAl is prepared from ubiquitin (*Protocol 5*) basically as described by Hershko and Rose (33). Initially, trypsin is used to cleave the 74-Arg-▼-Gly-Gly-COOH-76 bond in the ubiquitin molecule. Since the cleavage reaction is partially reversible and the reaction is carried out in the presence of Gly-ethyl ester, the ester can be bound to des-Gly-Gly ubiquitin to generate des-Gly-Gly-ubiquitin-Gly-ethyl ester (DGG-U-GEE). In successive reactions, the ester bond is cleaved and the exposed carboxyl group is reduced to aldehyde (UbAl).

Protocol 5. Preparation of ubiquitin aldehyde

Equipment and reagents

- Ubiquitin (Sigma)
- Glycine-ethyl ester (GEE; Sigma)
- Trypsin treated with *N*-tosyl-L-phenylalanine chloromethylketone (trypsin-TPCK; Sigma); make a 100 mg/ml stock solution
- 100 mg/ml soybean trypsin inhibitor (STI; Sigma)
- Carboxymethyl cellulose column (CM-52; Whatman): equilibrate the resin with 50 mM ammonium acetate pH 4.5 and load a 1.5 cm × 7 cm column
- Hydrazine hydrate (Fluka): the stock solution is ~ 19.5 M; make a 1 M solution containing 0.5 M HCl
- 5 M $NaNO_2$ (Fluka), freshly prepared
- 98% aminoacetaldehyde diethylacetal (Fluka)
- 100 mg/ml ovalbumin (Sigma)
- 1 M NaOH
- Superdex 75 HR column (1.6 cm × 60 cm) linked to an FPLC apparatus (Amersham Pharmacia Biotech): equilibrate the column in 50 mM ammonium acetate pH 4.5

- 1 M ammonium acetate pH 4.5
- 50 mM ammonium acetate pH 4.5, pH 5.5, and pH 7.0
- Concentrated HCl
- 10 mM HCl
- Dialysis tubing (3.5 kDa M_r cut-off; Spectrum)
- Freeze-drier
- Circulating water-bath at –4 °C (Amersham Pharmacia Biotech; with ethylene glycol)
- 3 M Na_2CO_3
- 20 mM Tris–HCl pH 7.2, 0.1 mM EDTA

A. *Preparation of DGG-U-GEE*

1. Dissolve 25 mg ubiquitin in 600 μl water.
2. Dissolve 670 mg GEE in 900 μl water. Then add 900 μl of 1 M NaOH (the final pH should be ~ 7.2).
3. Mix the ubiquitin solution (step 1) with the GEE solution (step 2).
4. Add 25 μl of 100 mg/ml trypsin-TPCK and incubate overnight at 37 °C.
5. Add 50 μl of 100 mg/ml STI. Incubate for 3 min at 37 °C.

From this step on, carry out all procedures at 4 °C.

6. Add 0.05 vol. of 1 M ammonium acetate pH 4.5.
7. Load the reaction mixture on the Superdex 75 HR column equilibrated with 50 mM ammonium acetate pH 4.5 and monitor the A_{280}. Two protein peaks will be detected: an early peak of trypsin and STI and a smaller peak containing unreacted ubiquitin and DGG-U-GEE.[a]
8. Load the second peak onto the CM-52 column equilibrated with 50 mM ammonium acetate pH 4.5. Wash the column with approx. 3 column volumes of 50 mM ammonium acetate pH 4.5 until the A_{280} returns to baseline.
9. Elute the ubiquitin with 50 mM ammonium acetate pH 5.5. Collect at least 40 fractions of 3 ml each. Make sure the A_{280} returns to baseline.
10. Elute the DGG-U-GEE with 50 mM ammonium acetate pH 7.0. Collect at least 20 fractions of 3 ml each until all the protein is eluted.
11. Dialyse the DGG-U-GEE eluate extensively using low molecular weight cut-off dialysis tubing against 10 mM HCl. Lyophilize the dialysed DGG-U-GEE.
12. Dissolve the lyophilized protein in 2 ml of 1 M hydrazine hydrate containing 0.5 M HCl. The pH of the solution should be 8.1–8.5.
13. Incubate the solution for 4.5 h at 37 °C and then dialyse it extensively against water (48 h at 4 °C, four changes of 2 litres each).
14. Measure the protein concentration against a ubiquitin standard.[b]
15. Lyophilize aliquots of ~ 2 mg. Store these at –70 °C.

B. *Preparation of Pre-UbAl*

1. Take a 2 mg aliquot of DGG-U-DEE and add 800 μl water. Then add 80 μl of 5 M $NaNO_2$ followed by 34.5 μl conc. HCl. Incubate for 10 min at –4 °C.

Protocol 5. *Continued*

2. Add 145 μl of 98% aminoacetaldehyde diethylacetal followed by 55 μl ice-cold water. Mix and incubate at –4 °C for 45 min.
3. Dialyse the mixture against water (two changes of 2.5 litres each). Freeze the dialysed pre-UbAl at –70 °C in ten aliquots of 200 μg (~ 200 μl).

C. *Conversion of Pre-UbAl to UbAl*

1. To each aliquot of pre-UbAl add 18.9 μl conc. HCl (final concentration 1 M). Mix gently and incubate 30 min at room temperature.
2. Titrate the acid by adding 3 M Na_2CO_3 until the pH is 7.0–8.5 (about 40 μl of the sodium carbonate solution will be required). Transfer the solution to ice.
3. Add 2.6 μl of 100 mg/ml ovalbumin and dialyse extensively against a buffer containing 20 mM Tris–HCl pH 7.2, 0.1 mM EDTA. Store dialysed aliquots of 10 μl (~ 10 μg) at –70 °C.[c]

[a] If desired, the second peak may be frozen at –70 °C.
[b] Because of the small molecular mass and atypical amino acid composition of ubiquitin, it is impossible to calculate its amount based on the absorbency or colour reaction of a typical standard protein such as BSA. Therefore, accurate measurement can be based only on comparison to the absorbency or colour reaction of a known weighed standard of ubiquitin.
[c] Thaw each aliquot only once.

Protocol 6 describes the method used to investigate the conjugation of ubiquitin to a test protein. Methylated ubiquitin (MeUb) is prepared as described in *Protocol 7*.

Protocol 6. Conjugation of ubiquitin to a test protein substrate in a cell-free system

The volume of the reaction mixture can vary from 12.5–50 μl and may contain ATP and an ATP-generating system, or ATP-γ-S, or (for depletion of ATP) 2-deoxyglucose and hexokinase. The assay procedure for the conjugates produced will also depend on the test protein used. For all these reasons, only an outline protocol rather than precise details can be given.

Method

1. Add the following reagents to a 1.5 ml microcentrifuge tube. The volume of the reaction mixture can vary from 12.5–50 μl. Addition of all the reagents should be carried out on ice:
   - 50 mM Tris–HCl pH 7.6 (final concentration)
   - 5 mM $MgCl_2$ (final concentration)

- 2 mM DTT (final concentration)
- 7.5–30 μl reticulocyte lysate or 50–200 μg complete cell extract protein
- 2.5–10 μg ubiquitin
- 0.5–2 μg UbAl
- ATP and ATP-regenerating system (0.5 mM ATP, 10 mM creatine phosphate, and 2.5–10 μg creatine phosphokinase)[a] or 2 mM ATP-γ-S[b] (final concentration)
- Substrate: use either a labelled protein (25 000–100 000 c.p.m.) or an unlabelled substrate in an amount sufficient for subsequent detection by Western blot analysis (20–2000 ng)

2. Incubate the mixture for 30–60 min at 37 °C.

3. Fractionate the proteins by SDS–PAGE using a 7.5% polyacrylamide or 10% polyacrylamide gel.

4. For labelled proteins, detect high molecular mass conjugates by fluorography or PhosphorImager analysis. For unlabelled protein substrates, detect high molecular mass conjugates by Western blot analysis using a specific primary antibody against the test protein and a secondary tagged antibody.

[a] For depletion of endogenous ATP, the system should contain 10 mM 2-deoxyglucose and 0.25–1 μg hexokinase ***instead*** of ATP and the ATP-regenerating system. Obtain the hexokinase from Boehringer Mannheim as an ammonium sulfate slurry. Centrifuge the slurry and resuspend the pellet at 10 mg/ml in 20 mM Tris–HCl buffer pH 7.6. Dilute this stock solution in the same buffer for use. The 10 mg/ml stock solution can be stored at 4°C for at least one month.

[b] Where it is suspected that phosphorylation of the substrate is required for its recognition and conjugation by E3, it is necessary to utilize ATP in the reaction.

There are several ways to demonstrate that the high molecular mass adducts generated are indeed ubiquitin conjugates of the test protein:

(a) Adducts should not be generated in an ATP-depleted system. To deplete the system of ATP, incubate the lysate with 2-deoxyglucose and hexokinase before the conjugation reaction is carried out (i.e. prior to addition of the substrate).

(b) Generation of the conjugates should be inhibited reversibly by the addition of increasing amounts of methylated ubiquitin (MeUb; see *Protocol 7*) (34). This reductively methylated derivative of ubiquitin lacks free amino groups and therefore cannot generate polyubiquitin chains. It serves therefore as a chain terminator in the polyubiquitination reaction, and consequently, as an inhibitor in this reaction (31, 34, 35).

(c) Conjugates can be precipitated from the reaction mixture with an antibody directed against the test protein, subjected to SDS–PAGE, and following transfer to nitrocellulose membrane probed with an anti-ubiquitin antibody to demonstrate the presence of ubiquitin in the adduct.

(d) A cell-free system can be reconstituted from purified or isolated components of the ubiquitin system and the formation of conjugates can be followed and shown to depend upon the addition of these components. Instead of adding a complete cell extract, it is possible to add *Fraction II* derived from ATP-depleted cells (see Section 1.3) and free ubiquitin. Since *Fraction II* is devoid of ubiquitin, formation of conjugates that is dependent upon the addition of exogenous ubiquitin will strongly suggest that the high molecular mass derivatives generated are indeed ubiquitin adducts of the test substrate. Since not all E2 enzymes are present in *Fraction II*, it may be necessary, at times, to add to the reconstituted system either crude *Fraction I*, or purified E2 enzymes contained in *Fraction II* (10, 36, 37).

Methylated ubiquitin (MeUb) is prepared as described in *Protocol 7*.

Protocol 7. Preparation of methylated ubiquitin

Equipment and reagents

- Ubiquitin (Sigma)
- Formaldehyde
- Cyanoborohydride (Fluka)
- 0.6 mg/ml fluorescamine (Sigma) in acetone
- 0.1 M Hepes–NaOH pH 7.4, 6 M urea
- Dialysis tubing (3.5 kDa M_r cut-off; Spectrum)
- 0.2 M $NaBO_4$ pH 9.0

Method

1. Dissolve 10 mg ubiquitin in 10 ml of 0.1 M Hepes–NaOH pH 7.4, 6 M urea.
2. Add formaldehyde and cyanoborohydride to final concentrations of 12 mM and 20 mM, respectively. Incubate the reaction at room temperature for 24 h. Then repeat the addition of formaldehyde and cyanoborohydride and continue the incubation for an additional hour.
3. Dialyse the mixture extensively against water.
4. Check that all the amino groups in the ubiquitin molecule have been methylated. To do this, use the fluorescamine assay:
   (a) Prepare duplicate samples of protein (2–20 μg) each in 700 μl H_2O.
   (b) To each sample, add 200 μl of 0.2 M $NaBO_4$ pH 9.
   (c) Add, while mixing, 0.1 ml of 0.6 mg/ml fluorescamine in acetone.
   (d) Measure the fluorescence at 475 nm following excitation at 390 nm. Fluorescamine reacts with amino groups and hence the fluorescence signal obtained titrates the amino groups remaining in the protein sample.
5. If necessary, repeat steps 2 and 3 to ensure methylation of all the amino groups.

1.6 Degradation of proteolytic substrates

With several exceptions, cell-free systems for monitoring degradation of proteolytic substrates are similar to those used for monitoring their conjugation. In proteolytic assays, ATP must be used as activity of the 26S proteasome complex is dependent upon cleavage of the high energy β-γ phosphoanhydride bond (see Section 1.5). ATP is added along with an ATP-regenerating system (see *Protocol 6*). Following incubation of 2-3 h at 37 °C, the disappearance of the substrate can be monitored by SDS–PAGE followed by either fluorography or PhosphorImager analysis (in the case of a labelled substrate), or via Western blot analysis (if the substrate is not labelled). Control reactions are mixtures that have been incubated on ice, and at 37 °C in the absence of ATP. At times, degradation rates are low, and it is extremely difficult to follow the reduction in the amount of a protein band in SDS–PAGE analysis. In these cases, it will be necessary to monitor the appearance of radioactivity in the TCA soluble fraction. Here, only radioactive substrates can be used. Radioiodinated proteins (*Protocol 4*) can be used directly. *In vitro* translated proteins must undergo DEAE fractionation or dialysis in order to remove excess unincorporated labelled methionine (see Section 1.4). At the end of the incubation, 0.5 ml ice-cold TCA (20%) is added and, following mixing, the reaction is incubated on ice for 10 min. To ensure complete precipitation of the labelled, undigested protein, it is recommended to add to the reaction prior to the addition of TCA a carrier protein (10–25 μl of 100 mg/ml BSA). Following centrifugation (5 min at 15 000 *g*), the supernatant is collected and the radioactivity is determined using either a β-scintillation counter (for [^{35}S]methionine labelled products) or a γ-counter (for iodine labelled protein).

1.7 Use of inhibitors to study proteasome function

All the known substrates of the ubiquitin system are degraded, following conjugation, by the 26S proteasome complex. The converse, that all substrates of the 26S proteasome must be ubiquitinated prior to their recognition by the proteasome enzyme, is true in all but one known case, that of ornithine decarboxylase (27). A non-covalent association with another protein, antizyme, renders ornithine decarboxylase susceptible to degradation by the proteasome.

The core catalytic subunit of the 26S proteasome is the 20S proteasome complex and inhibition of this complex inhibits all proteolytic activities of the 26S proteasome. To test whether a protein substrate is degraded by the 20S/26S proteasome, it is possible to inhibit proteolysis, both *in vitro* and *in vivo* (see *Protocol 8*). Stabilization of a protein under such conditions is a strong indication that the protein is indeed degraded by the 26S proteasome. Furthermore, inhibition of the 20S proteasome may lead to accumulation of ubiquitin adducts of the test protein that otherwise cannot be detected under conditions of rapid degradation when the proteasome is active. Detection of such intermediates will strongly suggest that the protein is degraded by the 26S proteasome complex following conjugation with ubiquitin.

There are several potent inhibitors of the proteasome. One is the *Streptomyces* metabolite lactacystin that is highly specific towards the proteasome and does not inhibit calpains (38). It is cell permeable and can be used at a concentration range of 1 μM (in a cell-free system) to 10 μM (*in vivo*) (39). An analogue that penetrates the cell even more efficiently, clasto-lactacystin β lactone, has been developed recently. The other inhibitor, MG132 (*N*-Carbobenzoxyl-Leu-Leu-Leucinal), is a derivative of the calpain inhibitor I, *N*-Acetyl-Leu-Leu-Norleucinal, and it has also a weak activity against calpain. It is also cell permeable and can be used in cells at a concentration range of 10 μM to 100 μM (26). For unknown reasons, we have found that it is much less effective in a cell-free system (unpublished observations). Both inhibitors can be dissolved in DMSO and then diluted in H_2O. In cell-free systems, it is recommended to pre-incubate the reaction mixture for 5 min in the presence of the inhibitor prior to the addition of the substrate. In pulse-chase experiments in intact cells, the best effects are attained if the inhibitor is added during the labelling period, preferably 60 min prior to the beginning of the chase. These inhibitors are now available from Calbiochem (Cat. Nos lactacystin, 426100; clasto-lactacystin β-lactone, 426102; MG132, 474790). As mentioned in Section 1.5, when following the accumulation of ubiquitin adducts in a cell-free system, it is also possible to inhibit the activity of the proteasome by the utilization of the non-hydrolysable ATP analogue adenosine-5′-*O*-(3-thiotriphosphate), ATP-γ-S.

Protocol 8. Involvement of the ubiquitin system in the degradation of proteins *in vivo*: effect of specific proteasomal inhibitors on the half-life of proteins in intact cells

Equipment and reagents

- Cultured cells in a monolayer or in suspension
- [^{35}S]methionine (1000 Ci/mmol at 10 mCi/ml; Amersham Pharmacia Biotech, ICN, or NEN)
- Lactacystin (or its lactone homologue) or MG132 dissolved in DMSO: stock solutions are 1–10 mM and 10–100 mM for lactacystin (or its lactone homologue) and MG132, respectively
- Methionine-free medium, preferably the same medium in which the cells are growing (however, for the short labelling period this may not be essential and any methionine-free medium with similar composition can be used)
- Dialysed serum (preferably the same source of serum in which the cells are grown)

Method

1. Wash the cells twice in methionine-free medium.
2. Add methionine-free medium containing dialysed serum (serum is

added in the concentration used for growing the cells) and incubate for 1 h.

3. Remove the medium (by aspiration for adherent cells and following centrifugation at 800 *g* for 10 min for cells in suspension) and add fresh methionine-free medium that contains dialysed serum. To save on labelled methionine, for adherent cells add medium to barely cover the cells. For cells in suspension, resuspend the cells to 2×10^6/ml.
4. Add labelled methionine (25–250 μCi/ml) and continue the incubation for 1–2 h (pulse).
5. Add the proteasome inhibitor to the appropriate dishes/tubes. Lactacystin or its lactone homologue should be added at a final concentration of 5–20 μM, whereas MG132 is added at a final concentration of 20–50 μM. The inhibitors should be added 1 h prior to the beginning of the chase, i.e. along with the labelled methionine (step 4) when labelling for 1 h, or following 1 h of labelling when labelling for 2 h.
6. Remove the labelling medium (which also contains the inhibitor for some of the samples) and wash the cells twice with complete medium containing 2 mM unlabelled methionine.
7. Add pre-warmed complete medium to the cells and continue the incubation for the desired time period (chase). The relevant proteasome inhibitors should be present in the chase medium as they were during the pulse (see step 5 for guide-lines).
8. Withdraw samples at various time points and monitor the fate of the target protein by immunoprecipitation followed by fluorography or phosphorimaging. For immunoprecipitation, use aliquots of the samples that contain equal amounts of radioactivity incorporated into proteins. To avoid proteolysis of the conjugates in cell lysates by ubiquitin C-terminal hydrolases, lyse the cells with a detergent-containing lysis buffer at 100 °C (40). Determination of total incorporated radioactivity is carried out by spotting an aliquot of the labelled cell extract onto a GF/C filter, washing the filter three times in 5% ice-cold TCA, and counting the filter in a β-scintillation counter. High molecular mass conjugates of the labelled protein with ubiquitin should be precipitated by the specific antibody directed against the target protein under study and can be visualized following SDS–PAGE and fluorography. The high molecular mass conjugates migrate in SDS–PAGE more slowly then the immunoprecipitated protein, and in many cases lie at the top of the gel. To corroborate that these compounds are indeed ubiquitin conjugates of the target protein, the gel can be blotted onto a nitrocellulose membrane to form a Western blot and then probed with anti-ubiquitin antibody (Sigma, Zymed, Chemicon, Santa-Cruz).

2. Proteolysis in mammalian lysosomes

2.1 Introduction

Subcellular compartmentalization is the main way in which a cell protects itself against the destructive potential of its proteolytic enzymes. To become degraded, proteins either have to enter a channel equipped with catalytic centres as in proteasomes and tricorn proteases (41), or they have to cross the membrane of an organelle containing peptidases, mostly by macro- or microautophagy (42).

In vitro, lysosomes show a very high proteolytic potential to degrade cell-derived proteins (43), and indeed, lysosomes are important organelles for cellular protein breakdown. For example, after 48 hours of starvation, rats and mice can rapidly mobilize up to 40% of their liver proteins by the lysosomal pathway to fill the amino acid pools for *de novo* protein synthesis and for gluconeogenesis (44). Numerous studies by Mortimore and Pösö (44) have clarified how this process is regulated in the perfused mammalian liver as well as in isolated hepatocytes. The lysosomal aspect of cellular protein breakdown in liver can be changed dramatically under the influence of amino acids and hormones (45). High serum levels of leucine and insulin are the essential inhibitors of macroautophagy, whereas amino acid deprivation and glucagon induce this process (45). In contrast, in muscle, the largest body protein reservoir, lysosomal protein degradation plays a minor role, and additional proteolytic systems are involved when accelerated protein breakdown is induced (46).

There are several mechanisms of protecting cells and their surrounding proteins against deleterious actions of lysosomal peptidases, e.g. after tissue injury. Protein-type endogenous inhibitors—the stefins, cystatins, and high molecular weight kininogens (47)—show a high affinity for lysosomal cysteine peptidases. These inhibitors act extralysosomally and, therefore, do not disturb the intralysosomal protein breakdown. However, due to the stable, but reversible, interactions of the inhibitors with the enzymes, the inhibitor:enzyme complexes represent a potential pool of active enzymes outside of lysosomes. No such cell-derived inhibitors against aspartic peptidases have been described yet.

The many similar lysosomal peptidases which have been preserved during evolution are able to digest almost all the peptide bonds in any protein. Therefore, once it has reached the lysosome, a substrate protein becomes completely degraded to amino acids and small peptides which leave the lysosomes (48) to be finally degraded by cytosolic exopeptidases.

It remains a question whether intralysosomal proteolysis is regulated. The propeptide regions of lysosomal peptidases often exhibit an inhibitory potential against their parental peptidase and can perhaps exert a regulatory role on lysosomal protein turnover. Another open question is whether all substrate proteins enter the lysosomes by micro- or macroautophagy. Some proteins carry recognition sequences, such as the KFERQ motif, which interact with receptor-like domains at the lysosomal membrane. This 'sorting' route of substrate proteins to the lysosomal compartment is under active study (49).

2.2 Isolation of mammalian lysosomes

The isolation of primary and secondary lysosomes from cell or tissue homogenates is difficult due to the great number of cellular vacuoles with various functions but similar physico-chemical properties. Nevertheless, it is often necessary to isolate the organelles for special purposes, e.g. as a first step in the purification procedure for lysosomal enzymes. The enzymes are manyfold concentrated within the lysosomes compared to the cell as a whole, and the gain of specific activity can be 50-fold by such a step. Furthermore, the interaction of lysosomal enzymes with enzyme inhibitors present in other cell compartments can be prevented by isolating the lysosomes first.

Unfortunately, the separation of lysosomes from other organelles is made more difficult by a number of factors:

- The heterogeneity of secondary lysosomes is exaggerated by entrapped molecules and parts of other organelles with different physico-chemical properties making the buoyant density of lysosomes far from uniform.
- The size of lysosomes varies in different cell types.
- Many reports have demonstrated variations in the enzyme content of lysosomes within a homogeneous cell population indicating the existence of different kinds of lysosomes in the same cell. This explains why there is no common isolation protocol for lysosomes. Special features of cells and tissues whose lysosomes should be isolated have to be considered. The following protocols, therefore, give only a general guide on how to plan the isolation procedure.

2.2.1 Tissue and cell homogenization

The object of tissue or cell homogenization is the selective disruption of plasma membranes of cells while retaining the lysosomal membranes intact. Gentle physical cell disruption methods are well suited to prepare such a homogenate. Avoid the use of detergents! For the isolation of peptidases, organ or tissue samples should be free of blood contamination since blood serum contains proteinase inhibitors that can interact with lysosomal peptidases. Whole organs should be perfused with 0.15 M NaCl before blood clotting occurs. If perfusion is not possible, careful washing of the tissue pieces is recommended.

A detailed discussion of other homogenization procedures is given by Dingle (50). *Protocol 9* describes a standard homogenization procedure. The success of the homogenization is judged by assaying marker enzymes, β-hexosaminidase for the lysosomes and lactate dehydrogenase (LDH) for the cytosol and unbroken cells. The aim is to adjust the homogenization conditions to minimize the LDH activity in the pellet which contains the lysosomes (but also unbroken cells) and minimize the β-hexosaminidase activity in the supernantants (cytosol). More than 10% of the total LDH activity in the pellet indicates insufficient cell disruption. More than 10% of the total β-hexosaminidase activity in the supernatant indicates too extensive disruption of the lysosomes. If this occurs, homogenize the cells or tissue to a greater or lesser extent according to the results of the marker enzyme assays.

Protocol 9. Homogenization of tissues and cells

Equipment and reagents

- Tissue or cultured cells for the isolation of lysosomes
- Scissors
- Sieve (kitchen pattern)
- Potter-Elvejehm homogenizer: Teflon-glass pattern with pestles giving 40 μm and 100 μm clearances at 4 °C
- Low speed centrifuge
- 0.25 M sucrose (for protection of cysteine proteinases the addition of 5 mM Na EDTA is recommended)
- PBS: 150 mM NaCl, 10 mM sodium phosphate buffer pH 7.2
- Assay kit for LDH (final concentrations): 50 mM sodium phosphate buffer pH 7.5, 600 μM pyruvate, 180 μM NADH
- Assay kit for β-hexosaminidase (final concentrations): 10 mM sodium acetate pH 4.4, 25 mM methyl-umbelliferyl-2-acetamido-2-deoxy-D-glucopyranoside, 50 mM sucrose
- Refrigerated centrifuge, e.g. Sorvall RC-5B (DuPont), GSA rotor, 500 ml tubes

Method

Note: perform all steps at 4 °C in a cold room, using pre-cooled solutions and instruments. Perform the homogenization in an ice-bath to keep the clearance of the homogenizer constant (Teflon and glass have different thermal expansions).

1. (a) For tissue pieces: mince the tissue finely with scissors and remove connective tissue and blood vessels by pressing the tissue through a kitchen sieve. Add 1–2 vol. of 0.25 M sucrose.
   (b) For cell cultures: harvest the cells with a rubber policeman. Wash them twice with 1 vol. PBS, sedimenting the cells at 1500 *g* for 3 min. Resuspend the washed cells in 2–3 vol. of 0.25 M sucrose.
2. Homogenize the tissue or cell suspension by five to ten strokes with a 100 μm clearance Teflon-glass homogenizer, followed by five to ten strokes with a 40 μm clearance homogenizer.
3. Centrifuge the homogenate at 10 000 *g* for 3 min.
4. Remove and retain the supernatant. Resuspend the lysosomal pellet in the original volume of 0.25 M sucrose.
5. Assay the activity of LDH and of β-hexosaminidase in the resuspended pellet and in the supernatant.
   (a) For LDH, mix the sample with reagents to give the following final concentrations: 50 mM sodium phosphate buffer pH 7.5, 600 μM

pyruvate, 180 μM NADH. Incubate at 25 °C for 10 min and then measure the change in absorbance at 340 nm.

(b) For β-hexosaminidase, mix the sample with reagents to give the following final concentrations: 10 mM sodium acetate pH 4.4, 25 mM methyl-umbelliferyl-2-acetamido-2-deoxy-D-glucopyranoside, 50 mM sucrose. Incubate at 37 °C for 10 min and then measure the change in fluorescence at 448 nm after excitation at 364 nm.

6. If cell disruption is insufficient based on the recovery of lysosomes, as judged by the marker enzymes, subject the cells or tissue to one to three freezing–thawing cycles and/or sonication (one to three bursts for 10 sec in an ice-bath) before the homogenization (step 2).

2.2.2 Preparation of lysosomes and lysosomal extracts

Homogenates from *Protocol 9* can be used to prepare either intact lysosomes or a lysosomal extract. *Protocol 10* results in a fraction containing lysosomes, mitochondria, and peroxisomes. Lysosomes are less resistant to hypo-osmotic solutions and so can be disrupted by simply resuspending them in water for some hours and then re-homogenizing using the 40 μm clearance homogenizer. The enrichment of specific activity of lysosomal enzymes in the resulting extract can be 50-fold above that of the original homogenate (51). For the preparation of intact lysosomes, use density gradient centrifugation (*Protocol 11*) or use Triton WR 1339 loading (*Protocol 12*). *In vivo*, Triton WR 1339 accumulates rapidly in liver lysosomes. The buoyant density of Triton WR 1339-filled lysosomes is lower than that of untreated lysosomes, thus allowing a rapid separation from mitochondria and peroxisomes. However, since the detergent interferes with normal lysosomal function, the use of Triton WR 1339-loaded lysosomes can be recommended only for studies of lysosomal membrane components. For more details and specialized isolation procedures for lysosomes and lysosomal extracts, see ref. 50.

Protocol 10. Preparation of a lysosomal fraction by differential centrifugation

Equipment and reagents

- Cell or tissue homogenate, 0.25 M sucrose, homogenizers, enzyme assay kits (see *Protocol 9*)
- 0.3 M sucrose
- Refrigerated centrifuge, e.g. Sorvall RC-5B (DuPont), GSA rotor plus 500 ml tubes, or SS-34 rotor plus 39 ml tubes

Method

1. Centrifuge the homogenate (from *Protocol 9*) at 5000 *g* for 6 min. Remove the supernatant and store it on ice.
2. Resuspend the pellet in 1 vol. of 0.25 M sucrose and centrifuge again.

Protocol 10. *Continued*

3. Discard the pellet and combine the second supernatant with the first.
4. Centrifuge the combined supernatants at 20 000 *g* for 5 min.
5. Resuspend the pellet (lysosomal fraction, LM) in 1 vol. of 0.3 M sucrose and centrifuge again. Discard the supernatant.
6. Check β-hexosaminidase activity of each fraction. To do this, mix one volume of any fraction with four volumes of the assay kit described in *Protocol 9*. Incubate for 10 min at 37 °C, and then measure the change in fluorescence at 448 nm after excitation at 364 nm.
7. Resuspend the pellet in 2.5 vol. water and allow it to stand at 4 °C for at least 1 h.
8. Homogenize the pellet in a 40 μm clearance homogenizer (ten strokes).
9. Centrifuge the homogenate at 20 000 *g* for 50 min.
10. Recover the supernatant (lysosomal extract) and store it at –20 °C.

Protocol 11. Preparation of lysosomes by density gradient centrifugation

Equipment and reagents

- Lysosomal fraction (LM) prepared as in *Protocol 10*, step 4
- Percoll® (density 1.13 g/ml as available from Pharmacia Biotech)
- 2.5 M sucrose in water
- 0.3 M imidazole in water
- Ultracentrifuge, vertical rotor (e.g. VTI 65 from Beckman), Quick-Seal tubes, and heat sealer
- Fraction recovery system (Beckman Instruments)
- Peristaltic pump allowing a flow of less than 0.5 ml/min

Method

1. Prepare 20 ml of 20% Percoll by mixing 4.6 ml Percoll, 2.5 ml of 2.5 M sucrose, 0.25 ml of 0.3 M imidazole, and water (to 20 ml). Adjust to pH 7.4 with 1 M HCl (~ one drop) and cool the Percoll mixture to 4 °C.
2. Nine-tenths fill a centrifuge tube with 20% Percoll (9 vol.) and load the LM fraction (1 vol.) on top. Heat seal the tube.
3. Allow the density gradient to form during 60 min centrifugation at 20 000 *g*, 4 °C. Stop the run slowly without the centrifuge brake.
4. Place the tube into the support of the fraction recovery system and fractionate the gradient slowly using one of the methods illustrated in *Figure 2*. For this purpose, connect the 'in' needle (see *Figure 2*) with a peristaltic pump. Make sure the selected method of recovery avoids disturbing any floating material in the tube. The number and volume of fractions collected depends on the size of the tube.

5. Add one-tenth volume of Triton X-100 to an aliquot of each fraction. Assay β-hexosaminidase (see *Protocol 10*, step 6). Combine fractions with the highest β-hexosaminidase activity as the lysosomal fraction. Lysosomes remain intact in the fractionation medium for up to 24 h when stored at 4 °C.

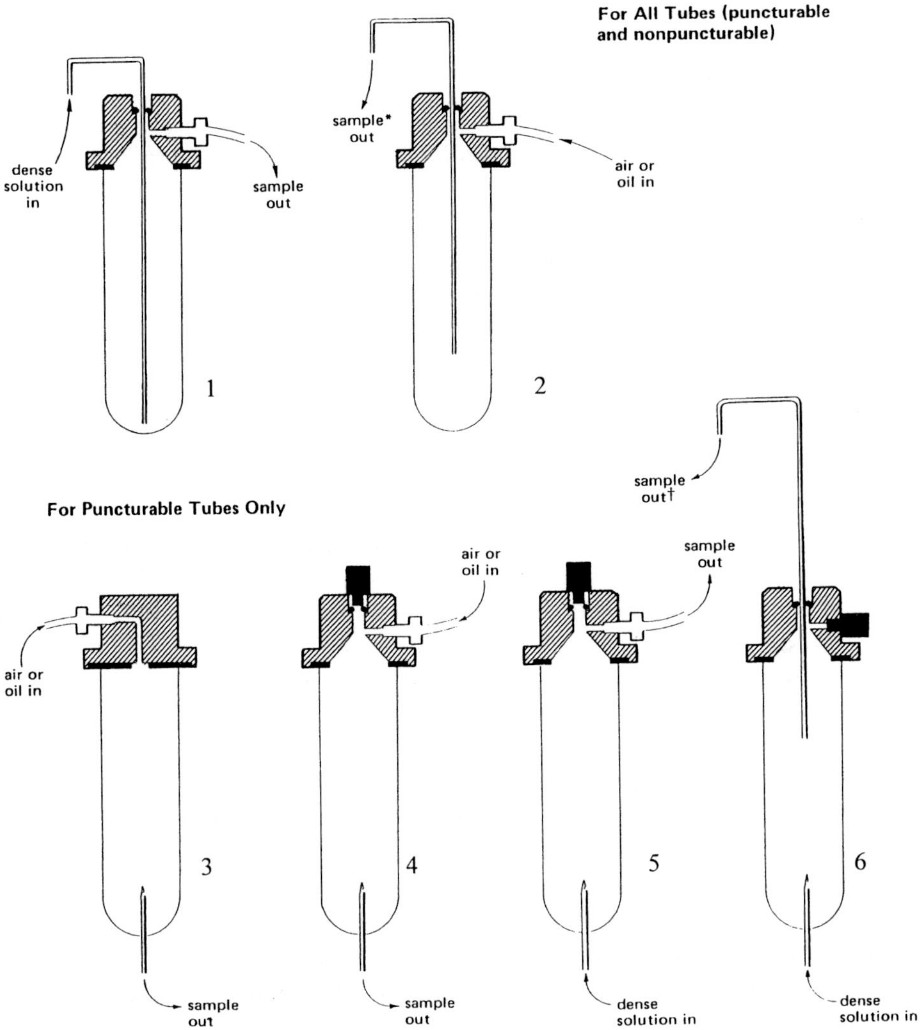

Figure 2. Methods for recovering fractions after density gradient centrifugation (reproduced with kind permission of Beckman Instruments). Methods 3–6 are only possible with puncturable tubes. Method 6 is recommended when only one definite fraction should be recovered. In method 2, the bottom of the drain tube* should be located just below the desired sample fraction. In method 6, the bottom of the drain tube† should be located just above the desired sample fraction.

Protocol 12. Preparation of rat liver lysosomes loaded with Triton WR 1339[a]

Equipment and reagents

- Equipment and reagents for preparing a tissue homogenate (see *Protocol 9*)
- Triton WR 1339
- 15%, 35%, 45%, and 60% (w/w) sucrose solutions
- Rat (at least 100 g body weight)

Method

Note: ensure all regulatory procedures and animal welfare measures specified by local and national authorities are strictly adhered to.

1. Inject a rat with 85 mg Triton WR 1339/100 g body weight via the intraperitoneal route.
2. Four days later kill the rat and remove the liver.
3. Prepare a liver homogenate as in *Protocol 10*, steps 1–3.
4. Centrifuge the homogenate at 3300 *g* for 3 min. Remove the supernatant and store it on ice.
5. Resuspend the pellet in 1 vol. of 0.25 M sucrose and centrifuge again. Discard the pellet of cellular debris and combine the two supernatants.
6. Centrifuge the combined supernatants at 20 000 *g* for 15 min. Discard the supernatant and resuspend the pellet in 45% (w/w) sucrose.
7. Place a cushion (one-quarter of the centrifuge tube volume) of 60% (w/w) sucrose in a centrifuge tube. Load the resuspended lysosomal pellet (step 6, another quarter of the tube volume) on top of the cushion. Gently pipette two further sucrose layers (another quarter of tube volume each) of 35% and 15% (w/w), respectively, on top of the lysosomal layer.
8. Centrifuge the gradient at 25 000 *g* for 20 min.
9. Remove the fractions from the bottom using a peristaltic pump as mentioned in *Protocol 11*. The Triton-loaded lysosomes float between the two upper sucrose layers, whereas most of the mitochondria and peroxisomes are found on top of the dense sucrose cushion.
10. Store the Triton-loaded lysosomes at 4°C. Storage overnight is possible, but it is recommended that the lysosomes be used immediately for further experimentation rather than be stored.

[a] Method from ref. 50.

2.3 Lysosomal peptidases

2.3.1 Properties of lysosomal peptidases

In 1929 Willstätter and Bamann introduced the term cathepsin (Greek origin, to digest) for cellular proteolytic enzymes of the stomach epithelium (52). Press, Porter, and Cebra (53) were the first to characterize an acid lysosomal peptidase from bovine spleen and named it cathepsin D. Cathepsin D was shown to be an aspartic peptidase. However, most lysosomal peptidases are cysteine proteinases. For a summary of the properties of cathepsins B, H, L, S, and C, see the review by Kirschke *et al.* (54). This summarizes the gene structures, the chromosomal localizations, purification procedures, main natural and synthetic substrates, molecular properties, and possible functions of the respective enzymes in normal mammalian physiology as well as in certain diseases, and includes a collection of 1237 citations. However, the recently discovered cathepsin O2, now called cathepsin K (55, 56), is not included. *Table 1* contains some selected properties of lysosomal peptidases.

2.3.2 Isolation of cathepsins

Most cathepsins were first detected in and isolated from animal organs. The original isolation protocol for cathepsin D started with a pH 3.5 treatment of the tissue homogenate overnight (53). Although this step is very effective for the isolation of cathepsin D, it delayed the discovery of the cysteine peptidases for a long time. Under these conditions, the cysteine peptidases are complexed with their endogenous inhibitors, and the yield of these enzymes was not improved until the protocols included the isolation of intact lysosomes as a first step (51). Only fresh tissue can be used to isolate lysosomes according to *Protocols 10* and *11*. Since access to human organs is extremely limited for experimental purposes (placenta is the only readily available source) alternative sources for human lysosomal cathepsins are being sought. Some established human tumour cell lines produce and secrete these enzymes or their precursor forms. Cell lines which can be used to isolate cathepsins are given in *Table 2* (57, 58).

The production of recombinant human cathepsins and pro-cathepsins in *E. coli*, yeast, and baculovirus-infected insect cells has been reported. Recombinant enzymes from these sources satisfy most experimental purposes, but incorrect glycosylation or even lack of glycosylation has to be considered when using recombinant enzymes. No significant differences in the kinetic constants between native and recombinant enzymes have been reported. Polyclonal antibodies raised against recombinant cathepsins recognize the native enzymes as well.

Purification schemes for lysosomal cathepsins are numerous and cannot be reviewed here. Instead, the following protocols describe general rapid and effective affinity purification procedures for the isolation of aspartic peptidases such as cathepsin D (59) and active cysteine peptidases (60). Whereas

Table 1. Summary of data for lysosomal peptidases[a,b]

	Cathepsin D	Cathepsin B	Cathepsin L	Cathepsin H	Cathepsin S	Cathepsin C	Cathepsin K
M_r of proenzyme form[a]	53 000	38 000	38 000	41 000	37 000	55 000	35 000[d]
M_r of mature enzyme[c]	48 000	30 000	28 000	28 000	24 000	25 000	23 500[d]
M_r of heavy chain[e]	33 000	25 000	24 000	23 000	–[f]	18 000	–[g]
M_r of light chain[e]	14 000	~ 5000	~ 4000	~ 5000	–[f]	~ 7000	–[g]
Accession number[h]	P07339	P07858	P07711	P09668	P25774	P80067[i]	HSOC2RNA[j]

[a] Data refer to the human enzymes except where indicated.
[b] Data from ref. 54.
[c] Mature enzyme form.
[d] Deduced from cDNA sequence.
[e] Enzymes can exist either in single chain form or in a two chain form, the latter originating from proteolytic cleavage. Heavy and light chains are connected by disulfide bridges.
[f] Enzyme does not exist in heavy and light chain forms.
[g] Data not available.
[h] Accession number for searches in the SWISSPROT database for the human enzymes.
[i] Accession number for rat enzyme.
[j] Accession number for the EMBL database.

Table 2. Cell lines expressing and secreting cathepsins and their precursors

Enzyme	Cell line(s)	Reference
Cathepsin B	HS-24	57
	MCF-10[a]	Unpublished observations; B. Wiederanders
	103H	58
Cathepsin L	103H	58
	MCF-7[a]	Unpublished observations; B. Wiederanders
Cathepsin S	97TM1	58

[a] Available from the American Cell Culture Collection.

pepstatin A-Agarose® (*Protocol 13*) can be used for most aspartic peptidases, the Thiopropyl-Sepharose® procedure of *Protocol 14* has been used for cathepsins S, B, H, and plant cysteine peptidases, but not for cathepsin L.

Protocol 13. Affinity purification of active aspartic peptidases[a]

Equipment and reagents

- Solid NaCl
- 0.1 M trisodium citrate pH 3.5, 1 M NaCl
- 0.1 M sodium phosphate pH 5.0, 1 M NaCl
- 0.1 M Tris–HCl pH 8.0, 1 M NaCl
- Neutralizing buffer: 0.3 M sodium phosphate pH 6.5
- 1 M citric acid
- Pepstatin A-Agarose® (pre-swollen suspension from Sigma)
- Chromatography column (1–5 ml capacity)
- Cathepsin D assay kit (6): 2% azocasein solution (2 g/100 ml water), urea buffer (0.4 M trisodium citrate pH 5.0, 6 M urea), 20% TCA (20 g/100 ml water)
- UV spectrophotometer
- Lysosomal preparation containing cathepsin D (e.g. lysosomal extract from *Protocol 10*)

Method

Note: perform all steps at 4°C. Purified cathepsin D inactivates rapidly at pH 3.5.

1. Equilibrate 1–5 ml pepstatin A-Agarose with 0.1 M trisodium citrate pH 3.5, 1 M NaCl and fill the chromatography column with this matrix.
2. Adjust the lysosomal extract to pH 3.5 with 1 M citric acid. From this point on work quickly to minimize loss of activity.
3. Adjust the molarity of the extract to 1 M with solid NaCl and then centrifuge the solution at 10 000 *g* for 2 min. Discard the pellet of precipitated protein. Cathepsin D is soluble at this pH.
4. Quickly load the supernatant onto the pepstatin A-Agarose column and wash the column with 0.1 M sodium phosphate pH 5.0, 1 M NaCl until all the unbound protein is eluted from the affinity matrix. Monitor the eluate until the A_{280} reaches the level of protein-free elution buffer.
5. Elute active cathepsin D with 0.1 M Tris–HCl pH 8.0, 1 M NaCl. Collect

Protocol 13. *Continued*

0.5 ml fractions directly into tubes containing 0.5 ml neutralizing buffer.

6. Assay the fractions for cathepsin D activity as follows:[b]
   (a) Mix 50 μl of the fraction with 50 μl of 2% azocasein solution and 100 μl urea buffer.
   (b) Incubate at 37 °C for 10 min.
   (c) Cool on ice, add 200 μl of ice-cold 20% TCA, and centrifuge at 10 000 *g* for 10 min.
   (d) Recover the supernatant and measure its absorbance at 366 nm.
7. Pool those fractions containing cathepsin D and store at −20 °C.

[a] Method of Lapresle *et al.* (59).
[b] Method of Wiederanders *et al.* (61).

Protocol 14. Affinity purification of active cysteine proteinases[a]

Equipment and reagents

- Lysosomal extract (from *Protocol 10*)
- Thiopropyl-Sepharose 6B® (Pharmacia Biotech)
- 0.1 M sodium acetate pH 4.5
- 0.1 M Tris–HCl pH 8.0
- 0.1 M Tris–HCl pH 6.8
- 30 mM cysteine, 0.1 M Tris–HCl pH 6.8
- 8 mM EDTA, 4 mM $HgCl_2$ in cysteine-free 0.1 M Tris–HCl pH 6.8
- Chromatography column (1–5 ml capacity)
- Ultrafiltration unit (Amicon)

Method

1. Resuspend an appropriate amount of Thiopropyl-Sepharose in 0.1 M sodium acetate pH 4.5 (1 g dry powder results in 3 ml gel; 1 ml gel contains ~ 20 μmol 2-pyridyl-disulfide groups sufficient to bind 3–5 mg active cysteine peptidase).
2. Adjust the lysosomal extract to pH 4.5 using 0.1 M sodium acetate pH 4.5.
3. Mix the extract with the Thiopropyl-Sepharose and incubate overnight at 4 °C using an end-over-end stirrer.
4. Pour the matrix into the chromatography column and wash it with 20 column volumes of 0.1 M sodium acetate pH 4.5.
5. Wash the column with 20 column volumes of 0.1 M Tris–HCl pH 8.0.
6. Elute the column with a linear gradient of 0.1 M Tris–HCl buffer pH 6.8, to 30 mM cysteine in 0.1 M Tris–HCl pH 6.8. Collect 0.5 ml fractions.
7. Assay the fractions for cysteine proteinases (see *Protocol 15*).

8. Pool the active fractions and concentrate them to about 1 mg enzyme protein/ml using an ultrafiltration unit (e.g. Amicon, use at least YM-10 membranes).
9. Equilibrate the concentrated solution with 8 mM EDTA, 4 mM $HgCl_2$ in cysteine-free 0.1 M Tris–HCl pH 6.8. Store at −20 °C.

[a] Method of Brömme *et al.* (60).
[b] Omit EDTA/$HgCl_2$ when isolating cathepsin S.

2.3.3 Selective determination of cathepsins

Although some peptidases seem to show high specificity towards certain substrates *in vivo*, the activity of one peptidase is hard to detect selectively with artificial substrates in the presence of a great number of similar enzymes. In lysosomes, we find exactly this situation, and the measurement of a defined enzyme activity *in vitro* needs a careful choice of experimental conditions. Methods for the selective determination of single cathepsins are given in *Protocols 15* and *16*. The recommended substrates are preferentially, but not exclusively, cleaved by the respective enzymes; other peptidases are also able to hydrolyse the substrates. Therefore, specific inhibitors or inactivating conditions have to be included in order to make an assay more selective. However, an absolute selectivity cannot be guaranteed for any peptidase, even if you follow the protocol exactly. If the assay according to *Protocol 15* fails to provide clear cut results because of other enzymes in the preparation, use one of the inhibitor protocols in *Protocol 16*.

Protocol 15. Assay of cysteine proteinase activity

Equipment and reagents

- Enzyme preparation for assay (from *Protocol 14*)
- 0.1 M EDTA stock solutions, adjusted to final pH values of 5.5, 6.0, 6.5, and 7.0 (store at −20 °C)
- 0.1 M DTT stock solution (store at −20 °C)
- Fluorimeter, for continuous or single point assays, λ_{exc}: ~ 360 nm, λ_{emm}: ~ 460 nm
- Activation buffer: 8 mM EDTA, 16 mM DTT, buffered to the pH appropriate for the enzyme being assayed with the relevant 50 mM sodium phosphate buffer
- 0.1 M substrate stock solutions for cathepsins[a,b] (Bachem):
  Z-F-R-AMC for cathepsin L (K_m at pH 5.5 ~ 2 μM)
  Z-R-R-AMC for cathepsin B (K_m at pH 6 ~ 200 μM)
  Z-V-V-R-AMC for cathepsin S (K_m at pH 6.5 ~ 18 μM)
  R-AMC for cathepsin H (K_m at pH 6.8 ~ 100 μM)
  Make a 0.1 M substrate stock solution of each in DMSO. Store these at −20 °C.
- 50 mM sodium phosphate buffers, containing 4 mM EDTA and 8 mM DTT, at final pH values of 5.5, 6.0, 6.5, and 7.0
- Buffered substrate solutions: using the 0.1 M substrate stock solutions, dilute each substrate to 20 μM in 50 mM phosphate buffer pH 5.5 for cathepsin L, pH 6.5 for cathepsin B, pH 6.5 for cathepsin H, and pH 7.0 for cathepsin S
- AMC (7-amido-4-methylcoumarine) standard solution: 1 nmol AMC (Bachem) in 200 μl of the appropriate sodium phosphate buffer
- Stop solution: 0.1 M monochloro acetic acid (Merck), 0.1 M sodium acetate pH 4.3

Protocol 15. *Continued*

Method

1. Mix 50 μl of enzyme solution with 50 μl of the appropriate activation buffer and incubate at 25 °C for 10 min.[c]
2. Add 100 μl of the required buffered substrate solution to each sample. Set up a blank sample (control) with the appropriate sodium phosphate buffer replacing the enzyme sample.
3. Incubate the mixture for 10 min at 25 °C.[d]
4. Stop the reaction by adding four volumes of stop solution to the enzyme sample and the control sample.
5. Read the fluorescence at 460 nm.[e] Subtract the fluorescence of the blank from that of the enzyme sample.
6. Calculate the amount of AMC released by comparison with the fluorescence of the standard AMC solution.

[a] Abbreviations: Z, benzyloxycarbonyl-; AMC, 7-amido-4-methylcoumarine; R, arginine; F, phenylalanine; V, valine.
[b] K_m data from Kirschke *et al.* (54).
[c] The volume of the assay has to be adapted according to the fluorimeter being used.
[d] Adjust the time of incubation if necessary.
[e] The fluorescence can also be measured continuously if the appropriate fluorimeter is available.

Protocol 16. Assay of individual peptidases using selective inhibitors

Equipment and reagents

- Equipment and reagents as in *Protocol 15*
- Inhibitors: pepstatin® A (Calbiochem), *N*-(L-3-*trans*-carboxirane-2-carbonyl)-L-leucyl]-agmatine (E-64®; Calbiochem), *N*-(L-3-*trans*-propylcarbamoyloxirane-2-carbonyl)-L-isoleucyl-L-prolyl-methyl ester (CA-074Me®; Bachem), Z-L-phenylalanyl-L-phenylalanyl-diazomethylketone (Z-F-F-CHN_2; Bachem), Bestatin® (Calbiochem). Make up 10 mM stock solutions of each inhibitor and store them at –20 °C.
- Activation buffer: 8 mM EDTA, 16 mM DTT buffered to the pH of the enzyme being assayed with the relevant sodium phosphate buffer (see *Protocol 15*)

A. *Differentiation of cysteine peptidases from other peptidases*

1. Make up activation buffer with and without 10 μM E-64.
2. Incubate the samples with and without E-64 as in *Protocol 15*, step 1.
3. Then follow *Protocol 15*, steps 2–6. The E-64-sensitive fluorescence is due to cysteine peptidases.[a]

B. *Assay of cathepsin B*

1. Make up activation buffer with and without 10 μM CA-074Me.
2. Incubate the samples with and without CA-074Me as in *Protocol 15*, step 1.
3. Then follow *Protocol 15*, steps 2–6. The CA-074Me-sensitive fluorescence is due to cathepsin B.

C. *Assay of cathepsin L*

1. Make up activation buffer with and without 10 μM Z-F-F-CHN_2.
2. Incubate the samples with and without Z-F-F-CHN_2 as in *Protocol 15*, step 1.
3. Then follow *Protocol 15*, steps 2–6. The Z-F-F-CHN_2-sensitive fluorescence is due to cathepsin L.[b]

D. *Assay of cathepsin S*

1. Incubate the sample for 1 h at pH 7.5 and 37 °C. This inactivates most other cysteine proteinases but cathepsin S remains active.
2. Assay the cathepsin S as described in part A.

E. *Assay of cathepsin H*

1. Make up activation buffer with and without 10 μM bestatin.
2. Incubate the samples with and without bestatin as in *Protocol 15*, step 1.
3. Then follow *Protocol 15*, steps 2–6. Non-lysosomal aminopeptidases are inhibited by bestatin, but cathepsin H is not. The remaining activity is due to cathepsin H.

[a] Serine peptidases of the β-trypsin type are also sensitive to the action of E-64 (62).
[b] Cathepsin B is also inhibited by Z-F-F-CHN_2, but the k_2 value is nearly three orders of magnitude higher for cathepsin L than for cathepsin B (54).

2.4 Measurement of lysosomal protein degradation

The lysosomal part of the cellular proteolysis cannot be measured directly. Generally, the overall protein breakdown in a cell culture, an organ culture, or in a perfused organ is quantified in the presence and absence of inhibitors of the lysosomal proteolysis in order to calculate the lysosomal part from the differences in the two rates. In whole animals, the administration of inhibitors for some lysosomal peptidases can only partially block the lysosomal degradative pathway. E-64 and CA-074 (*Protocol 16*) have been tested in mice with some effect, e.g. on bone resorption (63). Reports on active inhibitor concentrations within lysosomes and of their intralysosomal action are lacking. Leupeptin, a tripeptide aldehyde is rapidly degraded by a cytosolic enzyme

(64). Oral or intraperitoneal administration of pepstatin and derivatives of pepstatin has some influence on the overall proteolysis, but this effect cannot be interpreted as due to inhibition of cathepsin D alone, since other aspartic peptidases (cathepsin E, renin) are also blocked by this compound (65).

For these reasons, studies of lysosomal protein degradation in whole animals or even in humans remain limited. Most reports deal with concentrations, activities, and mRNA expression levels of lysosomal peptidases under the influence of regulating factors.

2.4.1 Measurement of cellular protein degradation

The general approach to the detection of protein degradation rates consists of two stages:

(a) Radioactive labelling of all the proteins in a given experimental system, either by a pulse or by continuous labelling.
(b) Measurement of the decay of radioactivity from the protein(s) of interest.

Since the half-lives of cellular proteins vary from minutes to several days, the labelling protocol has to take into account whether the turnover of long-lived and/or short-lived proteins should be followed. Short-lived proteins can be labelled with [^{35}S]methionine. Labelling of long-lived proteins in animals has to be followed by chase periods of some days. Therefore, the experiment has to be performed with radioactive precursor amino acids where reutilization of the label is minimal, e.g. [^{14}C]guanidino-arginine. Decay rates of individual proteins are usually followed by selective immunoprecipitation and subsequent analysis by high resolution electrophoresis (e.g. SDS–PAGE).

For studying the regulation of liver protein catabolism in rats, Mortimore and Pösö (44) employed a refined perfusion system which allowed the modification of numerous parameters. They followed the loss of [^{14}C]valine from liver after various labelling periods. Myofibrillar protein breakdown can be followed by the urinary excretion of 3-methyl histidine (66) but this degradation seems to be mainly non-lysosomal (67). Infusion of the non-radioactive amino acid, [^{13}C]leucine, and subsequent gas chromatography-mass spectrometry analysis of [^{13}C]leucine and its degradation product [^{13}C]α-ketocaproate, allows protein turnover studies in humans (46). Another approach is to study the degradation of *in vitro* labelled individual proteins administered to cell cultures, perfused organs, and whole animals. The intracellular fate of the labelled proteins can be followed after isolation of appropriate organelles. The contribution of lysosomal degradation to the process can be calculated by the addition of lysosomotropic agents to the experimental system (see Section 2.4.3).

2.4.2 Inhibition of lysosomal protein degradation

In cell cultures, as well as in organ cultures, lysosomal proteolysis can be completely blocked by the addition of so-called lysosomotropic substances (such as ammonium ions, 3-methyl adenine, colchicine, chloroquine) (68) to the culture or perfusion medium. These substances accumulate within the lysosomes, thus raising the intralysosomal pH, inhibiting mannose 6-phosphate receptor recycling, and interrupting lysosomal protein breakdown. However, treatments like this always influence other metabolic parameters as well. In muscle, colchicine inhibits mainly the organelle-specific protein degradation but non-lysosomal degradation mechanisms are also affected (69). *Protocol 17* describes how a first experiment can be planned if lysosomal protein degradation in a cell culture is to be studied. For detailed information on how to study lysosomal protein degradation in perfused organs, see Mortimore and Ward (70).

In general, researchers are much more interested in the degradation pathway of individual proteins. In these cases, immunoprecipitation of the specific proteins has to be used instead of the TCA precipitation described in *Protocol 17*. The procedures for washing the immunoprecipitates and subsequent analyses are tricky. However, we have successfully used the protocol recommended by von Figura *et al.* (71).

Protocol 17. Measurement of protein breakdown in cell cultures via the lysosomal route

Equipment and reagents

- Tissue culture cells expressing the protein of interest
- Cell culture plates (6 cm diameter are recommended since this size allows one to work with about 10^6 cells)
- Cell culture medium: usually RPMI or DMEM (Biochrom), however, some cells need special culture conditions
- Met⁻ medium: cell culture medium lacking methionine (Biochrom) but containing 10% dialysed FCS
- 1 M NH_4Cl stock solution
- 10% fetal calf serum (FCS): heat inactivated for 30 min at 56°C
- 10% dialysed FCS: heat inactivated for 30 min at 56°C, dialysed against PBS
- [^{35}S]methionine: *in vivo* cell labelling grade, > 37 TBq/mmol (Amersham)
- Rubber policeman (length 25 cm; Greiner)
- Growth medium: cell culture medium containing 10% FCS
- Met$^-$ medium containing 10 mM NH_4Cl
- Pulse medium: met$^-$ medium containing 1 MBq/ml [^{35}S]methionine (if labelling of the proteins is too weak, enhance the concentration of [^{35}S]methionine up to 15 MBq/ml)
- Pulse medium containing 10 mM NH_4Cl
- Chase medium: growth medium containing a fivefold higher concentration of methionine than in normal growth medium
- Chase medium containing 10 mM NH_4Cl
- PBS (see *Protocol 9*)
- 10% (w/w) Triton X-100
- 20% (w/v) TCA
- 1 M NaOH
- Scintillation counter, tubes, and scintillation cocktail
- Refrigerated cell centrifuge (controlled slow speed), plus a bench centrifuge

Method

1. Set up the cell culture plates; you will need sufficient cells to take several time point samples during the chase period to construct an accurate decay curve. In addition, to calculate the lysosomal contribution to the degradation rate, you will need to examine the decay in the presence and absence of NH_4Cl (the lysosomal contribution will

Protocol 17. *Continued*

be the difference between these rates). Label the dishes Am^+ (for NH_4Cl) and CON (control, no NH_4Cl).

2. Place 2.5 ml growth medium in each plate, seed the plates with the cells, and grow the cultures at 37°C to confluence (~ 10^6 cells/plate).
3. Wash cell the plates twice with 2.5 ml met^- medium.
4. Add 2.5 ml met^- medium to the CON plates and 2.5 ml met^- medium containing 10 mM NH_4Cl to the Am^+ plates. Incubate all the plates at 37°C for 1 h.
5. Remove the medium and carry out the pulse labelling using 0.5 ml pulse medium for the con plates and using 0.5 ml pulse medium with 10 mM NH_4Cl for the Am^+ plates. Incubate for 2–30 min at 37°C. The pulse time you will need is dependent on the cellular concentration of the proteins of interest and on the biosynthetic rate of these proteins. Therefore, no general recommendation can be given.
6. Remove the medium carefully and dispose it safely according to local regulations.
7. Carry out the chase by adding 2.5 ml chase medium (CON plates) or 2.5 ml chase medium containing 10 mM NH_4Cl (Am^+ plates).
8. Incubate the plates at 37 °C. At predetermined time intervals remove a CON plate and an Am^+ plate for analysis.
9. Wash the cells twice with 2.5 ml PBS at 4 °C. Harvest the cells in 2.5 ml PBS using a rubber policeman.
10. Centrifuge the cell suspensions at 800 *g* for 4 min (4 °C) to recover the cells.
11. Resuspend the cells in 1 ml of ice-cold 0.1% Triton X-100.
12. Leave the cells for 5 min on ice to lyse.
13. Centrifuge the cell lysate at 10 000 *g* for 10 min and recover the supernatant.
14. Assay the supernatant for radioactive protein by placing duplicate 0.1 ml samples in tubes with an equal volume of 20% TCA. Mix well and leave on ice for 10 min.[a]
15. Centrifuge the samples at 10 000 *g* for 10 min, and discard the supernatant.
16. Resuspend each pellet in 10 ml scintillation cocktail and determine the radioactivity by scintillation counting. If the pellet is insoluble in the scintillation cocktail, dissolve it in as little 1 M NaOH as possible before mixing with the scintillation cocktail. In this case, a blank containing the same volume of NaOH and a defined amount of

[^{35}S]methionine (e.g. giving 1000 d.p.m.) must be measured in order to calculate the quenching effect of the NaOH.

17. Calculate the lysosomal protein degradation rate using the formulae given in the text.

[a] At this point, immunoprecipitation of individual proteins may be carried out (71).

Calculate the degradation rate for the protein population via the lysosomal route in the following way. At each time point during the chase, subtract the radioactivity of the TCA precipitate (*Protocol 17*, step 17) from the Am^+ cells from the radioactivity of the CON cells. Plot the difference in radioactivity versus time and use a suitable software program to calculate the k_d (fractional degradation rate) and $t_{1/2}$ (half-life), e.g. *Scientist for Windows* (Micromath).

Use the following formula:

$$N = N_0 \times e^{-kt}$$

$$t_{1/2} = \frac{0.693}{k_d}$$

where N_0 is the radioactivity at the start of the chase period, and N is the radioactivity at time t in the chase period. For a detailed discussion of the determination of $t_{1/2}$ see ref. 72.

Acknowledgements

Research in the laboratory of Dr Aaron Ciechanover is supported by grants from the Israeli–German Foundation for Scientific Research and Development (G. I. F.; with Dr Dieter Wolf), European Community (T. M. R.; with Dr Dieter Wolf, among other researchers), Israel Science Foundation founded by the Israeli Academy of Sciences and Humanities - Centers of Excellence Program, US–Israel Binational Science Foundation (BSF), Israeli Ministry of Sciences, DKFZ, UK–Israel Science and Technology Research Fund, the Foundation for Promotion of Research at the Technion, and a research grant administered by the Vice President for research at the Technion.

References

1. Staub, O., Dho, S., Henry, P. C., Correa, J., Ishikawa, T., McGlade, J., *et al.* (1996). *EMBO J.*, **15**, 2371.
2. Kishino, T., Lalande, M., and Wagstaff, J. (1997). *Nature Genet.*, **15**, 70.
3. Hershko, A. and Ciechanover, A. (1998). *Annu. Rev. Biochem.*, **67**, 425.
4. Hochstrasser, M. (1995). *Curr. Opin. Cell. Biol.*, **7**, 215.
5. Jentsch, S. and Schlenker, S. (1995). *Cell*, **82**, 881.
6. Deshaies, R. J. (1995). *Trends Cell Biol.*, **5**, 428.

7. Ciechanover, A. (1994). *Cell*, **79**, 13.
8. Varshavsky, A. (1992). *Cell*, **69**, 725.
9. Jentsch, S. (1992). *Annu. Rev. Genet.*, **26**, 179.
10. Nuber, U., Schwarz, S., Kaiser, P., Schneider, R., and Scheffner, M. (1996). *J. Biol. Chem.*, **271**, 2795.
11. Pickart, C. M. and Rose, I. A. (1985). *J. Biol. Chem.*, **280**, 1573.
12. Stancovski, I., Gonen, H., Orian, A., Schwartz, A. L., and Ciechanover, A. (1995). *Mol. Cell. Biol.*, **15**, 7106.
13. Scheffner, M., Huibregtse, J. M., Vierstra, R. D., and Howley, P. M. (1993). *Cell*, **75**, 495.
14. Huibregtse, J. M., Scheffner, M., Beaudenon, S., and Howley, P. M. (1995). *Proc. Natl. Acad. Sci. USA*, **92**, 2563.
15. Nefsky, B. and Beach, D. (1996). *EMBO J.*, **15**, 1301.
16. Chen, Z., Hagler, J., Palombella, V. J., Melandri, F., Scherer, D., Ballard, D., *et al.* (1995). *Genes Dev.*, **9**, 1586.
17. Alkalay, I., Yaron, A., Hatzubai, A., Orian, A., Ciechanover, A., and Ben Neriah, Y. (1995). *Proc. Natl. Acad. Sci. USA*, **92**, 10599.
18. Scheffner, M., Werness, B. A., Huibregtse, J. M., Levine, A. J., and Howley, P. M. (1990). *Cell*, **63**, 1129.
19. Hee Lee, D. H., Sherman, M. Y., and Goldberg, A. L. (1996). *Mol. Cell. Biol.*, **16**, 4773.
20. Bercovich, B., Stancovski, I., Mayer, A., Blumenfeld, N., Laszlo, A., Schwartz, A. L., *et al.* (1997). *J. Biol. Chem.*, **272**, 9002.
21. Orlowski, M. (1990). *Biochemistry*, **29**, 10289.
22. Rechsteiner, M., Hoffman, L., and Dubiel, W. (1993). *J. Biol. Chem.*, **268**, 6065.
23. Peters, J. M. (1994). *Trends Biochem. Sci.*, **19**, 377.
24. Coux, O., Tanaka, K., and Goldberg, A. L. (1996). *Annu. Rev. Biochem.*, **65**, 801.
25. Hilt, W. and Wolf, D. H. (1996). *Trends Biochem. Sci.*, **21**, 96.
26. Rock, K. L., Gramm, C., Rothstein, L., Clark, K., Stein, R., Dick, L., *et al.* (1994). *Cell*, **78**, 761.
27. Murakami, Y., Matsufuji, S., Kameji, T., Hayashi, S.-I., Igarashi, K., Tamura, T., *et al.* (1992). *Nature*, **380**, 597.
28. Rabinowitz, M. and Fisher, J. M. (1964). *Biochim. Biophys. Acta*, **91**, 313.
29. Ciechanover, A., Hod, Y., and Hershko, A. (1978). *Biochem. Biophys. Res. Commun.*, **81**, 1100.
30. Ken Doyle, E. (1996). *Protocols and applications guide*, 3rd edn, pp. 259–76. Promega Corp. Cat. No. P1610 ISBN 1-882274-57-1.
31. Orian, A., Whiteside, S., Israël, A., Stancovski, I., Schwartz, A. L., and Ciechanover, A. (1995). *J. Biol. Chem.*, **270**, 21707.
32. Johnston, N. L. and Cohen, R. E. (1991). *Biochemistry*, **30**, 7514.
33. Hershko, A. and Rose, I. A. (1987). *Proc. Natl. Acad. Sci. USA*, **84**, 1829.
34. Hershko, A. and Heller, H. (1985). *Biochem. Biophys. Res. Commun.*, **128**, 1079.
35. Hershko, A., Ganoth, D., Pehrson, J., Palazzo, R. E., and Cohen, L. H. (1991). *J. Biol. Chem.*, **266**, 16376.
36. Blumenfeld, N., Gonen, H., Mayer, A., Smith, C. E., Siegel, N. R., Schwartz, A. L., *et al.* (1994). *J. Biol. Chem.*, **269**, 9574.
37. Scheffner, M., Huibregtse, J. M., and Howley, P. M. (1994). *Proc. Natl. Acad. Sci. USA*, **91**, 8797.

38. Fenteany, G., Standaert, R. F., Lane, W. S., Choi, S., Corey, E. J., and Schreiber, S. L. (1995). *Science*, **268**, 726.
39. Jensen, T. J., Loo, M. A., Pind, S., Williams, D. B., Goldberg, A. L., and Riordan, J. R. (1995). *Cell*, **83**, 129.
40. Strous, G. J., van Kerkhof, P., Govers, R., Ciechanover, A., and Schwartz, A. L. (1996). *EMBO J.*, **15**, 3608.
41. Tamura, T., Tamura, N., Cejka, Z., Hegerl, R., Lottspeich, F., and Baumeister, W. (1996). *Science*, **274**, 1385.
42. Seglen, P. O. and Bohley, P. (1992). *Experientia*, **48**, 158.
43. Bohley, P., Kirscke, H., Langner, J., Ansorge, S., and Hanson, H. (1971). *Acta Biol. Med. Germ.*, **27**, 229.
44. Mortimore, G. E. and Pösö, A. R. (1988). In *Methods in enzymology* (ed. R. A. Harris and J. R. Sokatch), Vol. 166, p. 461. Academic Press.
45. Mortimore, G. E. and Pösö, A. R. (1986). *Adv. Enzyme Regul.*, **25**, 257.
46. Mansoor, O., Beaufrere, B., Boirie, Y., Ralliere, C., Taillandier, D., Aurousseau, E., *et al.* (1996). *Proc. Natl. Acad. Sci. USA*, **93**, 2714.
47. Turk, V. E. (1986). *Cysteine proteinases and their inhibitors*. Walter de Gryuter, Berlin, New York.
48. Bird, S. J. and Lloyd, J. B. (1990). *Biochim. Biophys. Acta*, **1024**, 267.
49. Terlecky, S. R. and Dice, J. F. (1993). *J. Biol. Chem.*, **268**, 23490.
50. Dingle, J. T. E. (1972). *Lysosomes. A laboratory handbook*. North-Holland Publishing Co. Amsterdam, London.
51. Bohley, P., Kirschke, H., Langner, J., and Ansorge, S. (1969). *FEBS Lett.*, **5**, 233.
52. Willstätter, R. and Bamann, E. (1929). *Z. Physiol. Chem.*, **180**, 127.
53. Press, E. M., Porter, R. R., and Cebra, J. (1960). *Biochem. J.*, **74**, 501.
54. Kirschke, H., Barrett, A. J., and Rawlings, N. D. (1995). *Protein Profile*, **2**, 1581.
55. Brömme, D. and Okamoto, K. (1995). *Biol. Chem. Hoppe-Seyler*, **376**, 379.
56. Inaoka, T., Bilbe, G., Ishibashi, O., Tezuka, K., Kumegawa, M., and Kokubo, T. (1995). *Biochem. Biophys. Res. Commun.*, **206**, 89.
57. Spiess, E., Brüning, A., Gack, S., Ulbricht, B., Spring, H., Trefz, G., *et al.* (1994). *J. Histochem. Cytochem.*, **42**, 917.
58. Heidtmann, H. H., Salge, U., Havemann, K., Kirschke, H., and Wiederanders, B. (1993). *Oncol. Res.*, **5**, 441.
59. Lapresle, C., Puizdar, V., Porchon-Bertolotto, C., Joukoff, E., and Turk, V. (1986). *Biol. Chem. Hoppe-Seyler*, **367**, 523.
60. Brömme, D., Bonneau, P. R., Lachance, P., Wiederanders, B., Kirschke, H., Peters, C., *et al.* (1993). *J. Biol. Chem.*, **268**, 4832.
61. Wiederanders, B., Kirschke, H., and Schaber, S. (1986). *Biomed. Biochim. Acta*, **45**, 1477.
62. Sreedharan, S. K., Verma, C., Caves, L. S. D., Brocklehurst, S. M., Gharbia, S. E., Shah, H. N., *et al.* (1996). *Biochem. J.*, **316**, 777.
63. Towatari, T., Nikawa, T., Murata, M., Yokoo, C., Hanada, K., and Katunuma, N. (1991). *FEBS Lett.*, **280**, 311.
64. Brown, C. P. and Beynon, R. J. (1983). *Biosci. Rep.*, **3**, 179.
65. Lin, C. and Frishman, W. H. (1995). *Am. Heart J.*, **131**, 1024.
66. Young, V. R. and Munro, H. N. (1978). *Fed. Proc.*, **37**, 2291.
67. Furuno, K., Goodman, M. N., and Goldberg, A. L. (1990). *J. Biol. Chem.*, **265**, 8550.

68. Hendil, K. B., Lauridsen, A. M., and Seglen, P. O. (1990). *Biochem. J.*, **272**, 577.
69. Crie, J. S., Ord, J. M., Wakeland, J. R., and Wildenthal, K. (1983). *Biochem. J.*, **210**, 63.
70. Mortimore, G. E. and Ward, W. F. (1976). In *Lysosomes in biology and pathology 5* (ed. J. T. Dingle and R. T. Dean), p. 157. North Holland Publishing Co., Amsterdam, Oxford.
71. von Figura, K., Steckel, F., and Hasilik, A. (1983). *Proc. Natl. Acad. Sci. USA*, **80**, 6066.
72. Waterlow, J. C., Garlick, P. J., and Millward, D. J. E. (1978). In *Protein turnover in mammalian tissues and in the whole body*. North Holland Publishing Co., Amsterdam, London.

8

Protein degradation and proteinases in yeast

WOLFGANG HILT and DIETER H. WOLF

1. Introduction

Degradation of intracellular proteins has been proven to be essential for cellular life. Adjustment of protein levels by degradation initiated by extracellular or intracellular signals is a central mechanism to allow cells to adapt to changing environmental conditions or intracellular programmes. In eukaryotic cells two major proteolytic systems exist to exert these functions: the proteasome and the lysosome.

Proteasomes are multisubunit protein complexes localized in the cytoplasm and nucleus of all eukaryotic cells (1–4). The 26S proteasome is a complex structure consisting of two 19S regulatory complexes and the 20S proteasome (*Figure 1*). The proteolytic activity of the 20S proteasome depends on an amino terminal threonine residue located in the inner space of the cylinder in three of the seven β-subunits (5). These three active site β-subunits confer three different activities to the proteasome, when peptides are used as test substrates: Pup1 confers a trypsin-like activity, Pre2 a chymotrypsin-like activity, and Pre3 a peptidylglutamyl-peptide (PGPH) hydrolysing activity (6, 7). Accordingly, the proteasome can easily be tested *in vitro* for these enzyme activities using appropriate chromogenic or fluorogenic peptides (8, 9). Two, most likely identical, 19S cap complexes containing about 16–20 different subunits dock onto the outer α-rings of the cylinder core to form the 26S proteasome (*Figure 1*) which is thought to be the functional complex *in vivo* (3, 10, 11). Among the 19S cap subunits are six ATP binding proteins of the AAA type and a ubiquitin binding subunit (11–13). *In vivo* most of the substrates of the proteasome are marked for proteolysis via the 76 amino acid protein ubiquitin. Polyubiquitination is necessary for proteolysis to occur (14). The proteasome is vital for cellular life (2). Substrates of the proteasome are the rapidly turning over proteins of the cytoplasm and the nucleus. For instance, the enzyme complex is involved in the regulation of metabolic enzymes, transcription factors, cyclins, or malfolded proteins. As a result the

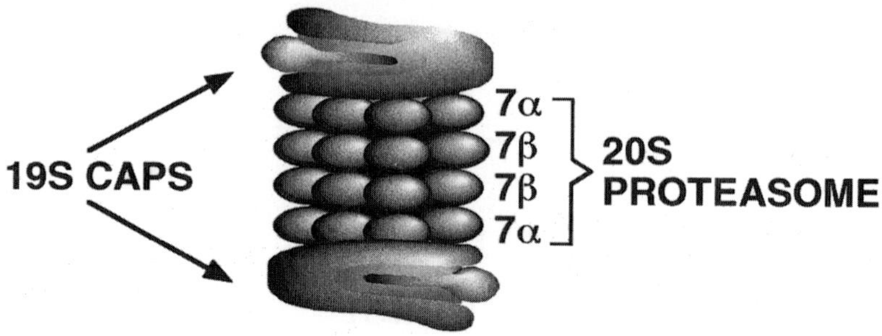

Figure 1. Structure of the 26S proteasome. The 26S proteasome consists of two 19S regulatory cap complexes and the 20S proteasome which acts as the proteolytically active core module. 20S proteasomes are hollow cylindrical structures built up from a stack of four rings each containing seven subunits. In eukaryotes, both inner rings consist of seven different β-type subunits, whereas the two outer rings are built up from seven different α-type subunits. The active sites are hidden in the central channel spanning the longitudinal axis of the 20S cylinder complex.

proteasome is a central regulatory tool in cell metabolism, gene transcription, the cell cycle, and the endoplasmic reticulum (2, 15–19).

In contrast to the proteasome, which acts as a highly specific proteolytic complex, the lysosome (or vacuole) of all eukaryotic cells is a membrane-surrounded 'bag' containing a variety of non-specific proteases that are able to digest all proteins to which they get access (20). Entry into the lysosome (or vacuole) via autophagocytosis (21, 22) or selective uptake (23, 24) delivers the proteins to a complete degradation pathway. A multitude of long-lived proteins are degraded via the lysosomal (vacuolar) pathway. Nutritional stress conditions initiate this degradation process (25). Under these conditions lysosomal (vacuolar) protein degradation is necessary for generation of carbon backbones from amino acids of proteins to allow energy production and gluconeogenesis. Consequently, cells depleted of vacuolar proteinases by mutation die very early under starvation conditions (25–27).

1.1 Advantages of yeast for studying eukaryotic cell biology

Due to its amenability to biochemical, genetic, and molecular biological manipulation, the yeast *Saccharomyces cerevisiae* has been used for many years as a model organism for studying the basic housekeeping functions of the eukaroytic cell. The belief that analysis of central biological problems in yeast is directly relevant to the biology of more complex eukaryotes has become an irrefutable guiding principle over the last decade. The value of yeast as a model organism has been dramatically increased due to the complete sequencing of its entire genome of 5885 genes, an important milestone in the biology of eukaryotes (28). Protein sequences from higher eukaryotes can be used in a similarity search to find their most closely related counterparts in yeast and

Table 1. Growth media for *Saccharomyces cerevisiae*

A. YPD (yeast extract, peptone, dextrose) medium

- 1% (w/v) yeast extract (Difco)
- 2% (w/v) peptone (Difco)
- 2% (w/v) D-glucose

Sterilize YPD medium by autoclaving for 30 min at 121 °C.

B. MV (mineral-vitamin) medium

- 0.67% (w/v) yeast nitrogen base without amino acids (Difco)
- 2% (w/v) D-glucose

Sterilize MV medium by autoclaving for 30 min at 121 °C. When yeast strains are used which exhibit deficiencies in biosynthetic pathways (i.e. auxotrophic strains), the product of the pathway must be added to allow auxothrophic growth. Aseptically add 0.1 vol. of the appropriate 100 × supplement stock solution (see Section G).

C. Synthetic complete medium

- 0.67% (w/v) yeast nitrogen base without amino acids (Difco)
- 2% (w/v) D-glucose
- 0.13% (w/v) dropout powder (see Section F)

Sterilize the medium by autoclaving for 30 min at 121 °C. When cool, aseptically add 0.01 vol. of the appropriate 100 × auxotrophic supplement stock solution (see Section G).

D. Starvation medium

- 0.67% (w/v) yeast nitrogen base without amino acids (Difco)
- 2% (w/v) potassium acetate

Sterilize starvation medium by autoclaving for 30 min at 121 °C.

E. Agar plates

When the above media are used as solid media, 2% (w/v) agar (Difco) is added before autoclaving; 600 ml agar medium is sufficient for 20 plates (10 cm diameter Petri dishes).

F. Dropout powder

To prepare dropout, powder, mix the following:

• L-alanine	2 g	• L-isoleucine	2 g
• L-arginine	2 g	• L-methionine	2 g
• L-asparagine	2 g	• *p*-aminobenzoic acid	0.29 g
• L-aspartic acid	2 g	• L-phenylalanine	2 g
• L-cysteine	2 g	• L-proline	2 g
• L-glutamine	2 g	• L-serine	2 g
• L-glutamic acid	2 g	• L-threonine	2 g
• glycine	2 g	• L-tyrosine	2 g
• *myo*-inositol	2 g	• L-valine	2 g

G. Auxotrophic supplements

For auxotrophic growth use the following 100 × stock solutions:

• adenine	30 mM
• uracil	20 mM
• L-leucine	100 mM
• L-lysine	10 mM
• L-histidine	60 mM
• L-tryptophan	40 mM

Prepare these solutions in water and autoclave.

then the biological information associated with homologous yeast genes can immediately suggest similar functions for the homologue in higher multicellular eukaryotes. In addition, the function of any new gene can easily be tested in yeast. The elucidation of the function of proteolysis in the cellular regulation of eukaryotic cells, the analysis of the proteasome as well as the lysosome (vacuole), and their intracellular actions is an example of the unique value of the yeast *S. cerevisiae* in eukaryotic cell biology. Mutants defective in proteolytic enzymes have enabled researchers to unravel many functions of these enzymes and to discover the variety of their substrates. Furthermore, the purification of proteins in a pure and unmodified state has been greatly facilitated by the use of proteinase-defective yeast mutants. Further studies on cellular regulation and intracellular signalling via proteolysis, as well as protein purification, should profit tremendously from the further use of proteinase mutants.

2. Growth of yeast cells and preparation of cell extracts

For the growth of yeast cultures for the protocols in this chapter, a 0.1% (v/v) dilution of a 24 h pre-culture, in YPD medium (*Table 1*) is generally used to inoculate the main culture. Cultivation of cells should be done at 30 °C (optimum growth temperature for wild-type yeast cells) or at 23 °C for temperature-sensitive strains. Growth time depends on the growth performance of the strain used and on the scientific question to be answered. The most common growth phases in which yeast cells are harvested are early, mid, or late logarithmic growth phases, diauxic growth phase, and early or late stationary growth phases. These growth phases have to be elucidated for each strain by determining a growth curve.

For any biochemical procedure (testing, purification, etc. of cellular contents) the preparation of small or large scale cell extracts (*Protocol 1*) is necessary. If the goal is the study of the protein in denatured form by pulse-chase experiments, or SDS–PAGE, small scale extracts of denatured protein (*Protocol 1*, part C) can be used. Because of the danger of proteolytic artefacts when using this method, peptidase inhibitors have to be included.

Protocol 1. Preparation of yeast cell extracts

A. Large-scale extracts (from 10–300 g cells)

Equipment and reagents

- Ultracentrifuge (Beckman Instruments or Centricon, Kontron-Hermle Instruments)
- Appropriate large scale culture of *S. cerevisiae*

- Cell breakage buffer: 0.1 M Tris–HCl pH 7.0, or 0.1 M potassium phosphate pH 7.0, or another buffer which suits the desired experiments best
- French pressure cell (SLM; Aminco Inc.)
- Low speed centrifuge (Centricon, Kontron-Hermle Instruments or Sorvall, Du Pont Instruments)

Method

Carry out all steps at 4°C using pre-cooled solutions and equipment.

1. Harvest the cells by low speed centrifugation at 5000 *g* for 15 min.
2. Wash the cells twice in cell breakage buffer. Use a 1:1 (w/v) cell suspension.
3. Break the cells in a French pressure cell by passing the cell suspension three times through the pressure cell at 1.4×10^8 Pascal. Cool the suspension on ice between passages.
4. Clear the homogenate of cell debris by low speed centrifugation for 10–20 min at 12 000 *g*.
5. If necessary, and a clear extract is desired, ultracentrifuge the extract for 30 min at 100 000 *g*.
6. Store the extract at 4°C.

B. Small scale extracts (100–500 μg cells)

Equipment and reagents

- Appropriate small scale culture of *S. cerevisiae*
- Cell breakage buffer (see part A)
- Glass beads (0.5 mm diameter)
- Vortex mixer
- Microcentrifuge

Method

1. Mix the cells suspended in cell breakage buffer 1:1 (w/v) in a 1.5 ml microcentrifuge tube with an equal volume of glass beads.
2. Break the cells by vortexing the tube for 1 min with intermittent cooling on ice. Repeat this vortexing two to four times.
3. Clear the homogenate of cell debris by centrifugation in a microcentrifuge for 5 min at maximum speed.

C. Small scale extracts of denatured protein

Equipment and reagents

- Appropriate small scale culture of *S. cerevisiae*
- Cell breakage buffer (see part A)
- Stock solution of protease inhibitor cocktail tablets complete (Boehringer Mannheim 1697 498): dissolve one tablet in 1 ml water

Protocol 1. *Continued*

- Glass beads (0.5 mm diameter)
- Boiling water-bath
- Vortex mixer
- Boiling solution: 4.5% (w/v) SDS, 2.25 mM EDTA, 2.25 mM EGTA pH 7.4

Method

1. Take 300 μl of a 20–50% (w/v) cell suspension in cell breakage buffer, add 6 μl protease inhibitor stock solution, and warm to 30 °C.
2. Heat 600 mg glass beads to 95–100 °C in a 1.5 ml microcentrifuge tube.
3. Pour the warmed cell suspension over the glass beads and heat at 95 °C for 10 min. Vortex for 1 min.
4. Add 600 μl of boiling solution and incubate for 10 min at 95 °C.
5. Freeze the samples at –20 °C.

3. Analysis of protein turnover

3.1 Protein degradation *in vivo*

Protein degradation *in vivo* is followed by measuring the liberation of TCA soluble peptides and amino acids in cells grown on a radioactively labelled amino acid. Frequently used radioactive amino acids are [^{3}H]leucine and [^{35}S]methionine (9, 25, 29, 30). *Protocol 2* is used when total protein turnover (no individual proteins) in cells is followed. Growth temperatures used depend on the choice of experiment and cell type. For temperature-sensitive strains a growth temperature of 23 °C is usually the best choice, for turnover measurements of non-temperature defective cells growth at 30 °C is appropriate, and turnover measurements under moderate heat stress conditions require growth at 37 °C. However, it might also be desirable to grow cells at their growth temperature optimum (23 °C or 30 °C) and then shift to temperatures under which the study of protein turnover is of interest. The labelling period depends on the type of experiment and the genetic background of the strain used. Some strains do not incorporate the radioactive amino acid well. This then requires a longer labelling period. For measurement of the degradation of long-lived proteins, a labelling period of several hours is required, for measurement of the degradation of short-lived proteins the labelling period should be a few minutes only. For turnover measurements of short-lived proteins, use [^{35}S]methionine as the radioactive amino acid to provide higher sensitivity. Calculate protein breakdown as the increase of TCA soluble activity in samples taken at different time points divided by the TCA insoluble radioactivity at time point zero. Correct for the different quenching effects of TCA and NCS-II using an internal standard.

Protocol 2. Measurement of protein turnover

Equipment and reagents

- Appropriate starter culture of *S. cerevisiae*
- MV medium (*Table 1*)
- [^{3}H]leucine or [^{35}S]methionine (Amersham Pharmacia Biotech)
- Chase medium: MV medium containing 5 mM leucine or 5 mM methionine (depending on the radioactive amino acid used for labelling)
- 110% (w/v) TCA
- Chase medium with 10% TCA
- Ethanol:ether (1:1, v/v)
- Liquid scintillant (Roth)
- NCS-II containing 10% (v/v) water (Amersham International plc)
- Scintillation counter (e.g. Wallac or Beckman)

Method

1. Grow the yeast cells at the desired temperature in MV medium up to A_{600} of about 1.
2. Harvest the cells by centrifugation, concentrate them in fresh MV medium (A_{600} = 2.5), and pulse label the cells with [^{3}H]leucine or [^{35}S]methionine for the desired time period. Use between 2–10 ml MV medium and add about 2 μCi/ml [^{3}H]leucine or about 60 μCi/ml [^{35}S]methionine.
3. After the desired labelling period, centrifuge the cells and wash them twice with 10 ml of chase medium.
4. Start the chase by resuspending the cells at the desired chase temperature in the desired volume (usually between 2–5 ml) of chase medium.
5. Take samples of the cell suspension (between 100 μl and 1 ml) at the desired time points (usually every 30 min for about 5 h), and add 110% TCA to a final concentration of 10% TCA. Mix and leave the samples on ice for at least 4 h.
6. Centrifuge the TCA samples for 5 min at 14 000 *g*.
7. For the determination of the *released TCA soluble radioactivity*, mix up to 900 μl of the TCA supernatant with 5 ml liquid scintillant. Measure the radioactivity in a scintillation counter.
8. For the determination of *total incorporated radioactivity*, wash the pellet of each TCA sample five times with chase medium containing 10% TCA and twice with ethanol:ether (1:1). To do this, vortex the pellet in the wash solution, recover it by centrifugation (step 6), and discard the washings (supernatant). Air dry the washed pellets and dissolve each in 1 ml of NCS-II (containing 10% water) at 40°C. Mix 900 μl of the NCS solution with 5 ml liquid scintillant. Measure the radioactivity in a scintillation counter.

3.2 Degradation of individual proteins

In many cases the turnover of the bulk protein in the cell is not of interest, but instead the object of the experiment is to study the degradation of individual proteins (31, 32). To follow the degradation of individual proteins a specific antibody directed against the chosen protein is required. *Protocol 3* is a general procedure for the pulse-chase measurement of protein degradation. *Protocol 4* is a variation applicable to membrane proteins.

Protocol 3. Measurement of degradation of individual proteins using a pulse-chase procedure

Equipment and reagents

- Appropriate starter culture of *S. cerevisiae*
- MV medium (*Table 1*)
- Vortex mixer
- Labelling medium: MV medium (*Table 1*) minus methionine
- [^{35}S]methionine (Amersham)
- Chase medium: MV medium (*Table 1*) plus 40 mM methionine
- 50% (w/v) TCA
- Liquid nitrogen
- Microcentrifuge
- Ethanol (absolute, at −20 °C)
- Glass beads (0.5 mm diameter)
- SDS breakage buffer: 50 mM Tris–HCl pH 7.5, 6 M urea, 1% (w/v) SDS, 1 mM EDTA
- Stock solution of protease inhibitor cocktail (see *Protocol 1*, part C)
- Boiling water-bath
- Buffer A: 50 mM Tris–HCl pH 7.5, 1.25% (v/v) Triton X-100, 6 mM EDTA, 0.19 M NaCl
- Antibody against the protein under investigation
- Buffer B: 50 mM Tris–HCl pH 7.5, 6 mM EDTA, 0.19 M NaCl
- Protein A–Sepharose® (Amersham Pharmacia Biotech): make a 7% (v/v) slurry in buffer B 1 h before use
- End-over-end shaker
- Electrophoresis buffer: 8 M urea, 5% (w/v) SDS, 0.1 mM EDTA, 1.5 mM DTT, 0.2 M Tris–HCl pH 6.8, 1 mg/ml bromophenol blue
- NCS-II, scintillation counter (see *Protocol 2*)
- SDS–PAGE gel, electrophoresis buffers, and equipment (Bio-Rad Protean II or Mini-Protean II)
- 5% (v/v) propan-2-ol, 5% (v/v) acetic acid in water
- Whatman filter paper (0.01 mm thick)
- Cellophane
- Gel drier and vacuum source
- Imaging cassette (Molecular Dynamics) or X-ray cassette and X-ray film (Biomax MR-1, Kodak)
- Imaging system (PhosphoImager, e.g. from Molecular Dynamics)

Method

1. Grow the cells in MV medium (with auxotrophic supplements, if required) to A_{600} = 2–6. Harvest the cells by centrifugation for 5 min at 500 *g*. Wash the cells two or three times in labelling medium by centrifugation.
2. Resuspend three A_{600} units of cells (about 1×10^8 cells) per ml in labelling medium (usually 5–10 ml). Add 50 μCi/ml [^{35}S]methionine to the cell suspension.[a] Pulse label the cells at the desired temperature (generally 30 °C) for 60–120 min, depending on the protein being studied.
3. Start the chase by adding an equal volume of chase medium and continue the incubation.

4. Take samples (usually 1 ml) at time zero and at least three additional time points during the chase into 0.1 vol. 50% TCA in microcentrifuge tubes. Mix and immediately freeze the samples in liquid nitrogen.
5. After all the samples are collected, thaw them, vortex each thoroughly, and incubate them for 10 min on ice. Centrifuge for 15 min at maximum speed in a microcentrifuge.
6. Remove the supernatants carefully. Add 1 ml ethanol (–20 °C) to each pellet and vortex to detach the pellet completely from the wall of the tube. Centrifuge for 5 min at maximum speed in a microcentrifuge. Repeat the ethanol wash and air dry the pellets at 50 °C for 2–3 min.
7. Add about 60 μl glass beads and 100 μl SDS breakage buffer to each pellet. Add 2 μl proteinase inhibitor cocktail. Break the cells by vigorous vortex mixing. The pellet should be completely dissolved. Incubate for 3 min at 95°C and then vortex the hot samples for an additional 2 min. Repeat the heat treatment and vortex mix twice. Add 1 ml buffer A plus the protease inhibitor cocktail as before and mix vigorously for 1 min. Centrifuge for 15 min at maximum speed in a microcentrifuge and transfer 1 ml of the supernatant into a new microcentrifuge tube.
8. Add the antibody.[b] Mix the samples gently on an end-over-end shaker for 90 min. Add 80 μl of 7% Protein A–Sepharose® in buffer B. Mix the samples again for 90 min. Centrifuge for 20 sec at 2000 r.p.m. in a microcentrifuge. Remove the supernatant very carefully without removing any Protein A–Sepharose®. Resuspend the Protein A–Sepharose® pellet in 1 ml fresh buffer A at 4°C. Mix about ten times using the end-over-end shaker. Centrifuge for 20 sec at 2000 r.p.m. in a microcentrifuge. Remove the supernatant very carefully and repeat this washing procedure four times.
9. Resuspend the Protein A–Sepharose® pellet in 100 μl electrophoresis buffer. Heat it for 5 min at 95°C and microcentrifuge it as before. Remove the supernatant to a fresh microcentrifuge tube. Take 5 μl of the supernatant from the zero time sample and measure the radioactivity in the scintillation counter. Load an aliquot of this sample containing 10 000–20 000 c.p.m. onto the SDS–PAGE gel and the same volumes of the other time samples. Run the gel.
10. Fix the gel for 30 min with a mixture of 5% propan-2-ol and 5% acetic acid in distilled water with gentle shaking. Wash the gel three times for 5 min in distilled water. Transfer the gel onto a Whatman filter. Cover with wet cellophane. Transfer the gel with filter and cellophane onto the gel drier. Dry it for 1–2 h at 60 °C under vacuum until the gel is completely dried. Transfer the dried gel into an imaging cassette or place the gel in an X-ray cassette and expose it to Kodak X-ray film.

Protocol 3. *Continued*

11. Quantify the radioactivity in each lane of the gel using an imaging system.

[a] This concentration should be varied according to the strain used and the cell density of the medium.
[b] The amount of antibody required has to be tested in pilot experiments.

Protocol 4. Pulse-chase analysis of membrane proteins

Equipment and reagents

- Appropriate culture of *S. cerevisiae*
- Vortex mixer, antibody, glass beads, Protein A–Sepharose, protease inhibitor cocktail, SDS–PAGE materials, scintillation counting, and imaging facilities (see *Protocol 3*)
- 30 mM and 10 mM solutions of NaN_3
- Breakage buffer: 50 mM Tris–HCl pH 7.5, 10 mM EDTA
- Buffer C: 0.15 M NaCl, 50 mM Tris–HCl pH 7.5, 0.1% (w/v) SDS
- Buffer D: 1% (v/v) Triton X-100, 0.1% (w/v) SDS, 0.15 M NaCl, 50 mM Tris–HCl pH 7.5, 10 mM EDTA

Method

1. Harvest samples of cells containing 10 A_{600} units of cells (about 3.3×10^8 cells) from a growing culture. Take as many samples as the time points that will be measured during the pulse-chase experiment. Concentrate the cells in each sample by centrifugation to 10 A_{600} units/ml. Follow *Protocol 3*, steps 2 and 3, to pulse label the cells and to start the chase.
2. Harvest 10 A_{600} units of cells (about 3.3×10^8 cells) for each time point during the chase period. Pipette 1 ml of cells into microcentrifuge tubes containing 0.5 ml of pre-cooled 30 mM NaN_3. Store the samples at –70 °C until all chase samples are collected. Thaw the cells at room temperature. Microcentrifuge to obtain the cell pellet. Wash the cells once with 1 ml of 10 mM NaN_3. Resuspend the cells in 1 ml breakage buffer. Disrupt the cells with 1 vol. of glass beads by four repeated cycles of vigorous vortex mixing for 30 sec at maximum speed with 30 sec incubation on ice between cycles. Clear the suspension by centrifugation at 370 *g* for 5 min at 4 °C. Remove the supernatant and centrifuge it at 16 000 *g* for 30 min at 4 °C to recover the membrane pellet.
3. Suspend the sedimented membranes in 100 μl of buffer C.
4. Add 0.9 ml of buffer D and let the membranes dissolve. Centrifuge at 16 000 *g* for 30 min at 4 °C. Take the supernatant and add the antibody.[a] Incubate overnight at 4 °C to immunoprecipitate the protein.

5. Add 80 μl Protein A–Sepharose® suspension and incubate for 60 min at room temperature with end-over-end shaking. Collect the Protein A–Sepharose® by centrifugation (*Protocol 3*, step 8), and wash it twice with buffer D. Heat the Protein A–Sepharose for 15 min at 40 °C, separate the proteins by SDS–PAGE, and analyse the results as in *Protocol 3*, steps 10 and 11.

[a] The quantity of antibody required must be determined in pilot experiments.

4. The proteasome and protein degradation in yeast

The yeast 20S proteasome contains three different types of active sites. These can be measured independently by use of artificial peptide substrates. These substrates contain fluorogenic or chromogenic residues at the carboxy terminal end of a short peptide chain. Cleavage after the C-terminal amino acid results in release of the fluorophore or chromophore which can be subsequently measured by a fluorimeter (*in situ* assays; *Protocols 5–7*) or directly visualized after dye formation (*in vitro* assays; *Protocols 8–10*).

The chymotrypsin-like activity cleaves at the C-terminal side of a hydrophobic amino acid; substrates used are succinyl-Leu-Leu-Val-Tyr-aminomethylcoumarin (Suc-Leu-Leu-Val-Tyr-AMC) or carbobenzoxy-Gly-Gly-Leu-*p*-nitroanilide (Cbz-Gly-Gly-Leu-pNA). The PGPH activity specifically clips at the carboxy terminus of acidic amino acids; the substrate used is carbobenzoxy-Leu-Leu-Glu-β-naphthylamide (Cbz-Leu-Leu-Glu-βNA). The trypsin-like activity cleaves at the C-terminus of basic amino acids; substrate used is carbobenzoxy-Ala-Arg-Arg-4-methoxy-β-naphthylamide (Cbz-Ala-Arg-Arg-MoβNA). All substrates may be purchased from Bachem GmbH Switzerland.

Protocol 5. *In vitro* assay for the chymotrypsin-like activity of the proteasome

Equipment and reagents

- Cell extract (*Protocol 1*) or proteasome preparation (*Protocols 12* and *13*) for assay
- Test buffer: 75 mM Tris–HCl pH 7.8, 3 mM ATP, 15 mM $MgCl_2$
- Substrate solution: 1.5 mM Suc-Leu-Leu-Val-Tyr-AMC (Bachem) in DMSO
- Ethanol (absolute, at 0 °C)
- Fluorimeter for single point registration(λ_{ex}: 380 nm, λ_{em}: 460 nm)
- 7-amino-4-methylcoumarin (Bachem)

Method

1. Mix 50 μl of the sample for assay, 100 μl test buffer, and 10 μl substrate solution. Also set up a control with 50 μl water in place of the sample.

Protocol 5. *Continued*

2. Incubate the tubes at 37 °C and stop the reactions after 15–60 min by adding 1 ml cold ethanol to each.
3. Measure the fluorescence in a quartz cuvette in a fluorimeter. Excite the sample at 380 nm.[a] Detect the emission at 460 nm and compare it to the emission of the control.
4. Determine the concentration of the fluorophore released (7-amino-4-methylcoumarine), using a calibration graph. To generate the calibration graph measure the fluorescence of a series of solutions containing increasing concentrations of the fluorophore. Calculate the saturation curve using the following formula:

$$I_F = C_1 \times \{1 - \exp(C_2 \times C_F\}.$$

The fluorophore concentration of test solutions can be calculated using the formula:

$$C_F = 1/C_2 \times \ln\{1 - I_F/C_1\}.$$

C_F = concentration of fluorophore; I_F = fluorescence; C_1/C_2 substrate-specific constants.

[a] A mercury lamp (366 nm) can be used instead.

Protocol 6. *In vitro* **assay for the PGPH activity of the proteasome**

Equipment and reagents

- Cell extracts (*Protocol 1*) or proteasome preparation (*Protocols 12* and *13*) for assay
- Test buffer: 1 M Tris pH 8.0
- Substrate solution: 10 mM Cbz-Leu-Leu-Glu-βNA (Bachem) in DMSO
- Stop solution: 0.2 M trisodium citrate pH 6.4, 4% (v/v) Tween 20
- Fluorimeter for single point registration(λ_{ex}: 340 nm, λ_{em}: 425 nm)
- β-Naphthylamine (Bachem)

Method

1. Mix 150 μl sample, 20 μl test buffer, 20 μl water, and 10 μl substrate solution. Also set up a control with 150 μl water in place of the sample.
2. Incubate the tubes at 37 °C. After 2–3 h incubation, stop the reactions by adding 200 μl stop solution.
3. Measure the fluorescence in a quartz cuvette in a fluorimeter. Excite at 340 nm.[a] Detect the emission at 410 nm and compare it to the emission of the control.
4. Determine the concentration of the released fluorophore using a calibration graph for β-naphthylamine (*Protocol 5*, step 4).

[a] A mercury lamp (366 nm) can be used instead.

Protocol 7. *In vitro* assay for the trypsin-like activity of the proteasome

Equipment and reagents

- Cell extracts (*Protocol 1*) or proteasome preparation (*Protocols 12* and *13*) for assay
- Test buffer: 1 M Tris–HCl pH 10.0
- Substrate solution: 10 mM Cbz-Ala-Arg-Arg-MoβNA (Bachem) in DMSO
- Fluorimeter and stop solution (*Protocol 6*)
- 4-methoxy-β-naphthylamine standard (Bachem)

Method

1. Mix 150 μl of the sample for assay, 20 μl test buffer, 20 μl water, and 10 μl substrate solution. Also set up a control with 150 μl water in place of the sample.
2. Follow *Protocol 6*, step 2.
3. Measure the fluorescence in a quartz cuvette in a fluorimeter. Excite at 340 nm.[a] Detect the emission at 425 nm and compare it to the emission of the control.
4. Determine concentration using a calibration graph for 4-methoxy-β-naphthylamine (*Protocol 5*, step 4).

[a] A mercury lamp (366 nm) can be used instead.

All three 20S proteasome activities can be determined directly and independently in single yeast colonies grown on agar plates. Release of the chromogenic residues from the peptide substrates by proteolytic cleavage is visualized by dye formation. The *p*-nitroaniline clipped off from the substrate used to measure chymotrypsin-like activity (Cbz-Gly-Gly-Leu-pNA) is detected by a coupling reaction with 1-naphthyl-ethylenediamine (33). The β-naphthylamine and MoβNA released from substrates used to measure respectively the PGPH activity (Cbz-Leu-Leu-Glu-βNA) or the trypsin-like activity (Cbz-Ala-Arg-Arg-MoβNA) are visualized by coupling with Fast Garnet GBC.

Protocol 8. *In situ* assay for chymotrypsin-like activity of the proteasome[a]

Equipment and reagents

- Appropriate *S. cerevisiae* strain for analysis
- YPD or CM agar plates (*Table 1*)
- YPD or CM agar plates covered with sterile filter dishes (8 cm diameter)
- Overlay agar: 1% agarose, 50 mM Tris–HCl pH 8.0; boil this to melt the agar
- Glass Petri dishes (9 cm diameter) and plastic Petri dishes (8.5 cm diameter) coated with 1% agarose. Add 10 ml overlay agar solution (at 50–60°C) to each Petri dish; distribute solution to moisten the whole bottom of the dish, and then pour out the surplus solution. Cool the dishes to solidify the agarose.

Protocol 8. *Continued*

- Chloroform
- Substrate solution: 50 mM Cbz-Gly-Gly-Leu-pNA (Bachem) in DMSO
- 0.1% (w/v) sodium nitrite in 1 M HCl
- 0.5% (w/v) ammonium sulfamate 1 M HCl
- 0.05% (w/v) 1-naphthyl-ethylenediamine in 47% (v/v) ethanol

Method

1. Grow the yeast cells at 30 °C on YPD agar plates or, if auxotrophic selection is needed, on CM agar plates, to a minimum colony size of 3 mm diameter.
2. Replica plate the colonies onto a fresh YPD or CM agar plate covered with a sterile filter dish.
3. Incubate the replica plate for 24 h at 30 °C.
4. Remove the filter and transfer it to a glass Petri dish coated with 1% agar (the agar coating is needed to fix the filter to the bottom of the glass Petri dish).
5. To permeabilize the cells, pour 10 ml chloroform onto the filter. Cover the Petri dish and incubate for 15 min.
6. Remove the filter, dry it, and transfer it to a plastic Petri dish coated with 1% agarose.
7. Melt 9.7 ml of overlay agar by boiling, cool down the solution to 45 °C, add 300 μl substrate solution, and mix thoroughly by stirring.[b]
8. Immediately overlay the filter (from step 6) with the substrate agar mixture (step 7), allow it to solidify, and then incubate the overlayed filter for 3 h at 37 °C.
9. Cover the overlayed filter with 10 ml of 0.1% sodium nitrite solution and incubate for 5 min.
10. Remove the nitrite solution, add 10 ml of 0.5% ammonium sulfamate solution, and incubate for 5 min.[c]
11. Remove the ammonium sulfamate solution, add 10 ml of 0.05% 1-naphthyl-ethylendiamine solution, and incubate until positive colonies stain pink.

[a] From ref. 9.
[b] Higher temperatures cause rapid decomposition of the substrate; work rapidly to avoid solidification of the melted agar.
[c] This step is needed to destroy surplus sodium nitrite; this is indicated by the appearance of nitrogen bubbles.

Protocol 9. *In situ* assay for PGPH activity of the proteasome[a]

Equipment and reagents

- Appropriate *S. cerevisiae* strain for analysis
- YPD or CM agar plates, sterile filter dishes, glass and plastic Petri dishes coated with 1% agarose, chloroform (*Protocol 8*)
- Overlay agar: 1% agarose, 0.1 M Tris–HCl pH 8.0
- Substrate solution: 50 mM Cbz-Leu-Leu-Glu-βNA in DMSO
- Coupling buffer: 0.2 M trisodium citrate pH 4.4, 4% (v/v) Tween
- Coupling reagent: Fast Garnet GBC, 50 mg/ml in DMSO

Method

1. Follow *Protocol 8*, steps 1–6.
2. Melt 9.7 ml overlay agar by boiling, cool the solution to 45°C, add 100 μl substrate solution, and mix thoroughly by stirring.[b]
3. Immediately overlay the filter with the substrate agar mixture, allow it to solidify, and then incubate the overlayed filter for 3 h at 37°C.
4. Dissolve 100 μl coupling reagent in 9.9 ml of coupling buffer.
5. Cover the overlayed filter with coupling mixture until positive colonies stain red.

[a] From ref. 34.
[b] Higher temperatures cause rapid decomposition of the substrate; work rapidly to avoid solidification of the melted agar.

Protocol 10. *In situ* assay for the trypsin-like activity of the proteasome[a]

Equipment and reagents

- Appropriate *S. cerevisiae* strain for analysis
- YPD or CM agar plates, sterile filter dishes, plastic Petri dishes coated with 1% agarose (*Protocol 8*)
- Permeabilization solution: 1% (v/v) toluene, 4% (v/v) ethanol, 0.5 mM EDTA, 0.5 mM EGTA, 20 mM Tris–HCl pH 9.0
- Overlay agar: 1% agarose, 20 mM Tris–HCl pH 9.3
- Substrate solution: 0.25 M Cbz-Ala-Arg-Arg-β-MoNA in DMSO
- Coupling buffer and coupling reagent (*Protocol 9*)

Method

1. Follow *Protocol 8*, steps 1–5.
2. Permeabilize the cells by transferring the filter to a second filter soaked with permeabilization solution. Incubate for 20 min.
3. Transfer the filter with the yeast colonies to a plastic Petri dish coated with 1% agarose.

Protocol 10. *Continued*

4. Melt 9.9 ml overlay agar by boiling, cool the solution to 45°C, add 100 μl substrate solution, and mix thoroughly by stirring.[b]
5. Follow *Protocol 9*, steps 3-5, until positive colonies stain red.

[a] From ref. 35.
[b] Higher temperatures cause rapid decomposition of the substrate; work rapidly to avoid solidification of the molten agar.

Protocol 11. Screening for proteasome mutants showing defects in 20S proteasomal activities

Equipment and reagents

- Appropriate *S. cerevisiae* cells for screening
- YPD medium, YPD agar plates (*Table 1*)
- 0.1 M sodium phosphate buffer pH 8.0
- Ethylmethane sulfonate
- Glycerol

Method

1. Grow the yeast cells at an appropriate temperature in 5 ml YPD medium to logarithmic growth phase (A_{600} = 1–3).
2. Centrifuge the culture for 10 min at 2000 *g* and resuspend the cell pellet in 5 ml sodium phosphate buffer pH 8.0.
3. Start the mutagenesis reaction by adding 150 μl of ethylmethane sulfonate.
4. Vortex the mixture vigorously to dissolve the ethylmethane sulfonate and then incubate the cells at 28°C.
5. Take 500 μl samples at different times of incubation (between 0.5–2.5 h).
6. Add glycerol to a concentration of 15% (v/v) and store the samples at –70°C.
7. Determine the survival rates of the mutagenized cells in the individual samples by counting colonies formed after plating aliquots of the cell suspension onto YPD agar plates.
8. Prepare YPD agar plates each with 300–500 distinct colonies using those samples which have 5–10% surviving cells.
9. Determine the proteasomal peptide-cleaving activities of the mutagenized colonies (*Protocols 8–10*).

Protocol 12 can be used for the preparation of 20S proteasomes from yeast but large amounts of highly purified 20S proteasomes are most easily obtained by the method described in *Protocol 13*. This procedure has been used for crystallization of yeast 20S proteasomes (5).

Protocol 12. Purification of 20S proteasomes from yeast[a]

Equipment and reagents

- Yeast cells grown to stationary phase in YPD medium (*Table 1*)
- Tris-EDTA buffer: 0.1 M Tris–HCl pH 7.2, 1 mM EDTA
- French press and centrifuges (*Protocol 1*)
- Ammonium sulfate (crystalline)
- Tris-EDTA buffer with 40% ammonium sulfate
- Tris-EDTA buffer with 15% ammonium sulfate
- Octyl–Sepharose® CL-4B column (Amersham Pharmacia Biotech) (3 × 14 cm) equilibrated in Tris-EDTA buffer containing 40% (w/v) ammonium sulfate
- Materials for peptide activity assays (*Protocols 5–7*)
- Tris-EDTA buffer with 0.15 M NaCl
- Tris-EDTA buffer with 0.35 M NaCl
- DEAE–Sepharose® CL-6B column (Amersham Pharmacia Biotech) (1.6 × 15 cm) equilibrated in Tris-EDTA buffer containing 50 mM NaCl
- Storage buffer: 20 mM Tris–HCl pH 7.2, 0.15 M NaCl, 10% (v/v) glycerol, 1 mM EDTA

Method

Carry out all steps at 4 °C.

1. Suspend 160 g (wet weight) of stationary phase yeast cells in 160 ml Tris-EDTA buffer.
2. Break the cells by two passages through the French pressure cell at 20 000 p.s.i.
3. Centrifuge the homogenate for 10 min at 12 000 *g* and recover the supernatant.
4. Centrifuge the supernatant for 30 min at 134 000 *g* and recover the cleared supernatant.[b]
5. Stir the cleared supernatant slowly while adding crystalline ammonium sulfate to 40% (w/v) saturation.[c]
6. After stirring for 1 h, sediment the precipitate at 90 000 *g* for 35 min.
7. Load the supernatant onto the octyl–Sepharose® CL-4B column.
8. Elute the proteins from the column with a linear gradient of ammonium sulfate (decreasing from 40–15%) in Tris-EDTA buffer, at a flow rate of 19 ml/h.
9. Collect fractions of 1.7 ml and assay the desired proteasomal activity (*Protocols 5–7*). Combine the fractions containing the highest proteasomal activity.
10. Precipitate the protein from the pooled fractions by addition of solid ammonium sulfate to 70% (w/v) saturation.
11. After stirring for 1 h, collect the precipitate by centrifugation at 90 000 *g* for 35 min.
12. Dissolve the precipitate in Tris-EDTA buffer containing 0.15 M NaCl. Use a volume one-tenth that of the original pooled fractions. Dialyse the solution overnight against the same buffer (500 ml, twice).

Protocol 12. *Continued*

13. Apply the solution to the DEAE–Sepharose® CL-6B column.
14. Elute the protein with a linear gradient of 0.15–0.3 M NaCl in Tris-EDTA buffer, at a flow rate of 16 ml/h.
15. Collect fractions of 1.5 ml, assay them for proteasomal activity (step 9), and pool the active fractions.[d]
16. Precipitate the protein by adding ammonium sulfate to 70% saturation. Centrifuge for 30 min at 90 000 *g*, and dissolve the precipitate in storage buffer (one-tenth to one-fifth volume of the original pooled fraction).

[a] From ref. 36.
[b] Protein concentrations obtained are usually about 30 mg/ml.
[c] The solution usually reaches a final pH of 6.
[d] The 20S proteasome elutes from the column in two activity peaks at 0.27 M NaCl and 0.28 M NaCl.

Protocol 13. Large scale purification of 20S proteasomes from yeast[a]

Equipment and reagents

- Yeast (fresh baker's yeast)
- Buffer A: 20 mM Tris–HCl pH 7.5, 1 mM EDTA, 1 mM NaN_3
- Cell disintegrator (Biomatik; B Euler)
- Glass beads (0.5 mm diameter)
- Light microscope (200–400-fold magnification)
- Sintered glass filter funnel
- 0.28 M NaCl in buffer A
- 0.8 M NaCl in buffer A
- Q-Sepharose (Sigma): pack a 5 × 20 cm column and equilibrate in 0.25 M NaCl in buffer A
- Materials for the assay of the relevant proteasomal enzyme activity (*Protocols 5–7*)
- Buffer B: 60 mM potassium phosphate pH 7.5
- Hydroxylapatite (Bio-Rad): pack a 3 × 10 cm column and equilibrate it in buffer B
- 0.3 M potassium phosphate pH 7.5
- Ultrafiltration device (Amicon with YM30 membrane)
- Superose-6 column (Amersham Pharmacia Biotech): pack a 1 × 30 cm column and equilibrate it in buffer A
- FPLC equipment (Amersham Pharmacia Biotech)

Method

Carry out all steps at 4 °C except the FPLC chromatography (step 16).

1. Wash 500 g (wet weight) yeast cells twice with ice-cold water, recovering the cells by centrifugation at 2500 *g* for 15 min.
2. Resuspend the cells in buffer A at a ratio of cells to buffer of 2:3 (wet w/v).
3. Break the cells for 10 min in a disintegrator with glass beads using a ratio of beads to cell suspension of 3:2 (v/v). Monitor the cell breakage microscopically.
4. Filter the cell homogenate through a sintered glass filter funnel to

remove the broken glass beads. Centrifuge the filtered homogenate for 45 min at 10 000 *g*.

5. Centrifuge the supernatant at 134 000 *g* for 60 min. Remove the lipids from the top layer using a Pasteur pipette. Recover the remaining (yellow) supernatant.[b]
6. Immediately apply the supernatant to the Q-Sepharose column.
7. Wash the column with 0.28 M NaCl in buffer A.
8. Elute the protein with a 2 litre gradient of 0.28–0.8 M NaCl in buffer A at a flow rate of 120 ml/h.
9. Collect 12 ml fractions and determine the relevant proteasomal peptide-cleaving activities in all the fractions.[c]
10. Combine the fractions containing the proteasome and dilute the solution threefold with water.
11. Apply the diluted solution to the hydroxylapatite column.
12. Wash the column with two column volumes of buffer B.
13. Elute the protein with a 1 litre gradient of buffer B to 300 mM potassium phosphate pH 7.5 at a flow rate of 60 ml/h.
14. Collect 12 ml fractions and determine the relevant peptide-cleaving activities in the fractions.[d]
15. Combine the active fractions. Concentrate the solution 20-fold by ultrafiltration using an Amicon YM30 membrane.
16. Apply the concentrated sample to the Superose-6 column connected to the FPLC apparatus and elute the column with buffer A at a flow rate of 18 ml/h. The proteasome elutes at high purity with an elution time of 37 min.

[a] From ref. 5.
[b] Protein concentrations are normally about 50 mg/ml.
[c] The proteasome normally elutes at 0.4–0.45 M NaCl.
[d] The proteasome normally elutes at 0.12 M phosphate.

5. Proteinase yscD, a major cytoplasmic peptidase

5.1 Introduction

Proteinase yscD is a major enzyme, mainly cytoplasmic, capable of splitting *p*-nitroaniline derivatives of peptides such as benzoyl-prolyl-phenylalanyl-arginyl-*p*-nitroanilide (Bz-Pro-Phe-Arg-pNA), acetyl-alanyl-alanyl-prolyl-alanyl-*p*-nitroanilide (Ac-Ala-Ala-Pro-Ala-pNA), methoxy-succinyl-alanyl-alanyl-prolyl-methionyl-*p*-nitroanilide (Mo-Suc-Ala-Ala-Pro-Met-pNA), or acetyl-alanyl-alanyl-prolyl-phenylalanyl-*p*-nitroanilide (Ac-Ala-Ala-Pro-Phe-pNA) (37). Cleavage of these substrates, however, does not occur at the *p*-nitroanilide bond but at an internal peptide bond. For instance, cleavage of Bz-Pro-Phe-Arg-pNA occurs at the Pro–Phe bond (37). The enzyme is suspected to be the equivalent of the mammalian metalloendopeptidase EP 24.15 (38) and, like the mammalian enzyme, proteinase yscD cleaves bradykinin. It has been shown to cleave peptides *in vivo* (38). The enzyme might function in late stages of protein degradation in the cytoplasm by further degrading peptides released from the proteasome.

5.2 Assay of proteinase yscD

Proteinase yscD uses the substrate Bz-Pro-Phe-Arg-pNA. The enzyme splits the substrate into a dipeptidyl-*p*-nitroanilide and this can be completely cleaved by a helper enzyme, aminopeptidase M, liberating *p*-nitroaniline as shown below:

$$\text{Bz-Pro-Phe-Arg-pNA} \xrightarrow{\text{Proteinase yscD}} \text{Bz-Pro} + \text{Phe-Arg-pNA}$$

$$\text{Phe-Arg-pNA} \xrightarrow{\text{Aminopeptidase M}} \text{Phe, Arg, } p\text{-nitroaniline}$$

A procedure for the proteinase yscD assay is given in *Protocol 14*.

Protocol 14. Assay of proteinase yscD

Reagents

- 1 M Tris–HCl pH 7.5
- Aminopeptidase M (Boehringer Mannheim): 1 mg in 500 μl of 0.1 M Tris–HCl pH 7.5
- Extract (see *Protocol 1*)
- Substrate solution: 10 mM Bz-Pro-Phe-Arg-pNA (Bachem) in DMSO

Method

1. Set up two tubes (sample and control) with 30 μl of 1 M Tris–HCl pH 7.5 in each.
2. To the sample tube, add 245 μl distilled water and add 265 μl distilled water to the control tube.
3. Add 20 μl of aminopeptidase M to each tube.
4. Add 1–20 μl cell extract to the sample tube. If less than 20 μl is used, make up the volume to 20 μl with distilled water.
5. Add 15 μl of substrate solution to each tube.
6. Measure the A_{405} continuously in each tube using a recording spectrophotometer. Subtract the readings of the control tube from those of the sample tube. One unit of proteinase yscD activity corresponds to 1 μmol *p*-nitroaniline produced per min under these conditions (ϵ_{405} = 9500 litre/mol.cm).

6. Protein degradation in the vacuole

6.1 Introduction

In *S. cerevisiae* the vacuole is the equivalent of the mammalian lysosome. The organelle contains a variety of endopeptidases, carboxypeptidases, aminopeptidases, and a dipeptidyl aminopeptidase (*Table 2*) (for reviews see refs 20, 39–46). The two endopeptidases, proteinase yscA and proteinase yscB, are the major peptidases involved in intracellular protein degradation under nutritional stress (25). Under these conditions the degradation rate is about 3.3% per hour and both proteinases catalyse the degradation of about 80% of radioactively labelled yeast protein. Under normal growth conditions, the degradation rate is about 1.1% per hour and about 40% of labelled intra-

Table 2. Peptidases present in the vacuole of *Saccharomyces cerevisiae*

Enzyme[a]	Gene	Peptidase type	Molecular mass (kDa)	Inhibitors
Proteinase yscA	*PRA1/PEP4*	Aspartic acid endopeptidase	42	Pepstatin; diazo-acetyl-DL-norleucine methylester; epoxy-3-4-nitrophenoxypropane
Proteinase yscB	*PBR1*	Serine endopeptidase	33	Chymostatin; DFP; PMSF; antipain; pCMB
Aminopeptidase yscl	*LAP4*	Metallo (Zn^{2+}) exopeptidase	640	EDTA; bestatin
Aminopeptidase yscCo(Y)	*APY1*		70	EDTA
Dipeptidyl amino-peptidase B (yscV)	*DAP2*	Serine exopeptidase		DFP; PMSF
Carboxypeptidase yscY	*PRC1*	Serine exopeptidase	61	DFP; PMSF, TPCK
Carboxypeptidase yscS	*CPS1*	Metallo (Zn^{2+}) exopeptidase	64	EDTA

[a] For each enzyme activity listed, *S. cerevisiae* mutants are available as follows:

- Mutants defective in proteinases yscA and/or yscB or carboxypeptidase yscY can be obtained from The Yeast Genetic Stock Center, Department of Molecular and Cell Biology, 229 Stanley Hall, University of California, Berkeley, CA 94720-3206; or Dr Elizabeth W. Jones, Biological Sciences, Carnegie Mellon University, 4400 Fifth Avenue, Pittsburgh, PA 15213, USA; or Dr Dieter H. Wolf, Institut für Biochemie, Universität Stuttgart, Pfaffenwaldring 55, D-70569 Stuttgart, Germany.
- Mutants defective in proteasome core subunits, mutants defective in carboxypeptidase yscS, and proteinase yscD mutants can be obtained from Dr Dieter H. Wolf, Institut für Biochemie, Universität Stuttgart, Pfaffenwaldring 55, D-70569 Stuttgart, Germany.
- Mutants defective in 19S cap subunits of the proteasome can be obtained from Dr Carl Mann, Commissariat a l'Energie Atomique DSV/DBCM, Service de Biochimie, et de Genetique Moleculaire (BAT 142), Centre d'Etudes de Saclay, F-91191 Gif-sur-Yvette Cedex, France; and Dr Daniel Finley, Harvard Medical School, Department of Cell Biology, 240 Longwood Avenue, Boston, MA 02115, USA.
- Aminopeptidase yscl and dipeptidylaminopeptidase B (yscV) mutants can be obtained from Dr Paz Suarez-Rendueles, Departamento de Bioquimica y Biologia Molecular, Universidad de Oviedo, C/Julian Claveria S/N, E-33006 Oviedo, Spain.
- Aminopeptidase yscCo(Y) mutants can be obtained from Dr Toshimasa Yasuhara, Bayer Yakuhin Pharmacology Institute, Nabetani, Kande-cyo, Nishi-ku, Kobe 651-23, Japan.

cellular protein is degraded by the combined action of proteinases yscA and yscB. The essential roles of proteinase yscA and proteinase yscB in yeast cells under nutritional stress conditions is documented by the fact that cells devoid of these two enzymes fail to survive a long starvation period (25, 27). None of the other vacuolar peptidases have been found to be vital for yeast cells, nor does a defect in any of them have any effect on the growth rate of cells under normal nutritional supply (e.g. in YPD or MV media). Only when yeast is grown on certain carboxy terminally blocked dipeptides as sole nitrogen source (e.g. Cbz-Gly-Leu, Cbz-Phe-Leu, or Cbz-Leu-Leu) do the activities of carboxypeptidase yscY or carboxpeptidase yscS become essential for growth. The availability of mutants defective in each of the vacuolar peptidases (39–43, 46) allows the functional study of the peptidases. Such strains are useful for purifying proteins or analysing protein functions *in vitro* without proteolytic artefacts due to the action of these non-specific peptidases.

6.2 Assay of vacuolar peptidases

This section describes assays for the vacuolar peptidases as well as the purification of the most important vacuolar endopeptidases, proteinase yscA and proteinase yscB. Highest activities of the vacuolar proteinases yscA and yscB are reached in stationary growth phase or after incubation of cells (as a 1% cell suspension) for about 18 hours in acetate mineral medium (29, 47). However, protein inhibitors of the two endopeptidases, proteinase yscA and proteinase yscB, as well as carboxypeptidase yscY, exist in the cytoplasm of yeast cells (48, 49) and combine with the peptidases when cells and their organelles are broken. Therefore, after cell extracts are prepared, activation procedures are required to destroy the inhibitors or disassemble the peptidase–inhibitor complex before these peptidases can be assayed. Destruction of the inhibitors occurs via cross-activation by the proteinases yscA and yscB. Adjustment of cell extracts to pH 5 liberates some of the proteinase yscA from its inhibitor I^A. This results in digestion of the proteinase yscB–inhibitor I^B complex liberating active proteinase yscB. The activated proteinase yscB itself then digests the rest of the proteinase yscA–inhibitor I^A complex leading to fully active proteinase yscA. Proteinase yscA and proteinase yscB together degrade the carboxypeptidase yscY–inhibitor I^C complex liberating active carboxypeptidase yscY.

6.2.1 Activation of proteinases yscA and yscB

Protocol 15 describes a procedure for the activation of proteinases yscA and yscB in cell extracts using low pH.

6.2.2 Assay of proteinase yscA and its inhibitor I^A

Proteinase yscA cleaves acid denatured haemoglobin into tyrosine-containing acid soluble peptides at acidic pH. These peptides are measured using a protein determination procedure (49, 50). The assay is described in *Protocol 16*.

Protocol 15. Activation of proteinases yscA and yscB

Reagents

- 1 M acetic acid
- Potassium azide
- Cell extract (*Protocol 1*)

Method

1. Adjust the cell extract to pH 5.0 using 1 M acetic acid. Add potassium azide (final concentration 1 mM) to avoid growth of micro-organisms.
2. Incubate the extract at 30 °C for 24–96 h.
3. Measure the activity of proteinase yscA or yscB as appropriate (*Protocols 16* and *17*).

Protocol 16. Assay of proteinase yscA

Reagents

- Haemoglobin
- 1 M HCl
- Lactic acid
- 1 M NaOH
- Acid activated yeast cell extract for assay (*Protocol 15*)
- 10% (w/v) TCA
- Folin reagent
- 1% (w/v) $CuSO_4$
- 2.7% sodium potassium tartrate
- 24% (w/v) Na_2CO_3
- Folin solution A: mix 0.5 ml 1% $CuSO_4$ with 0.5 ml 2.7% sodium potassium tartrate; add 5 ml 24% Na_2CO_3 and 6 ml 1 M NaOH
- Folin solution B: mix 1 ml Folin reagent with 19 ml water

Method

1. Prepare the proteinase yscA substrate (1.2% acid denatured haemoglobin) by dissolving 240 mg haemoglobin in 5 ml water and adjusting to pH 1.0 using 1 M HCl. Incubate the haemoglobin solution for 2 h at room temperature. Add 150 μl lactic acid and adjust to pH 3.0 with 1 M NaOH. Add distilled water to 20 ml final volume.[a] Prepare this substrate solution fresh daily.
2. Add an appropriate volume of activated yeast cell extract to a test-tube and add distilled water to 100 μl final volume. Set up a control with water only.
3. Add 500 μl haemoglobin substrate solution to each tube.
4. Incubate the tubes for 30 min at 30 °C.
5. Add 500 μl 10% TCA to both tubes. Add the equivalent volume of activated extract used in the test reaction to the control.
6. Centrifuge the samples for 10 min in a microcentrifuge at full speed. Recover the supernatant and repeat the centrifugation.

Protocol 16. *Continued*

7. Take a sample (5–100 μl)[b] of each of the clear TCA supernatants and add distilled water to 100 μl.
8. Add 100 μl of Folin solution A. Incubate for 30 min at 37 °C.
9. Add 500 μl of Folin solution B and mix immediately. Incubate for 30 min at 37 °C.
10. Measure the A_{578} of the test sample and subtract from it the A_{578} of the control. Proteinase yscA activity is calculated as follows: one unit of enzyme activity is equivalent to a change in A_{578} of 0.185.

[a] This volume of haemoglobin substrate solution is sufficient for 40 assays.
[b] The volume of sample required will depend on the yscA activity present.

The proteinase yscA assay can be modified for the measurement of proteinase yscA–inhibitor by adding increasing amounts of I^A (for preparation see *Protocol 23*) to an extract containing proteinase yscA. Include a proteinase yscA control without I^A. Calculate the 50% inhibition value from the resulting graph of proteinase yscA activity.

6.2.3 Assay of proteinase yscB

Besides acidic activation in crude extracts (see Section 6.2.1 and *Protocol 15*), proteinase yscB activity can also be liberated by incubation of extracts with 0.1–0.5% SDS (51). The amount of SDS required will depend on the protein concentration. Note that in this procedure buffers must not contain potassium ions because of the insolubility of potassium dodecyl sulfate.

The principle of the yscB assay (described in *Protocol 17*) is that proteinase yscB cleaves Azocoll or Hide Powder Azure, two collagen derivatives modified with chromophores, to yield soluble peptides. The solubilized red (Azocoll) or blue (Hide Powder Azure) dye can be measured photometrically.

Protocol 17. Assay of proteinase yscB[a]

Equipment and reagents

- Azocoll (Sigma)[b]
- 0.1 M Sodium phosphate pH 7.0
- Acid activated (*Protocol 15*) or SDS activated yeast cell extract (see text) for assay
- Shaker (any shaker that keeps the particles suspended during the reaction)
- 10% (w/v) TCA
- Microcentrifuge

Method

1. Suspend 240 mg Azocoll in 10 ml of 0.1 M phosphate buffer pH 7.0.
2. Take an aliquot of activated cell extract and add 0.1 M phosphate buffer pH 7.0 to 100 μl final volume. Set up a control with 100 μl of 0.1 M phosphate buffer pH 7.0.

3. Add 500 μl Azocoll suspension to each tube. Use of a cut-off 1 ml pipette tip will allow the Azocoll particles to enter the tip.
4. Shake the tubes for 30 min at 30 °C.
5. Add 500 μl of 10% TCA to each tube.
6. Centrifuge the samples in a microcentrifuge at maximum speed to precipitate undigested Azocoll.
7. Measure the A_{520} nm of the supernatants.[c]
8. Substract the A_{520} of the control from the A_{520} of the test sample.
9. Calculate the enzyme activity from the corrected A_{520}. Enzyme activity is expressed as the change in A_{520}/min/mg extract protein.

[a] From ref. 49.
[b] Hide Powder Azure (Calbiochem) may also be used.
[c] Use 578 nm if Hide Powder Azure is the substrate.

6.2.4 Assay of carboxypeptidase yscY

The principle of the carboxypeptidase yscY assay (described in *Protocol 18*) is that carboxypeptidase yscY catalyses the hydrolysis of *N*-benzoyltyrosine-*p*-nitroanilide to yield the yellow *p*-nitroaniline product which can be detected by absorbance at 405 nm. The reaction can be followed kinetically or in a fixed time point assay. When the assay described in *Protocol 18* is used, acid activation of the cell extract (*Protocol 15*) is not necessary because dimethylformamide liberates the inhibitor I^C from carboxypeptidase yscY instantaneously.

Protocol 18. Assay of carboxypeptidase yscY[a]

Reagents

- Acid activated yeast cell extract (*Protocol 15*) or non-activated extract (*Protocol 1*)
- 0.1 M Sodium phosphate pH 7.0
- Dimethylformamide
- Substrate solution (3 mM *N*-benzoyl-L-tyrosine-*p*-nitroanilide in dimethylformamide): dissolve 12.16 mg substrate in 10 ml dimethylformamide (sufficient for 50 assays)
- 10 mM $HgCl_2$

Method

1. Take an aliquot of the extract and add 0.1 M phosphate buffer pH 7.0 to 1.2 ml final volume. Set up two controls each with 1.2 ml of 0.1 M phosphate buffer pH 7.0.
2. Add 200 μl substrate solution to the test sample and to one of the controls. Add 200 μl dimethylformamide to the other control.
3. Incubate the tubes at 30 °C and stop the reactions after 30 min by adding $HgCl_2$ to a final concentration of 1 mM.[b]
4. Measure the A_{405} of each solution.

Protocol 18. *Continued*

5. Subtract both control values from the test value. Calculate the enzyme activity: 1 unit = 1 μmol *p*-nitroaniline liberated per min (ϵ_{405} of *p*-nitroaniline = 9500 litre/mol.cm).

[a] From ref. 52.
[b] The reactions can also be followed continuously at 405 nm using a recording spectrophotometer.

6.2.5 Assay of carboxypeptidase yscS

While carboxypeptidase yscY (Section 6.2.4) can be tested specifically with *N*-benzoyl-L-tyrosine-*p*-nitroanilide in yeast cell extracts, both carboxypeptidase yscS and carboxypeptidase yscY react with the N-terminally blocked peptide substrate, *N*-carbobenzoxy-glycyl-L-leucine. To assay carboxypeptidase yscS, carboxypeptidase yscY must be selectively inhibited with 1 mM PMSF. As the assay involves the determination of the liberated amino acid L-leucine, cell extracts have to be dialysed extensively against 0.1 M imidazole chloride buffer pH 5.2, 0.1 mM $ZnCl_2$ before assay to remove free amino acids.

The principle of the assay (described in *Protocol 19*) is that carboxypeptidase yscS liberates L-leucine from *N*-carbobenzoxy-glycyl-L-leucine. L-Leucine is then oxidized with L-amino acid oxidase to the respective keto acid and hydrogen peroxide. Reduction of the hydrogen peroxide is coupled to the oxidation of *o*-dianisidine by peroxidase, yielding oxidized *o*-dianisidine (brown).

Protocol 19. Assay of carboxypeptidase yscS

Equipment and reagents

- Yeast cell extract dialysed for 24 h at 25°C in imidazole buffer
- Dialysis tubing
- Imidazole buffer: 0.1 M imidazole chloride pH 5.2, 0.1 mM $ZnCl_2$
- L-Amino acid oxidase (type I, Sigma)
- Horseradish peroxidase (Sigma)
- 0.1 M $MnCl_2$
- 0.1 M sodium phosphate pH 7.4
- *o*-Dianisidine (Sigma)
- *N*-carbobenzoxy-glycyl-L-leucine (Bachem)
- 0.2 M potassium phosphate pH 7.4
- Solution I: 0.25 mg/ml L-amino acid oxidase, 0.4 mg/ml horseradish peroxidase, 1 μl of 0.1 M $MnCl_2$ in 0.1 M sodium phosphate buffer pH 7.4
- Solution II: 2 mg/ml *o*-dianisidine in distilled water
- Solution III: 20 mM *N*-carbobenzoxy-glycyl-L-leucine in 0.2 M potassium phosphate buffer pH 7.4
- 20% (v/v) Triton X-100
- PMSF

Method

1. Dialyse 1–10 ml cell extract against 100 volumes of imidazole buffer to remove free amino acids. Repeat the dialysis. Add PMSF to the extract to 1 mM final concentration.

2. Set up three series of tubes and add the following in the order given:

	Test (extract)	Control 1 (no extract)	Control 2 (no peptide substrate)
Solution I	500 μl	500 μl	500 μl
Solution III	500 μl	500 μl	–
Solution II	50 μl	50 μl	50 μl
Extract[a]	*x* μl	–	*x* μl
0.1 M sodium phosphate buffer pH 7.4	50–*x* μl	50 μl	550–*x* μl

3. Mix the contents of each tube and incubate the tubes at 30°C for 30 min. Stop the reactions by adding 100 μl of 20% Triton X-100.[b]
4. Measure the A_{405} of each sample.
5. Subtract the A_{405} values of control 1 and control 2 from the test A_{405}. Calculate the enzyme activity from the corrected A_{405}. One unit of enzyme activity corresponds to 1 μmol L-leucine produced per min; 1 μmol L-leucine corresponds to a change in A_{405} nm of 7.25 under these conditions. The specific activity is expressed as nmol L-leucine/min/mg extract protein.

[a] The volume (*x* μl) of extract required will depend on the yscS activity of the extract. Make up the difference in volume with phosphate buffer as in the table.
[b] The reaction can be followed continuously at 405 nm using a recording spectrophotometer.

The assay described in *Protocol 19* for carboxypeptidase yscS may be used to measure the activity of carboxypeptidase yscY in the same extracts. In this case *N*-carbobenzoxy-L-phenylalanyl-L-leucine should be used as substrate and the activity of carboxypeptidase yscS against this substrate (5% of the activity of carboxypeptidase yscY) must be inhibited by addition of 5 mM EDTA to the extract.

6.2.6 Assays of aminopeptidase yscI and aminopeptidase yscCo(Y)

Aminopeptidase I and aminopeptidase II (a non-vacuolar enzyme) both use L-leucine-*p*-nitroanilide as a substrate. Aminopeptidase I is sensitive to 1 mM nitrilo triacetic acid and thus its activity can be calculated from the loss of activity after adding this agent. The non-vacuolar aminopeptidase yscII and the vacuolar aminopeptidase yscCo(Y) cleave lysine-*p*-nitroanilide. However, aminopeptidase yscCo(Y) activity is completely dependent on Co^{2+} ions. Thus activity of this enzyme can be determined after a one hour pre-incubation with 0.1 mM $CoCl_2$. The difference in activity with and without Co^{2+} ions is the aminopeptidase yscCo(Y) activity. The assays for yscI and yscCo(Y) are described in *Protocols 20* and *21*. Each assay involves the liberation of *p*-nitroaniline from the substrate; this product can be measured at 405 nm. Note: it is also possible to measure aminopeptidase II activity in these assays (see *Protocol 20*).

Protocol 20. Assay of aminopeptidase yscl

Reagents

- 0.2 M Tris–HCl pH 7.5
- 10 mM nitrilo triacetic acid
- 50 mM EDTA
- 4 mM H_2SO_4
- L-Leucine-*p*-nitroanilide (Bachem or Sigma)
- Yeast cell extract (see *Protocol 1*)
- Substrate solution: 2 mM L-leucine-*p*-nitroanilide in 4 mM H_2SO_4

Method

1. Set up two samples of extract (each 90 µl). To one (test extract) add 10 µl 10 mM nitrilo triacetic acid. To the other (control extract) add 10 µl water. Incubate both extracts for 1 h at 30 °C.
2. Set up three tubes and mix 500 µl of 0.2 M Tris–HCl pH 7.5 with 500 µl substrate solution in each tube.
3. Add 20 µl test extract to one tube, 20 µl control extract to the second, and 20 µl water to the third (blank).
4. Incubate the tubes at 30 °C. After 30 min, add 100 µl of 50 mM EDTA to each.[a]
5. Measure the A_{405} of each sample. Calculate the difference in A_{405} between the control (activities I and II) and test (activity II only) samples; this gives the aminopeptidase yscl activity. The difference between the test and the blank tubes gives the activity of aminopeptidase II. In each case, one unit of enzyme activity corresponds to 1 µmol *p*-nitroaniline produced per min under these conditions (ϵ_{405nm} *p*-nitroaniline = 9500 litre/mol.cm).

[a] The reaction may also be measured continuously at 405 nm using a recording spectrophotometer.

Protocol 21. Assay of aminopeptidase yscCo(Y)[a]

Reagents

- L-Lysine-*p*-nitroanilide
- Yeast cell extract for assay (*Protocol 10*)
- 4 mM H_2SO_4
- 50 mM EDTA
- Substrate solution: 2 mM L-lysine-*p*-nitroanilide in 4 mM H_2SO_4
- 0.1 mM $CoCl_2$
- 0.2 M Tris–HCl buffer pH 8.0

Method

1. Activate the extract by incubating 90 µl of yeast cell extract with 10 µl of 0.1 mM $CoCl_2$ at 25 °C for 60 min. At the same time prepare a control extract by carrying out a similar incubation with the $CoCl_2$ replaced by water.
2. Set up two assay tubes. To each tube add 500 µl of 0.2 M Tris–HCl buffer pH 8.0 and 500 µl substrate solution.

3. Add 20 μl activated extract to one tube and 20 μl control extract to the other.
4. Incubate both tubes at 30 °C.
5. After 30 min, stop the assay by adding 100 μl of 50 mM EDTA to each tube.[b] Calculate the difference in A_{405} between the test (yscCo and II activities) and the control (activity II only) to determine the yscCo(Y) activity (see *Protocol 20*, step 5).

[a] From ref. 53.
[b] The reaction may be followed continuously at 405 nm using a recording spectrophotometer.

6.2.7 Assay of dipeptidyl aminopeptidase B (yscV)

Two major dipeptidyl aminopeptidases are found in yeast: dipeptidyl aminopeptidase A (yscIV), in the Golgi complex, and the vacuolar dipeptidyl aminopeptidase B (yscV). Both enzymes can be tested with the substrate L-alanyl-prolyl-*p*-nitroanilide and following the production of *p*-nitroaniline at 405 nm. As the vacuolar dipeptidyl aminopeptidase B (yscV) is heat-sensitive, it can be inactivated by incubating cell extracts for 30 min at 60 °C. Thus the individual activities of dipeptidyl aminopeptidaseB (yscV) and dipeptidyl aminopeptidase A can be calculated from assays before and after heating the cell extract. The procedure is given in *Protocol 22*.

Protocol 22. Assay of dipeptidyl aminopeptidase B (yscV)[a]

Reagents

- Yeast extract for assay (see *Protocol 1*)
- Substrate solution: 3 mM L-alanyl-prolyl-*p*-nitroanilide in 25% methanol
- 5% $ZnSO_4$
- Hepes buffer: 0.4 M Hepes pH 7 with Tris base
- 7.5% $Ba(OH)_2$

Method

1. Heat treat 1 ml of the extract at 60 °C for 30 min.
2. Mix 10–100 μl of heated extract with distilled water in a final volume of 200 μl, and add 250 μl Hepes buffer. Mix an identical volume of non-heated extract with water and Hepes buffer. Also set up a control without extract by mixing 200 μl water with 250 μl Hepes buffer.
3. Start the reactions by adding 50 μl substrate solution to each tube.
4. Incubate the tubes at 37 °C for 30 min.[b]
5. Add 500 μl of 5% $ZnSO_4$ and 100 μl of 7.5% $Ba(OH)_2$ to stop the reactions.
6. Centrifuge the samples for 10 min at maximum speed in a microcentrifuge to remove the precipitate. Measure the A_{405} nm of each supernatant.

Protocol 22. *Continued*

7. Calculate the difference in A_{405} between the non-heated (activities A and B) and heated (activity A only) extracts to determine the aminopeptidase B (yscV) activity (see *Protocol 20*, step 5). One unit of dipeptidyl aminopeptidase B (yscV) is defined as the amount of enzyme which liberates 1 μmol *p*-nitroaniline from the substrate in 1 min under these assay conditions.

[a] From ref. 54.
[b] The reactions may be followed continuously at 405 nm using a recording spectrophotometer.

6.3 Purification of yeast vacuolar proteinases

This section describes the purification of vacuolar proteinases yscA and yscB. The procedure for yscA, described in *Protocol 24*, utilizes the specific inhibitor of this proteinase, I^A, to affinity purify the proteinase. *Protocol 23* describes the preparation of the proteinase yscA–inhibitor affinity matrix. Proteinase yscB is purified as described in *Protocol 25*.

Protocol 23. Preparation of the proteinase yscA–inhibitor affinity matrix[a]

Equipment and reagents

- Yeast cells grown into stationary phase (Section 2) or use fresh baker's yeast
- 1 M NaOH
- Glass spiral (about 16 m in length with 600 ml volume) placed in a boiling water-bath
- Büchner funnel with filter paper
- 1 M acetic acid
- Standard buffer: 50 mM sodium acetate buffer pH 4.5
- Acetate buffer: 50 mM sodium acetate pH 4.5
- SP-Sephadex (Amersham Pharmacia Biotech)
- SP-Sephadex® C-25 column A (5 × 17 cm) equilibrated in standard buffer
- SP-Sephadex® C-25 column B (2.6 × 15 cm) equilibrated in standard buffer
- Standard buffer containing 0.5 M NaCl
- Standard buffer containing 50 mM NaCl
- Materials for assay of proteinase yscA (*Protocol 16*)
- Freeze-dryer
- Dialysis tubing
- 0.1 M Tris–HCl pH 9.0
- Tris base (solid)
- QAE-Sepharose A-25 (Amersham Pharmacia Biotech)
- 50 mM Tris–HCl pH 9.0
- QAE-Sephadex® A-25 column A (2.6 × 12 cm) equilibrated in 50 mM Tris–HCl pH 9.0
- QAE-Sephadex® A-25 column B (1 × 16 cm) equilibrated in 50 mM Tris–HCl pH 9.0
- 50 mM Tris–HCl pH 7.0
- Triethanolamine buffer: 0.2 M triethanolamine–HCl pH 8.2
- CH-Sepharose® 4B (Amersham Pharmacia Biotech)
- 1 M NaCl
- Hydroxysuccinimide, 1-ethyl-3-(dimethylaminopropyl) carbodiimide
- 1 M HCl
- Shaker
- 0.5 M Tris–HCl pH 8.0
- Sodium azide

A. *Preparation of the specific proteinase yscA–inhibitor*

1. Suspend 500 g yeast cells in 5 litres of distilled water and adjust to pH 6.0 with 1 M NaOH. Heat the cell suspension to 100 °C for about

20 min by pumping it (200 ml/h) through the glass spiral immersed in a boiling water-bath. Cool the heated suspension immediately in ice and centrifuge it at 27 000 *g* for 30 min.

2. Filter the turbid yellow supernatant (about 6.7 litres) through filter paper on the Büchner funnel and adjust to pH 4.5 with 1 M acetic acid. Adjust the ionic strength of the filtrate to the equivalent of the standard buffer. To do this, dilute 6.7 litres filtrate with acetate buffer to 11.6 litres.
3. Load the solution onto the SP–Sephadex® C-25 column A at a flow rate of 180 ml/h. Wash the column with 8 litres standard buffer. Elute the protein with a 4 litre linear gradient of 0–0.5 M NaCl in standard buffer at a flow rate of 100 ml/h. Collect fractions of 20 ml.
4. Assay the fractions for proteinase inhibitor.[b] Pool and lyophilize the inhibitor-containing fractions. Dissolve the protein in distilled water and dialyse the solution overnight against standard buffer.
5. Adjust the dialysed solution to 320 ml with standard buffer. Apply the material to the SP–Sephadex® C-25 column B. Wash the column with standard buffer until the A_{280} of the washings is zero. Elute the inhibitor with a 1 litre linear gradient of 0–50 mM NaCl in standard buffer at a flow rate of 50 ml/h. Collect fractions of 10 ml.
6. Assay the fractions for proteinase inhibitor.[b] Pool the active fractions and lyophilize them. Dissolve the protein in distilled water and dialyse against distilled water.
7. Dilute the inhibitor solution (about 26 ml) to 160 ml with 0.1 M Tris–HCl pH 9.0. Adjust the solution carefully to pH 9.0 with Tris base and apply the solution to the QAE–Sephadex® A-25 column A at a flow rate of 50 ml/h. Elute the inhibitor with a 1 litre gradient of 50 mM Tris–HCl from pH 9.0 to pH 7.0. Collect fractions of 10 ml.
8. Repeat step 6.
9. Chromatograph the dialysed solution on QAE–Sephadex® A-25 column B as in steps 7 and 8.
10. Store the inhibitor preparation until the affinity matrix has been prepared (part B).

B. *Preparation of the inhibitor affinity matrix*

1. Equilibrate the lyophilized inhibitor with triethanolamine buffer by gel filtration.
2. Wash 5 g CH–Sepharose® 4B with 500 ml of 1 M NaCl and then with 1 litre distilled water.[c]

Protocol 23. *Continued*

3. Dissolve 0.5 g hydroxysuccinimide and 0.75 g 1-ethyl-3-(dimethylaminopropyl) carbodiimide in 100 ml distilled water. Add the swollen gel (part B, step 2) to this solution. Adjust the gel suspension to pH 5.0 with 1 M HCl, and keep the gel at this pH for 120 min adding 1 M NaOH as required to maintain the pH constant. Wash the gel with 1 litre of ice-cold water.
4. Dilute the inhibitor preparation (part A, step 10) to 0.5 mg/ml and add the washed gel to 20 ml diluted inhibitor. Shake the suspension overnight at 4 °C and then wash it successively with 1 litre of 0.5 M Tris–HCl pH 8.0, 500 ml distilled water, and 1 litre acetate buffer. Add 200 μg of sodium azide and store the gel at 4 °C.

[a] From ref. 55.
[b] The inhibitory action of I^A can be easily assayed by the inhibition of proteinase yscA in the proteinase yscA assay (*Protocol 16*).
[c] About 7 mg of inhibitor will be bound to 5 g gel.

Protocol 24. Purification of proteinase yscA[a]

Equipment and reagents

- Yeast cells grown to stationary phase (Section 2)
- Homogenizer (Manton Gaulin Mfg.)
- Glacial acetic acid
- Streptomycin sulfate
- Penicillin G
- Ammonium sulfate
- Dialysis tubing
- Phenyl–Sepharose® (Amersham Pharmacia Biotech)
- Sintered glass filter
- Buffer A: 0.1 M potassium phosphate pH 6
- Buffer A containing 2 M NaCl
- Ethylene glycol
- Buffer A containing 50% (v/v) ethylene glycol
- Buffer A containing 85% (v/v) ethylene glycol
- DEAE–Sephacel® (Amersham Pharmacia Biotech)
- Buffer B: 1 M potassium phosphate buffer pH 7.0
- Chromatography column (5 × 6 cm)
- Buffer C: 0.1 M potassium phosphate buffer pH 7.0
- Buffer C containing 0.8 M NaCl
- Reagents for assay of proteinase yscA (*Protocol 16*)
- $HgCl_2$
- Proteinase yscA inhibitor affinity matrix (*Protocol 23*)
- Buffer A containing 6.5 M urea
- Storage buffer: 50 mM ammonium acetate buffer pH 5.5
- Ultrafiltration equipment (*Protocol 13*)

Method

Carry out all operations at 0–5 °C, except where stated otherwise.

1. Suspend 1.2 kg yeast cells in 1200 ml cold water, and pass the suspension six times through the homogenizer at 500 kg/cm², cooling the suspension to 10 °C between cycles. Centrifuge the homogenate at 27 000 *g* for 30 min and adjust the supernatant to pH 5.0 with glacial

acetic acid. Add streptomycin and penicillin (60 mg each) and incubate the suspension at 25°C for 20 h to liberate proteinase yscA from its inhibitor.

2. Gently stir the activated crude extract and slowly add solid ammonium sulfate to 95% (w/v) saturation. Continue to stir the suspension for 2 h and then centrifuge it at 27 000 *g* for 30 min. Dissolve the precipitate in 400 ml distilled water and dialyse it overnight against 5 litres distilled water.
3. Add 500 ml phenyl–Sepharose® to the dialysate. Stir the gel suspension for 30 min. Transfer the gel suspension to a sintered glass filter and wash the gel with 400 ml buffer A containing 2 M NaCl, and then with 360 ml buffer A containing 50% (v/v) ethylene glycol. Elute the proteinase yscA from the gel by washing the gel with 500 ml buffer A containing 85% (v/v) ethylene glycol. Dialyse the eluate overnight against 15 litres distilled water.
4. Equilibrate 120 ml DEAE–Sephacel® overnight with buffer B. Transfer the gel into the chromatography column and wash the column extensively with buffer C.
5. Adjust the dialysed eluate (step 3) to pH 7.0 and apply the solution to the DEAE column. Wash the column with 300 ml buffer C and then elute the enzyme with a 1.5 litre linear gradient of 0–0.8 M NaCl in buffer C. Maintain a flow rate of 70 ml/h and collect fractions of 20 ml.
6. Assay the fractions for proteinase yscA (*Protocol 16*). Pool the active fractions.[b]
7. Dilute the pooled active fractions threefold with buffer A. Add $HgCl_2$ to a final concentration of 1 mM. Apply portions of extract containing about 30000 units proteinase yscA onto the inhibitor affinity matrix. Wash the affinity column with 200 ml buffer A containing 2 M NaCl. Elute the proteinase yscA from the gel with buffer A containing 6.5 M urea. Immediately dilute the eluate threefold with buffer A. Dialyse the eluate against 20 litres distilled water for several hours. Continue the dialysis overnight against storage buffer.
8. Concentrate the dialysed proteinase yscA by ultrafiltration and store it in 500 μl portions at –20°C.

[a] From ref. 56.

[b] This step yields a partly purified proteinase yscA preparation which will be sufficiently pure for biochemical studies. For a highly purified enzyme, affinity chromatography using the intrinsic peptide inhibitor I^A has to be carried out as described in the remaining steps of the protocol.

Protocol 25. Purification of proteinase yscB[a]

Equipment and reagents

- Yeast cells, dialysis tubing, homogenizer, ammonium sulfate, penicillin, streptomycin $HgCl_2$ (*Protocol 23*)
- Breakage buffer: 0.1 M potassium phosphate buffer pH 7.0
- 20% (v/v) acetic acid
- K_2HPO_4
- Standard buffer: 10 mM potassium phosphate pH 7.0
- DEAE–Sephadex® A-50 (Amersham Pharmacia Biotech)
- Chromatography column (5 × 70 cm)
- 0.4 M NaCl in standard buffer
- Reagents for the assay of proteinase yscB (*Protocol 17*)
- Equilibration buffer: 10 mM potassium phosphate buffer pH 7.0, 25% (w/v) ammonium sulfate
- Octyl–Sepharose® 4B (Amersham Pharmacia Biotech)
- Octyl–Sepharose® 4B column (1.6 × 25 cm) equilibrated in equilibration buffer
- Buffer A: 10 mM potassium phosphate pH 7.0, 50% (v/v) ethylene glycol
- Buffer B: 20 mM potassium phosphate pH 7.0, 5% (v/v) glycerol, 0.15 M NaCl
- Sephadex® G25 (Amersham Pharmacia Biotech)
- Sephadex® G25 column (5 × 40 cm) equilibrated in standard buffer
- DEAE–Sepharose® (Amersham Pharmacia Biotech)
- DEAE–Sepharose® column (1.6 × 10 cm) equilibrated in standard buffer
- 0.5 M NaCl in standard buffer
- PEG (flake; Aquacide II; Roth)
- Sephadex® G100 (Amersham Pharmacia Biotech)
- Sephadex® G100 column (2.6 × 90 cm) equilibrated in buffer B

Method

Carry out all steps at 4 °C unless otherwise stated.

1. Suspend 2 kg of yeast cells in 2 litres of breakage buffer and homogenize the cells by passing the suspension through the homogenizer five times at 500 kg/cm^2 with intermittent cooling to 10 °C. Cool the homogenate again to 10 °C and centrifuge it at 27 000 *g* for 30 min. Adjust the supernatant to pH 5.0 with 20% acetic acid. Add 100 mg penicillin G and 100 mg streptomycin sulfate to the extract, and incubate it at 25 °C for 20 h.
2. Adjust the extract to pH 7.0 with K_2HPO_4. Add $HgCl_2$ (10 mM final concentration)[b] and add ammonium sulfate to 90% saturation. Stir the suspension for 2 h and then centrifuge it at 27 000 *g* for 60 min. Dissolve the precipitate in standard buffer and dialyse the solution extensively against this buffer.
3. Swell and equilibrate 30 g DEAE–Sephadex® A-50 in standard buffer. Add the gel to the dialysate (step 2) and stir for 2 h. Pour the suspension into a chromatography column (5 × 70 cm), wash with standard buffer, and elute the protein with a 3 litre linear gradient of 0–0.4 M NaCl in standard buffer at a flow rate of 100 ml/h. Collect 5 ml fractions and assay them for proteinase yscB (*Protocol 17*). Pool the active fractions and concentrate the solution in a dialysis bag embedded in solid ammonium sulfate overnight. Centrifuge the dialysed solution at 27 000 *g* for 30 min. Dissolve the precipitate in standard

buffer and dialyse the solution against this buffer overnight. Continue the dialysis for 2 h against equilibration buffer.

4. Apply the dialysed solution to the octyl–Sepharose® 4B column. Wash the column extensively with equilibration buffer and elute the proteinase yscB with a linear gradient between equilibration buffer and buffer A. Collect 5 ml fractions and assay them for the proteinase activity (*Protocol 17*). Pool the proteinase yscB-containing fractions and remove the ethylene glycol by gel filtration on the Sephadex® G25 column run in standard buffer. Concentrate the protein (which elutes in the excluded volume) by adsorption on the DEAE–Sepharose® column. Elute the column with 0.5 M NaCl in standard buffer. Collect 2 ml fractions and assay them for proteinase (*Protocol 17*). Pool the active fractions and concentrate the solution in a dialysis bag embedded in flake PEG.

5. Apply the concentrated enzyme solution to the Sephadex® G100 column. Collect 1 ml fractions and assay them for the proteinase (*Protocol 17*). Pool the active fractions and concentrate the solution in a dialysis bag embedded in flake PEG. Store the proteinase in 100 μl fractions at –20 °C.

[a] From ref. 57.

[b] Note that inhibition of proteinase yscB during purification by $HgCl_2$ is essential to prevent self-digestion of the enzyme. Reactivation of the enzyme is achieved by adding an excess of 2-mercaptoethanol followed by gel filtration on a Sephadex® G25 column with 50 mM Tris–HCl pH 8.0.

6.4 Isolation of yeast vacuoles

Protocol 26 describes the isolation of vacuoles from yeast cells. These may be used for purification of a variety of proteinases or for studying protein degradation *in vitro*.

Protocol 26. Isolation of vacuoles[a]

Equipment and reagents

- JA10 and JA20 centrifuges, rotors, and tubes (Beckman Instruments)
- Ultracentrifuge SW41 rotor and tubes (Beckman Instruments)
- Buffer A: 0.1 M Pipes–KOH pH 9.4, 10 mM DTT
- Yeast cells in the desired growth phase (Section 2)
- YPD medium (*Table 1*)
- Sorbitol
- Dextrose
- Oxalylticase (Enzogenetics)
- Corex tubes (30 ml; DuPont-NEN)
- Buffer B: 50 mM potassium phosphate pH 7.5, 300 mM sorbitol in YPD medium, with 0.2% (w/v) dextrose, and 1800–3600 U/ml oxalyticase
- Buffer C: 10 mM Pipes–KOH pH 6.8, 200 mM sorbitol
- DEAE dextran (Amersham Pharmacia Biotech): just before use prepare a 0.4 mg/ml solution in 15% Ficoll 400 (Amersham Pharmacia Biotech) in buffer C
- 4%, 8%, 15% Ficoll 400 solutions in buffer C

Protocol 26. *Continued*

Method

1. Grow yeast cells in 1 litre of YPD medium at 30 °C until an A_{600} 0.6–0.8 is reached.
2. Centrifuge the culture for 5 min at 4000 *g* at 23 °C in a JA10 rotor and resuspend the cells in 50 ml buffer A.
3. Incubate the cell suspension for 10 min at 30 °C and then centrifuge it for 5 min at 4000 *g* at 2 °C in a JA10 rotor. Resuspend the pellet in 15 ml buffer B.
4. Transfer the cell suspension into 30 ml Corex tubes. Incubate the tubes for 25 min at 30 °C. Recover the cells by centrifugation for 1 min at 800 *g*. Resuspend the cells in buffer B and recover them by centrifugation for 1 min at 1500 *g* at 2 °C in a JA20 rotor.
5. Resuspend the cells in 2.5 ml of 15% Ficoll 400 in buffer C by gentle stirring with a glass rod.
6. Add 100–200 μl DEAE dextran in 15% Ficoll in buffer C. Incubate the cells for 2 min at 0 °C and then for 75 sec at 30 °C. Cool the suspension on ice.
7. Transfer the cell suspension to a SW41 tube and overlay it successively with 3 ml of 8% Ficoll in buffer C, 3 ml of 4% Ficoll in buffer C, and 1.5 ml buffer C.
8. Centrifuge the gradient for 90 min at 150 000 *g* in the SW41 rotor. Harvest the vacuoles from the buffer C–4% Ficoll interface using a pipette tip with a wide orifice.
9. Place the vacuoles at 4 °C and use them immediately for the desired experiments.[b]

[a] From ref. 58.
[b] The vacuole preparation may be tested by measuring the specific activity of any vacuolar proteinase in relation to a cytoplasmic enzyme (e.g. glucose 6-phosphate dehydrogenase) in the purified vacuoles and comparing this to the enzyme activity in the cell extract.

Acknowledgements

The work of the authors was funded by the Deutsche Forschungsgemeinschaft (Bonn), the Fond der Chemischen Industrie (Frankfurt), and by a Grant from the German–Israeli Foundation for Scientific Research and Development (G. I. F.) (Jerusalem).

References

1. Coux, O., Tanaka, K., and Goldberg, A. L. (1996). *Annu. Rev. Biochem.*, **65**, 801.
2. Hilt, W. and Wolf, D. H. (1996). *Trends Biochem. Sci.*, **21**, 96.
3. Peters, J. M. (1994). *Trends Biochem. Sci.*, **19**, 377.
4. Rubin, D. M. and Finley, D. (1995). *Curr. Biol.*, **5**, 854.
5. Groll, M., Ditzel, L., Löwe, J., Stock, D., Bochtler, M., Bartunik, H. D., *et al.* (1997). *Nature*, **386**, 463.
6. Arendt, C. S. and Hochstrasser, M. (1997). *Proc. Natl. Acad. Sci. USA*, **94**, 7156.
7. Heinemeyer, W., Fischer, M., Krimmer, T., Stachon, U., and Wolf, D. H. (1997). *J. Biol. Chem.*, **272**, 25200.
8. Orlowski, M. and Michaud, C. (1989). *Biochemistry*, **28**, 9270.
9. Heinemeyer, W., Kleinschmidt, J. A., Saidowsky, J., Escher, C., and Wolf, D. H. (1991). *EMBO J.*, **10**, 555.
10. Rechsteiner, M., Hoffman, L., and Dubiel, W. (1993). *J. Biol. Chem.*, **268**, 6065.
11. Glickman, H., Rubin, D., Fried, V. A., and Finley, D. (1998). *Mol. Cell. Biol.*, **18**, 3149.
12. Deveraux, Q., Ustrell, V., Pickart, C., and Rechsteiner, M. (1994). *J. Biol. Chem.*, **269**, 7059.
13. van Nocker, S., Deveraux, Q., Rechsteiner, M., and Vierstra, R. D. (1996). *Proc. Natl. Acad. Sci. USA*, **93**, 856.
14. Hochstrasser, M. (1996). *Annu. Rev. Genet.*, **30**, 405.
15. Brodsky, J. M. and McCracken, A. A. (1997). *Trends Cell Biol.*, **7**, 151.
16. Biederer, T., Volkwein, C., and Sommer, T. (1996). *EMBO J.*, **15**, 2069.
17. Hiller, M. M., Finger, A., Schweiger, M., and Wolf, D. H. (1996). *Science*, **273**, 1725.
18. Hochstrasser, M. (1995). *Curr. Opin. Cell. Biol.*, **7**, 215.
19. Sommer, T. and Wolf, D. H. (1997). *FASEB J.*, **11**, 1227.
20. Knop, M., Schiffer, H. H., Rupp, S., and Wolf, D. H. (1993). *Curr. Opin. Cell. Biol.*, **5**, 990.
21. Bohley, P. and Seglen, P. O. (1992). *Experientia*, **48**, 151.
22. Dunn, W. J. (1994). *Trends Cell Biol.*, **4**, 139.
23. Dice, J. F. (1990). *Trends Biochem. Sci.*, **15**, 305.
24. Cuervo, A. M. and Dice, J. F. (1996). *Science*, **273**, 501.
25. Teichert, U., Mechler, B., Müller, H., and Wolf, D. H. (1989). *J. Biol. Chem.*, **264**, 16037.
26. Egner, R., Thumm, M., Straub, M., Simeon, A., Schüller, H. J., and Wolf, D. H. (1993). *J. Biol. Chem.*, **268**, 27269.
27. Rupp, S. and Wolf, D. H. (1995). *Eur. J. Biochem.*, **231**, 115.
28. Goffeau, A. (1997). *Nature*, **387**, 1.
29. Betz, H. and Weiser, U. (1976). *Eur. J. Biochem.*, **62**, 65.
30. Straub, M., Bredschneider, M., and Thumm, M. (1997). *J. Bacteriol.*, **179**, 3975.
31. Schork, S. M., Thumm, M., and Wolf, D. H. (1995). *J. Biol. Chem.*, **270**, 26446.
32. Finger, A., Knop, M., and Wolf, D. H. (1993). *Eur. J. Biochem.*, **218**, 565.
33. Ohlsson, B. G., Weström, B. R., and Karlson, B. W. (1986). *Anal. Biochem.*, **152**, 239.
34. Hilt, W., Enenkel, C., Gruhler, A., Singer, T., and Wolf, D. H. (1993). *J. Biol. Chem.*, **268**, 3479.

35. Gueckel, R., Enenkel, C., Wolf, D. H., and Hilt, W. (1998). *J. Biol. Chem.*, **273**, 19443.
36. Achstetter, T., Ehmann, C., Osaki, A., and Wolf, D. H. (1984). *J. Biol. Chem.*, **259**, 13344.
37. Achstetter, T., Ehmann, C., and Wolf, D. H. (1985). *J. Biol. Chem.*, **260**, 4585.
38. Büchler, M., Tisljar, U., and Wolf, D. H. (1994). *Eur. J. Biochem.*, **219**, 627.
39. Jones, E. W. (1984). *Annu. Rev. Genet.*, **18**, 233.
40. Achstetter, T. and Wolf, D. H. (1985). *Yeast*, **1**, 139.
41. Suarez-Rendueles, P. and Wolf, D. H. (1988). *Fems. Microbiol. Rev.*, **4**, 17.
42. Hirsch, H. H., Suarez-Rendueles, P., and Wolf, D. H. (1989). In *Molecular and cell biology of yeasts* (ed. E. F. Walton and G. T. Yarranton), pp. 134–200. Blackie and Van Nostrand Reinhold.
43. Wolf, D. H. (1986). *Microbiol Sci.*, **3**, 107.
44. Klionsky, D. J., Herman, P. K., and Emr, S. D. (1990). *Microbiol. Rev.*, **54**, 266.
45. Thumm, M. and Wolf, D. H. (1998). *Adv. Mol. Cell Biol.*, **27**, 43.
46. Jones, E. W. (1991). *J. Biol. Chem.*, **266**, 7963.
47. Hansen, R. J., Switzer, R. L., Hinze, H., and Holzer, H. (1977). *Biochim. Biophys. Acta*, **496**, 103.
48. Müller, M. and Holzer, H. (1980). In *Enzyme regulation and mechansims of action* (ed. O. Mildner and B. Ries), pp. 339–49. Pergamon Press.
49. Saheki, T. and Holzer, H. (1974). *Eur. J. Biochem.*, **42**, 621.
50. Hata, T., Hayashi, R., and Doi, E. (1967). *Agric. Biol. Chem.*, **31**, 357.
51. Jones, E. W. (1977). *Genetics*, **85**, 23.
52. Aibara, S., Hayashi, R., and Hata, T. (1971). *Agric. Biol. Chem.*, **35**, 658.
53. Achstetter, T., Ehmann, C., and Wolf, D. H. (1982). *Biochem. Biophys. Res. Commun.*, **109**, 341.
54. Suarez-Rendueles, M. P., Schwenke, J., Garcia-Alvarez, N., and Gascon, S. (1981). *FEBS Lett.*, **131**, 296.
55. Nunez de Castro, I. and Holzer, H. (1976). *Hoppe-Seyler´s Z. Physiol. Chem.*, **357**, 727.
56. Meussdoerffer, F., Totora, P., and Holzer, H. (1980). *J. Biol. Chem.*, **255**, 12087.
57. Kominami, E., Hofschulte, H., and Holzer, H. (1981). *Biochim. Biophys. Acta*, **661**, 124.
58. Mayer, A., Wickner, W., and Haas, A. (1996). *Cell*, **85**, 83.

A1

List of suppliers

Aldrich Chemical Co., The Old Brickyard, Gillingham, Dorset SP8 4XT, UK.

Amersham

Amersham International plc., Lincoln Place, Green End, Aylesbury, Buckinghamshire HP20 2TP, UK.

Amersham, Amersham Place, Little Chalfont, Buckinghamshire HP7 9NA, UK.

Amersham Corporation, 2636 South Clearbrook Drive, Arlington Heights, IL 60005, USA.

Amersham Buchler GmbH & Co. KG, Gieselweg 1, 38110 Braunschweig, Germany.

Amicon, Millipore GmbH, Postfach 5647, D-65731 Eschborn, Germany.

Aminco, via Spectronic Instruments Ltd., 93/96 Chadwick Road, Astmoor Industrial Estate, Runcorn, Cheshire WA7 1PR, UK.

Anderman

Anderman and Co. Ltd., 145 London Road, Kingston-Upon-Thames, Surrey KT17 7NH, UK.

Bachem

Bachem Feinchemikalien AG, Hauptstrasse 14, CH-4416 Bubendorf, Switzerland.

Bachem (USA), 3700 Horizon Drive, Renaissance at Gulph Mills, King of Prussia, PA 19406, USA.

BDH Laboratory Supplies, Poole, Dorset BH15 1TD, UK.

Beckman Instruments

Beckman Instruments UK Ltd., Progress Road, Sands Industrial Estate, High Wycombe, Buckinghamshire HP12 4JL, UK.

Beckman Instruments Inc., PO Box 3100, 2500 Harbor Boulevard, Fullerton, CA 92634, USA.

Beckman Instruments GmbH, Frankfurter Ring 115, D-80807 München, Germany.

Becton Dickinson

Becton Dickinson and Co., Between Towns Road, Cowley, Oxford OX4 3LY, UK.

Becton Dickinson and Co., 2 Bridgewater Lane, Lincoln Park, NJ 07035, USA.

Bibby Sterelin Ltd., Customer Services dept., Tilling Drive, Stone, Staffordshire ST15 0SA, UK.

Bio

Bio 101 Inc., c/o Stratech Scientific Ltd., 61–63 Dudley Street, Luton, Bedfordshire LU2 0HP, UK.

Bio 101 Inc., PO Box 2284, La Jolla, CA 92038–2284, USA.

Biochrom, Biochrom GmbH & Co. KG, Leonorenstrasse 2–6, 12247 Berlin, Germany.

Bio-Rad Laboratories

Bio-Rad Laboratories Ltd., Bio-Rad House, Maylands Avenue, Hemel Hempstead HP2 7TD, UK.

Bio-Rad Laboratories, Division Headquarters, 3300 Regatta Boulevard, Richmond, CA 94804, USA.

Bio-Rad, 2000 Alfred Nobel, Hercules, CA 94547, USA.

Bio-Rad, Heidemann Strasse 164, D-80901 München, Germany.

Boehringer Mannheim

Boehringer Mannheim UK (Diagnostics and Biochemicals) Ltd., Bell Lane, Lewes, East Sussex BN17 1LG, UK.

Boehringer Mannheim Corporation, Biochemical Products, 9115 Hague Road, PO Box 504, Indianopolis, IN 46250–0414, USA.

Boehringer Mannheim Biochemica, GmbH, Sandhofer Str. 116, Postfach 310120, D-6800 Ma 31, Germany.

British Drug Houses (BDH) Ltd., Poole, Dorset, UK.

Calbiochem-Novabiochem

Calbiochem-Novabiochem, 3 Heathcoat Building, Highfields Science Park, University Boulevard, Nottingham NGT 2QJ, UK.

Calbiochem-Novabiochem, Lisztweg 1, D-65812 Bad Soden, Germany.

Cellmax, Germantown, MD, USA.

Corning Costar UK, 1 The Valley Centre, Gordon Road, High Wycombe, Buckinghamshire HP13 6EQ, UK.

Costar, 1 The Valley Centre, Gordon Road, High Wycombe, Buckinghamshire HP13 6EQ, UK.

Difco Laboratories

Difco Laboratories Ltd., PO Box 14B, Central Avenue, West Molesey, Surrey KT8 2SE, UK.

Difco Laboratories, PO Box 331058, Detroit, MI 48232–7058, USA.

Difco Laboratories, Postfach 101466, D-86004 Augsburg, Germany.

Du Pont

Dupont (UK) Ltd. (Industrial Products Division), Wedgwood Way, Stevenage, Hertfordshire SG1 4Q, UK.

Du Pont Co. (Biotechnology Systems Division), PO Box 80024, Wilmington, DE 19880–002, USA.

Dupont - NEN Life Science, PO Box 666, Hounslow TW5 9RT, UK.

Eli-Lilly, Indianapolis, USA.

Enzogenetics, PO Box 1715, Corvallis, OR 97330, USA.

Eppendorf, Hamburg, Germany.

B. Euler, Biotechnologie Mikrobiologie-Geraete und Verbauchsmaterialien, Starkenburger Strasse 55, D-60386 Frankfurt am Main, Germany.

European Collection of Animal Cell Culture, Division of Biologics, PHLS Centre for Applied Microbiology and Research, Porton Down, Salisbury, Wiltshire SP4 0JG, UK.

Falcon (Falcon is a registered trademark of Becton Dickinson and Co.)

Fisher Scientific

Fisher Scientific Co., 711 Forbest Avenue, Pittsburgh, PA 15219–4785, USA.

Fisher Scientific, 1600 West Glenlake Avenue, Itasca, IL 60143, USA.

Flow Laboratories, Woodcock Hill, Harefield Road, Rickmansworth, Hertfordshire WD3 1PQ, UK.

Fluka

Fluka-Chemie AG, CH-9470, Buchs, Switzerland.

Fluka Chemicals Ltd., The Old Brickyard, New Road, Gillingham, Dorset SP8 4JL, UK.

Forma Scientific Inc., PO Box 649, Millcreek Road, Marietta, OH 45750, USA.

Fugisawa Pharmaceutical, Osaka, Japan.

Gibco BRL

Gibco BRL (Life Technologies Ltd.), Trident House, Renfrew Road, Paisley PA3 4EF, UK.

Gibco BRL (Life Technologies Inc.), 3175 Staler Road, Grand Island, NY 14072–0068, USA.

Gibco BRL, Unit 4, Cowley Mill Trading Estate, Longbridge Way, Uxbridge, Middlesex UB8 2YG, UK.

Greiner, Greiner Labortechnik GmbH, Maybachstrasse 2, 12636 Frickenhausen, Germany.

Hereaus Equipment Ltd., Unit 9, Wates Way, Brentwood, Essex CM15 9TB, UK.

Heto-Holten, PO Box 31, Camberley, Surrey GU15 ITN, UK.

Hewlett Packard, Analytical Products Group, Heathside Park Road, Cheadle Heath, Stockport, Cheshire SK3 ORB, UK.

Hoefer, now part of Amersham Pharmacia Biotech

Arnold R. Horwell, 73 Maygrove Road, West Hampstead, London NW6 2BP, UK.

Hybaid

Hybaid Ltd., 111–113 Waldegrave Road, Teddington, Middlesex TW11 8LL, UK.

Hybaid, National Labnet Corporation, PO Box 841, Woodbridge, NJ 07095, USA.

HyClone Laboratories, 1725 South HyClone Road, Logan, UT 84321, USA.

ICN Pharmaceuticals Inc., 3300 Hyland Avenue, Costa Mesa, CA 92626, USA.

International Biotechnologies Inc., 25 Science Park, New Haven, Connecticut 06535, USA.

Invitrogen Corporation

Invitrogen Corporation, 3985 B Sorrenton Valley Building, San Diego, CA 92121, USA.

Invitrogen Corporation, c/o British Biotechnology Products Ltd., 4–10 The Quadrant, Barton Lane, Abingdon, Oxon OX14 3YS, UK.

Irvine Scientific Inc., 2511 Daimler Street, Santa Anna, CA 92705-558, USA.

Jencons Scientific, Cherrycourt Way Industrial Estate, Stanbridge Road, Leighton Buzzard LU7 8UA, UK.

Jones Chromatography Ltd., New Road, Hengoed, Mid-Glamorgan CF8 8AU, Wales.

Kodak

Kodak: Eastman Fine Chemicals, 343 State Street, Rochester, NY, USA.

Kodak, 25 Science Park, New Haven, CT 06511, USA.

Kontron Instruments Ltd., Blackmore Lane, Croxley Business Park, Watford, Hertfordshire WD1 8XQ, UK.

Life Technologies Inc., 8451 Helgerman Court, Gaithersburg, MN 20877, USA.

Merck

Merck Industries Inc., 5 Skyline Drive, Nawthorne, NY 10532, USA.

Merck, Frankfurter Strasse, 250, Postfach 4119, D-64293, Germany.

Micromath Scientific Software, PO Box 21550, Salt Lake City, Utah 84121, USA.

Millipore

Millipore (UK) Ltd., The Boulevard, Blackmoor Lane, Watford, Hertfordshire WD1 8YW, UK.

Millipore Corp./Biosearch, PO Box 255, 80 Ashby Road, Bedford, MA 01730, USA.

Molecular Probes

Molecular Probes Inc., Eugene, Oregon, USA.

c/o *Mo Bi Tec*, Wagenstieg 5, D-37077 Gottingen, Germany.

NEN (New England Nuclear)

Dupont Co., 549 Albany Street, Boston, MA 02118, USA.

Dupont (UK) Ltd., Wedgewood Way, Stevenage, Hertfordshire SG1 4QN, UK.

DuPont et Nemours International SA, Pumpwerk Strasse 15, CH-8105 Regensdorf, Switzerland.

New Brunswick Scientific, Edison House, 163 Dixons Hill Road, North Mymms, Hatfield, Hertfordshire AL9 2JE, UK.

New England Biolabs (NBL)

New England Biolabs (NBL), 32 Tozer Road, Beverley, MA 01915–5510, USA.

New England Biolabs (NBL), c/o CP Labs Ltd., PO Box 22, Bishops Stortford, Hertfordshire CM23 3DH, UK.

Nikon Corporation, Fuji Building, 2–3 Marunouchi 3-chome, Chiyoda-ku, Tokyo, Japan.

Oxford Glycosystems Ltd., Hitching Court, Blacklands Way, Abingdon OX14 1RG, UK.

Pall Biosupport, Europa House, Havant Street, Portsmouth PO1 3PD, UK.

PE Applied Biosystems, Kelvin Close, Birchwood Science Park North, Warrington, Cheshire WA3 7PB, UK.

Peptide International, 11621 Electron Drive, Louisville, KY 40299, USA.

Perkin-Elmer

Perkin-Elmer Ltd., Maxwell Road, Beaconsfield, Buckinghamshire HP9 1QA, UK.

Perkin Elmer Ltd., Post Office Lane, Beaconsfield, Buckinghamshire HP9 1QA, UK.

Perkin Elmer-Cetus (The Perkin-Elmer Corporation), 761 Main Avenue, Norwalk, CT 0689, USA.

PerSeptive Biosystems Inc., 500 Old Connecticut Path, Framingham, MA 01701, USA.

Pharmacia Biotech Europe, Procordia EuroCentre, Rue de la Fuse-e 62, B-1130 Brussels, Belgium.

Pharmacia Biosystems

Pharmacia Biosystems Ltd. (Biotechnology Division), Davy Avenue, Knowlhill, Milton Keynes MK5 8PH, UK.

Pharmacia LKB Biotechnology AB, Björngatan 30, S-75182 Uppsala, Sweden.

Pharmacia LKB Biotechnology, Midsummer Boulevard, Central Milton Keynes, Buckinghamshire MK9 3HP, UK.

Pierce & Warriner, 44 Upper Northgate Street, Chester CH1 4EF, UK.

Promega

Promega Ltd., Delta House, Enterprise Road, Chilworth Research Centre, Southampton SO1 7NS, UK.

Promega Corporation, 2800 Woods Hollow Road, Madison, WI 53711–5399, USA.

Qiagen

Qiagen Inc., c/o Hybaid, 111–113 Waldegrave Road, Teddington, Middlesex TW11 8LL, UK.

Qiagen Inc., 9259 Eton Avenue, Chatsworth, CA 91311, USA.

Research Biochemicals International, 1 Strathmore Road, Natick, MA 01760-2418, USA.

Roth, Carl Roth GmbH, Schoemperlen Strasse 1-5, D-76185 Karlsruhe, Germany.

Sandoz Pharmaceuticals, East Hanover, NJ, USA.

Savant (Stratech Scientific Ltd.), 61–63 Dudley Street, Luton, Bedfordshire LU2 OAP, UK.

Schleicher and Schuell

Schleicher and Schuell Inc., Keene, NH 03431A, USA.

Schleicher and Schuell Inc., D-3354 Dassel, Germany.
Schleicher and Schuell Inc., c/o Anderman and Co. Ltd.
Seikagaku Corporation
Seikagaku Corporation, Tokyo, Japan.
c/o *WAK-Chemie*, Hessenring 63, D-61348 Bad Homburg, Germany.
Serva (*see Universal Biologicals Ltd.*)
Shandon Scientific Ltd., Chadwick Road, Astmoor, Runcorn, Cheshire WA7 1PR, UK.
Sigma-Aldrich Chemie GmbH, Grünwalder Weg, D-82041 Deisenhofen, Germany.
Sigma Chemical Company
Sigma Chemical Company (UK), Fancy Road, Poole, Dorset BH17 7NH, UK.
Sigma Chemical Company, 3050 Spruce Street, PO Box 14508, St. Louis, MO 63178–9916, USA.
Sonics & Materials Inc., West Kenosia Avenue, Danbury, Connecticut, CT 06819, USA.
Sorvall DuPont Company
Sorvall DuPont Company, Biotechnology Division, PO Box 80022, Wilmington, DE 19880–0022, USA.
Sorvall, 31 Pecks Lane, Newton, CT 06470-2337, USA.
Sterilin (*see Bibby Sterilin Ltd.*)
Stratagene
Stratagene Ltd., Unit 140, Cambridge Innovation Centre, Milton Road, Cambridge CB4 4FG, UK.
Stratagene Inc., 11011 North Torrey Pines Road, La Jolla, CA 92037, USA.
United States Biochemical, PO Box 22400, Cleveland, OH 44122, USA.
Upstate Biotechnology Ltd., 89 Saranac Avenue, Lake Placid, NY 12946, USA.
Universal Biologicals Ltd., Fome Hall Mill, Lodgemoor Lane, Stroud, Gloucester GL5 3EH, UK.
VWR Scientific International, PO Box 1002, South Plainfield, NJ 07080, USA.
Wako Chemicals, 1600 Bellwood Road, Richmond, VA 23237, USA.
Waters Ltd., The Boulevard, Blackmoor Lane, Watford WD1 8YW, UK.
Wellcome Reagents, Langley Court, Beckenham, Kent BR3 3BS, UK.
Whatman, St. Leonard's Road, Maidstone, Kent ME16 OLS, UK.

Index